TECHNICAL RESCUE OPERATIONS

Volume II: COMMON EMERGENCIES

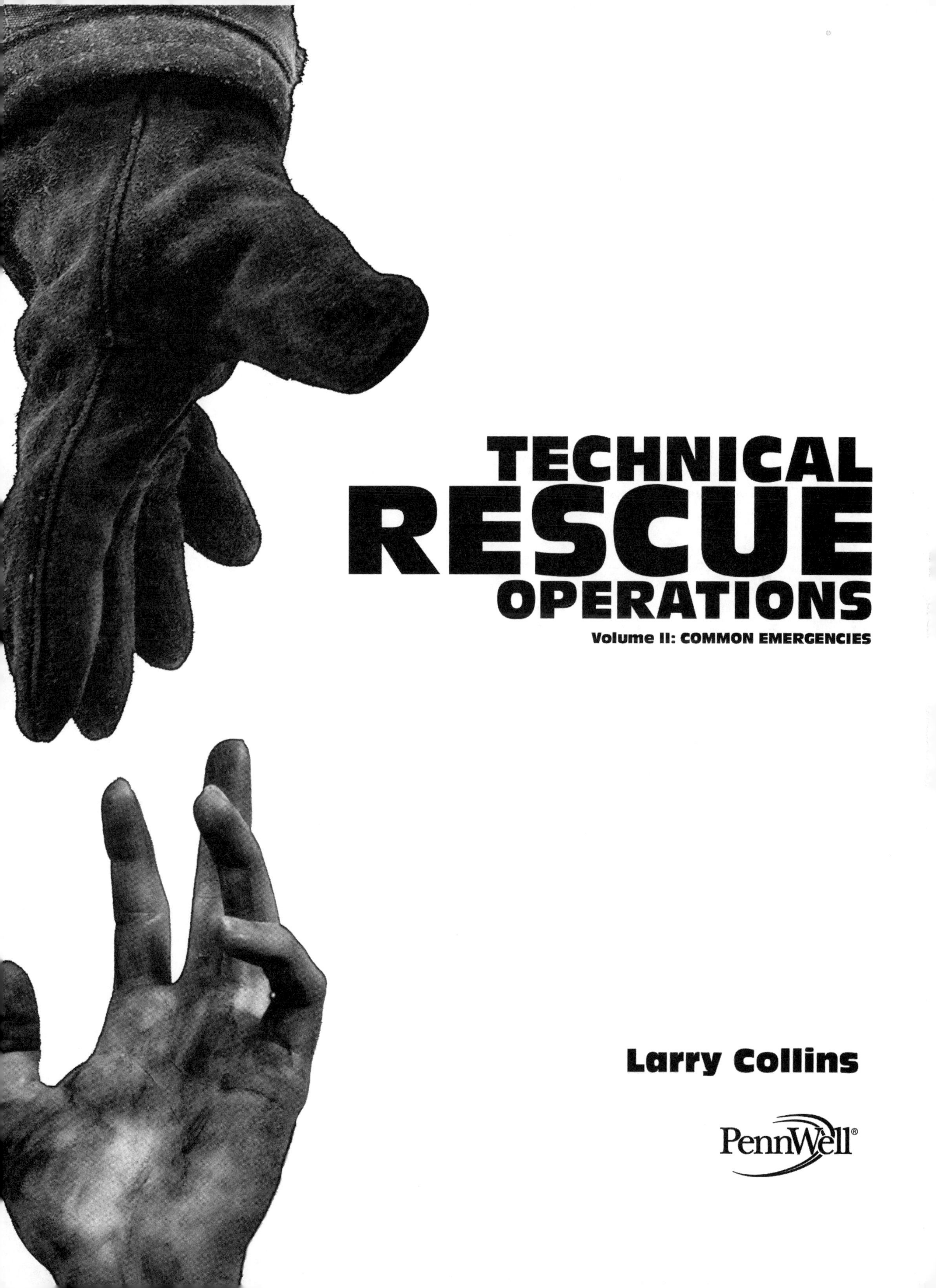

TECHNICAL RESCUE OPERATIONS

Volume II: COMMON EMERGENCIES

Larry Collins

PennWell®

PennWell Corporation
1421 South Sheridan Road
Tulsa, Oklahoma 74112-6600 USA
800.752.9764
+1.918.831.9421
sales@pennwell.com
www.pennwellbooks.com
www.pennwell.com
Managing Editor: Jared d'Orr Wicklund
Production Editor: Sue Rhodes Dodd
Cover Designer: Ken Wood
Book Designer: Wes Rowell

Library of Congress Cataloging-in-Publication Data

Collins, Larry, 1952–
Technical rescue operations / by Larry Collins.
p. cm.
Includes bibliographical references and index.
ISBN 1-59370-020-2 (2)
1. Emergency management--United States. 2. Rescue work--United States--Planning. 3. Search and rescue operations--United States. 4. Disaster relief--United States. I. Title.

HV551.3.C65 2004
363.34'81'0973--dc22

2004011227

Printed in the United States of America

1 2 3 4 5 09 08 07 06 05

This book is dedicated to
the first responders who gave their lives to save others
in the aftermath of the 9-11 terrorist attacks.

In the words of Han-Jochen Blatte,
Chief of a German fire brigade,
"The firefighters from FDNY had only a small chance.
But they used it completely for others and died themselves."

Contents
Volume II

List of Acronyms

ACFD (Arlington County Fire Department)
AFD (Alexandria Fire Department)
APR (air-purifying respirator)
A/SDP (air/sea disaster plan)
BoO (base of operations)
CAD (computer-assisted dispatch)
CATF-2 (California Task Force 2)
CDC (U.S. Center for Disease Control)
CFR (Code of Federal Regulations)
cfs (cubic feet per second)
CISD (critical incident stress debriefing)
CPR (cardiopulmonary resuscitation)
CSFA (California State Firefighters Association)
DMAT (disaster medical assistance team)
DMORG (disaster mortuary team)
EMS (emergency medical service)
ER (emergency room)
FCCF (fire command-and-control facility)
FDNY (Fire Department of New York City)
FLIR (forward-looking infrared)
FOG (field operations guide)
HERS (heavy equipment/rigging specialist)
HHR (helo/high-rise)
HHRT (helo/high-rise team)
HSW (helo/swift-water)
HSWR (Helo/swift-water rescue)
IAP (incident action plan)
IC (incident commander)
ICP (incident command post)
ICS (incident command system)
IMT (incident management team)
IST (incident support team)
L.A. (Los Angeles)
LACoFD (County of Los Angeles Fire Department)
LAFD (Los Angeles Fire Department)
LAPD (Los Angeles Police Department)
LCES (lookout, communications, escape routes, safe zones)
MASRS (multi-agency swift-water rescue system)
MDW (Military District of Washington)
NASAR (National Association for Search and Rescue)
NBC (nuclear/biological/chemical)
NFPA (National Fire Protection Association)
NIOSH (National Institute of Occupational Safety and Health)
NMTF-1 (New Mexico Task Force 1)
NTSB (National Transportation and Safety Bureau)
NYTF-1 (New York City USAR Task Force)
OCFD (Oklahoma City Fire Department)
OES (office of emergency services)
OIC (ocean incident command)
ONP (FEMA Office of National Preparedness)
OSHA (Occupational Safety and Health Administration)
PALS (personal alert devices)
PAR (personnel accountability report)
PFD (personal flotation device)
PLS (point last seen)
PM (paramedic)
PPE (personal protective equipment)
PWC (personal watercraft)
RIC (rapid intervention crew)
RIO (rapid intervention operation)
RIS (rapid intervention system)
SAR (search and rescue)
SCBA (self-contained breathing apparatus)
SEMS (standard emergency management systems)
SMC (federal search and rescue coordinator)
SOG (standard operating guidelines)
USAR (urban search and rescue)
USFA (U.S. Fire Administration)
VMAT (veterinary medical assistance team)
WMD (weapons of mass destruction)
WRP (waterway rescue preplan)
WTC (World Trade Center)

Introduction

Whether known by the term *urban search and rescue (USAR), technical rescue,* or some other common or regional term, *rescue* is among the most essential jobs of the fire service and other branches of public safety. One of the most time-honored duties of firefighters has been to rescue people trapped by fire. The goal of rescue is equally important in non-fireground situations: Whenever people are physically trapped or otherwise unable to reach safety on their own, they must be *rescued* and moved to a place of safety before we can say our job is done. And although we can provide certain types of pre-hospital care for injured people while they are still trapped, they must at some point be rescued before they can receive definitive treatment.

Rescue is a mission that requires dedication to protecting lives; a willingness to take calculated risks; an understanding of how things work (and, consequently, how things *fail*); and the ability to devise and implement solutions for complex problems under hazardous and rapidly changing conditions—without the luxury of time for reflection and deep contemplation. Broken down to its most basic elements, rescue is something performed in the most crucial moments of an emergency by firefighters and other rescuers guided by a combination of gut instinct, experience, knowledge of standard protocol, and the ability to improvise a solution from seemingly unsolvable problems.

This book, the second in a series, emphasizes the operations of rescue companies, USAR companies and task forces, and other specialized rescue units as well as various types of search and rescue teams that will typically respond to the full range of "daily" technical rescues and the major rescue operations that occur less frequently. But the principles contained herein also apply to the typical fire department truck company as well as to standard engine companies, brush patrols, fire/rescue helicopters, chief officers, disaster planners, rescue instructors, and others with responsibility for managing USAR emergencies and disasters (or those charged with teaching others how to do it.

Volume I (*Rescue Operational Planning, Training, Equipment, and Command*) laid out the basic building blocks for development and use of comprehensive, multi-tier rescue response systems. A history of rescue was discussed, along with trends for the future, taking into account the ever-changing rescue hazards we encounter in the real world. Effective planning based on accurate hazard assessment (and the training and experience needed to make precise hazard evaluations) was another important issue in the process of operational readiness for rescue operations. Primary and training and continuing education (and its twin cousins *research and development* and their application to multi-level rescue response systems) were main topics of Volume I. Naturally, equipment is an important part of effective response to rescues of every sort, and that also was covered in the first book. Finally, *command* is a topic that has an overarching effect on how rescue is done. Indeed, there was vigorous discussion in Volume I about how to manage technical rescues and disaster rescue operations with maximum effectiveness and adequate margins of safety for firefighters and rescuers.

Now in Volume II we move to the next stage: Responding to, managing, and conducting rescues in the daily setting of fire/rescue agencies. This includes the kind of technical rescues that confront firefighters and rescuers on practically a daily basis somewhere in North America. We will also consider how to handle more complex and large-scale rescue (or "sub-disaster") operations, where it may take many hours to extract the last victim. In some cases, this includes freeing the bodies of the dead, who must then be removed from the scene in a dignified manner and in a way that allows the local coroner's office to determine the cause of death as prescribed by law. We are not talking here about disasters, which by definition overwhelm the ability of local resources to manage them. Instead, we are discussing more complex or larger scale rescue operations that challenge local responders to apply solid rescue principles for longer periods of time, with assistance required of additional resources. These operations are under more strict Command and Control because of the scope of the incident, its newsworthiness (with the attendant news reporters, television cameras, media helicopters, etc.), crowds of people (possibly including relatives of the trapped and missing) arriving on the scene, and getting the immediate attention of local or regional elected officials.

In Volume II, we explore how applying the concepts taught in Volume I can lead to more efficient rescue operations during daily and major incidents. This, in turn, will let us move more easily to the management of unusual rescues that may be "once in a career" events for most firefighters and rescuers (the topic of Volume III).

About This Series

In writing an instructional text, an author attempts to transfer a body of knowledge to current practitioners of a selected discipline and to the next generation. The knowledge being transferred within a text may have been gleaned from pure research, field experiments, case studies, or the actual experiences of the author. Or it may be a consolidation of previously documented lessons, edited and reformatted to meet modern demands.

Just as successful living species tend to evolve—and thereby thrive—by selecting the most helpful genetic traits, the fire service can select valuable bits of knowledge from its body of experience, passed along in the form of written and spoken words supported by illustrations, photos, and other media. And so it goes with *Technical Rescue*. This series is the author's modest attempt to select the most helpful parts of a highly dynamic and ever-changing body of information in order to improve the chances of survival for victims and rescuers alike during the course of USAR emergencies. This includes rapid intervention situations where our own colleagues have themselves become trapped, lost, or injured in high-risk environments, and in need of our immediate assistance.

As a reminder, *Rescue* is not a "nuts and bolts" series of books. Although we will discuss rescue systems and their application as part of the response to challenging rescues and disasters, we are not learning to tie the knots required to establish systems, and we are not concentrating on topics like nail patterns for shoring system components. That kind of detailed information is widely available in a variety of other books, magazine articles, and training manuals, and it should be taught in a hands-on setting with qualified instructors and with all the appropriate safety precautions due to its high-risk nature.

The real intent of this series is to move beyond the "nuts and bolts" of rescue and begin putting together the pieces in a way that facilitates the development and efficient operation of multi-tiered response systems while affording maximum safety to the rescuers and victims alike. It is a process, and in some cases it requires a philosophical and even an organization change in the way rescue is approached. The effectiveness of this systems-based approach to rescue is clearly demonstrated by the number of successful rescues conducted every year by progressive fire and rescue agencies that use it. It is also seen in the effective operation of local, state, and regional search and rescue teams and in the utility of the state and federal USAR task forces (and their various components) that have evolved in the United States and other nations since the 1980s.

With the current emphasis on Homeland Security and effective management of the consequences of terrorism, a systematic approach to daily rescue and disaster search and rescue has become even more important to first responders and the public alike.

On The Evolution of Rescue

The fire and rescue services are creatures of evolution, however painfully slow that progress might seem to be at times. Our chosen profession (like so many others) is sometimes plagued with a form of institutional amnesia. Many of us tend to forget that modern fire and rescue service practice derives (in part) from knowledge and traditions inherited from long-deceased predecessors, knowledge and experience that's been hammered into contemporary form to suit the needs of modern fire and rescue culture.

Speaking of evolution, it is possible today in modern-day Rome to visit the subterranean remains of the ancient city that has been buried and built upon throughout the ages. Not far from the Spanish Steps, and some 16 meters below the surface, you can descend through passageways to explore the ruins of ancient structures, including a Roman fire station that existed in the time of Nero. This fire station, built more than 2,000 years ago, shares many characteristics that would be familiar to any modern firefighter, including dormitory areas for overnight shifts, apparatus rooms (complete with stables), offices, and a kitchen. Inscriptions made by members of the *Militia Vigilium* on a wall of the station relate some of the same concerns shared by modern-day firefighters. "Send my relief," wrote one member (presumably waiting for shift change after a particularly busy tour), "for I am weary."

The point is this: Although we are the benefactors of modern technology applied to fire and rescue work, it's also true that many modern rescue and firefighting concepts are actually rooted in principles developed by firefighting and military forces that existed long before the time of Christ, Mohammed, and Abraham. Others are the echoes of tactics employed by fire brigades that protected European, Greek, Middle Eastern, and Far Eastern civilizations *thousands of years ago*. Even some aspects of modern fire department organizational structure and management are rooted in ancient fire brigade models like the *Militia Vigilium*.

Consequently, the combination of evolution and tradition are constant factors in shaping the modern fire and rescue services. Over millennia, some of the more helpful principles have been selected and passed on by long-dead fire and rescue authorities in the form of traditions, sometimes supported by written lessons. Succeeding generations have built upon the existing body of knowledge that was their heritage—and, in turn, added their own interpretations and knowledge based on their more modern experiences. If not for this evolutionary process, each succeeding generation of firefighters and rescuers would be hamstrung by the need to learn the basics from experience alone.

This series is a humble attempt to support the evolutionary march of progress with regard to rescue by presenting a certain viewpoint on assessing, planning, and managing USAR technical rescue operations, which hopefully will stimulate conversation and thought about the topics herein. While some of the information may have a certain "West Coast" flavor (owing to the experiences of the author and his colleagues), it's also been deeply influenced by the shared experiences of members of active rescue companies and USAR units across North America, from New York to Seattle, from Albuquerque to British Columbia, from Florida to Phoenix, from L.A. to Oklahoma City. This book series is also based on the experiences of fire and rescue agencies on every continent. Clearly, the fire/rescue services are a truly global calling, and we continue to have much to learn from one another, regardless of where we come from.

It should also be noted that some of the most important information herein came from firefighters and rescuers across North America and other parts of the world who ask the following question every day: "What if?" Without intelligent people asking that question about the potential for unexpected events, the fire and rescue services would be left with lessons gained strictly through experience—much of which comes at a painful cost. It's important to keep asking the question "What if?" because the world is changing fast. Potential rescue problems—many of which would never have been contemplated just a few years ago—are developing with equal speed.

Cutting edge USAR capabilities remain an elusive goal for many fire/rescue agencies. Balanced with the need to properly train and equip its members for the myriad emergencies for which today's emergency services are responsible for managing, these agencies have the desire to develop effective rescue response. In too many places, the ability to implement a formal rescue or USAR program is considered a luxury, something to be done after legal mandates and other local priorities are addressed. It's not uncommon to find that rescue is treated as an afterthought in terms of dedication, training and continuing education, budgeting for equipment, and development of operational guidelines. And who suffers worst from that dynamic? In some cases, it's an innocent victim awaiting help. In other cases it's the firefighters and rescuers themselves, who may suffer for lack of advance rescue and rapid intervention capabilities.

Add to that the difficulty encountered by firefighters who lack the benefit of routine exposure to technical search and rescue (SAR) emergencies, mostly because true *working* rescue emergencies are relatively rare events in many regions and because realistic training is often hard to come by. And when it comes to disaster

SAR, it's even more difficult for local fire departments to develop the levels of experience necessary to effectively manage the multiple demands. Consequently, major rescues and disasters are sometimes marked by some (if not all) of the following:

- misappropriation of emergency resources
- mistaken priorities
- questionable strategies implemented by decision-makers lacking solid rescue experience
- excessively dangerous and time-consuming tactics by inexperienced, ill-trained, and improperly equipped rescuers
- lengthy delays in making the decision to request outside assistance even when the local jurisdictional agency is clearly overwhelmed by the scope or nature of the incident.

Yet despite these difficulties—-which are often regional in nature—the following statements are clearly true: (1) Never before in history has the fire service been better prepared to handle the wide range of technical SAR emergencies that occur every day across the United States; and the nation has never been better prepared to manage disaster SAR operations. (2) Recent international events have demonstrated that many other nations are exceedingly well prepared to manage the consequences of disasters and daily rescue emergencies; and the level of international response to a litany of recent disasters worldwide has never been equaled.

In short, many challenges remain for those charged with managing and conducting technical SAR operations, including firefighters assigned to engines, truck companies, rescue and USAR companies, paramedic squads, and other fire department units. Equal challenges remain for personnel assigned to local and regional rescue teams, as well as state and federal USAR task forces. But two decades of renewed emphasis in SAR has invigorated funding and other support for improvement of these capabilities. For the tailboard firefighter, the rescue company member, and the chief officers, this equates to a time of discovery and improvement in the field of SAR.

For those whose agencies aren't yet committed to developing rescue and USAR programs, this book offers realistic and field-tested guidelines for establishing viable response systems. For those whose agencies aren't capable of providing the most advanced levels of training, this may provide a sort of guide for developing and implementing effective training regimens. And for firefighters who don't benefit from daily exposure to rescue-related emergencies, this book is intended to convey valuable lessons gleaned from other firefighters whose daily role is to manage technical SAR operations.

Rescue in its modern form is in fact a diverse collection of disciplines, often related by common equipment, ruled by common methodology, and conflicted by different regional names and rules. Despite ever-greater attempts at national and international standardization by organizations such as the National Fire Protection Association (NFPA), the National Association for Search and Rescue (NASAR), several state offices of emergency services, and the Occupational Safety and Health Administration (OSHA), the vast majority of fire/rescue agencies still operate with wildly varying standards for training, equipment, and continuing education.

By practically any measure, both daily rescues and disaster SAR operations in North America are now conducted faster, more efficiently, and more safely than ever before. These revolutionary improvements aren't by any means limited to North America; in many parts of world, fire departments and rescue teams safely conclude SAR operations that would have been unthinkable just two decades ago. And revolutionary changes continue to occur at an evermore rapid pace as new technologies, funding sources, and other resources are directed at improving SAR capabilities. We all benefit from this revolution: From the trapped victim who is rescued faster and suffers less, to the rescuer who does his job knowing that he is better prepared to rescue trapped victims faster, with a greater chance of success, and with enhanced personal safety. Intentionally, these are the main goals of this book.

1

Terrorism and Rescue

"During the Crusades, a battle between the crusaders and the Muslim Saracens was announced with much pomp and circumstance. In the Third Crusade, the heralds of Richard the Lionhearted would blow their trumpets and parade a great banner of golden rampant lions. Saladin would begin with the pounding of massive kettledrums, often mounted on a camel, as his bravest emirs would carry forward the green flags of the Prophet (Fig. 1–1).

"In the new war between extremist Muslims and the West, there is no such medieval panoply and fanfare to announce the opening of a battle. Sometimes people become casualties before they even know that a battle has commenced. It was this way the day Osama bin Laden decided to bring his brand of Islamic holy war to the homeland of the United States on September 11, 2001."[1]

Fig. 1–1 September 11, 2001

The 9-11 Terrorist Attacks

At 8:20 A.M. Eastern Standard Time, American Airlines Flight 77 took off from Dulles Airport, en route from Washington D.C. to Los Angeles. There was little indication that the most deadly terrorist plan in modern times had just been set into motion.

Within 2 hours, thousands of innocent victims from more than 80 nations would die, along with hundreds of firefighters and dozens of police officers, who knowingly risked their lives attempting to rescue others. New York would experience the largest collapse of a single occupied structure in the history of the world. Within 1 hour, it would happen again, killing some of the most experienced rescuers and firefighters in the nation. The attack would result in the longest, most complex, and largest USAR operation in history. The changing face of radicalized religious (in this case Islamic) terrorism would be revealed once again, and the nation's fire and rescue services would be confronted by their most deadly challenge. There is every indication that the evolution of modern terrorism could lead to larger, more lethal terrorist attacks in the United States and/or its ally nations.

At 20 minutes before American Flight 77 departed from Dulles, American Flight 11 had departed Boston Airport for Los Angeles International Airport. Meanwhile, United Airlines Flight 175 had just lifted off from Boston, also headed toward Los Angeles. Neither airliner would arrive at its scheduled destination.

United Flight 93, scheduled to depart 22 minutes later,[2] was at that moment being boarded by passengers, including four terrorists who left behind a checklist later discovered by investigators.[3] The instructions on the terrorists' handwritten list mapped out a carefully choreographed plan that read: "Bring ID, clothes, knives, and a last will and testament." Also on the list was a final instruction: "When you board the plane, remember this is a battle in the sake of God, which is worth the whole world and all that is in it."

Once in the air, the crew of American Flight 77 had no way of knowing that air traffic controllers had already lost radio contact with a sister plane, Flight 11, over the skies of Massachusetts just six minutes earlier. FAA controllers watched the radar screen in disbelief as Flight 11 turned right and headed toward New York City. At 6 minutes before Flight 11 barreled into the North Tower of the World Trade Center (WTC), United Flight 175 inexplicably deviated from *its* flight plan, and within moments the plane discontinued radio contact with air traffic control.

At 8:47 A.M., Flight 11 barreled into the North Tower, the first outward sign of a chain of events that in quick succession would cost the lives of thousands of innocent civilians and hundreds of firefighters, emergency medical service (EMS) personnel, and police officers. The Fire Department of New York City (FDNY) mounted a massive response to the high-rise fire that now threatened thousands of people.

At 8:50 A.M., 3 minutes after Flight 11 crashed into the North Tower, air traffic controllers lost radio contact with American Flight 77. The plane had been heading west for 30 minutes, and now the pilot failed to answer urgent radio calls from the controllers. At 8:56, the plane's transponder was silenced. At that point, there was no obvious indication to signal where the plane was ultimately headed. But to air traffic controllers, it was lost, and no one seemed able to raise it on the radio or determine its precise location. And there were no reports of a plane crash in that region.

At 9:03 A.M., the South Tower of the WTC was set aflame by the crash of Flight 175. FDNY commanders knew they were under attack, and U.S. military fighter jets were scrambled to provide air cover over New York City. At 9:24, fighter jets were scrambled from Virginia to locate American Flight 77. Then, at 9:25, air traffic controllers spotted Flight 77 on a reverse track, headed toward Washington D.C.

At about 9:30 A.M., air traffic controllers overheard the first signs that Flight 93 was in trouble. According to the *Los Angeles Times*, the sound of a struggle was transmitted from the cockpit of Flight 93, and a voice in English was heard to say, "Hey, get out of here!"

Then a different voice was transmitted over the radio and (apparently) on the airplane intercom. The new voice said, "This is your captain. There is a bomb on board. Please remain seated. We are returning to the airport."

At that point, Flight 93 reversed its direction and a new flight plan was filed from the cockpit. The airliner was now heading back toward Washington D.C. From cellular phone calls made from the cabin, it is now known that at least one group of passengers eventually hatched a plan to rush the hijackers and regain control of the airplane. It's not known whether any of them knew how to fly a plane, but there's little doubt that they understood that they had suddenly been thrust into the middle of a larger terrorist scheme.

Naturally, many people at the Pentagon were watching or listening to live news media coverage of the New York attack. According to survivors, a group of military personnel and civilian workers were in a conference room in a section of the Pentagon (between structural landmarks known as Wedge 1 and Wedge 2) watching the FDNY facing its worst nightmare at the WTC. Others were no doubt watching or listening to the ongoing WTC disaster. In a tragic twist of fate, many of them were about to become victims of the third terrorist attack that was heading right toward them, even as they watched the consequences of the New York attacks unfold.

According to the *New York Times*, the same air traffic controller who handled Flight 77's takeoff 70 minutes earlier, then spotted an unidentified object on the radar screen racing toward Washington at 9:33. He immediately called Reagan National Airport controllers to report a "fast moving primary target" headed toward the Capital. At 9:36 A.M., air traffic controllers asked a military C-130 cargo plane flying out of nearby Andrews Air Force Base to intercept and identify the unknown target. Soon thereafter, the C-130's crew reported a Boeing 757 moving toward Washington "low and fast." Flight 77 crossed over the Pentagon and made a 360° descending turn to the right and came in nearly level with the ground (Fig. 1–2).

Fig. 1–2a to Fig. 1–2c Security Cameras Record Strike on the Pentagon

A taxi driver on a nearby access road saw the plane strafe the open grass area in front of the Pentagon. The next thing he saw was a street light standard hurtling through the air toward his car. The light pole slammed through the windshield of the taxi, barely missing the driver.

Two firefighters of the Fort Meyer Fire Department (staffing a crash/rescue unit assigned to protect flights landing at the heliport directly in front of the Pentagon) were shocked to see the big jetliner skimming across the grass toward them, just a few feet off the ground. A wing clipped the grass and the plane cartwheeled into the Pentagon, clipping the Fort Meyer unit as it struck the building. The airliner penetrated hundreds of feet into the Pentagon, setting it ablaze and burning the firefighters who still attempted to get water to fight the fire. The crash killed all of Flight 77's passengers instantly and trapped dozens of Pentagon personnel in the burning five-story reinforced concrete building. The time was 9:38 A.M.

At around 10:00 A.M., with Flight 93 headed toward the White House, cellular phone calls from within the plane indicated that a group of passengers had decided to make a dash to the cockpit to retake control of the airliner. One of the last statements transcribed from a passenger's cell phone call was: "Are you ready? OK, let's roll."

Moments later, United Flight 93 crashed into a field in Pennsylvania, effectively ending the sequence of events. The 9-11 attacks left thousands of casualties in their wake and prompted an unprecedented series of local, state, and federal USAR operations that would continue for weeks and months. The attacks also ignited the War on Terrorism, currently in progress.

Case Study 1: Statistics on WTC Attack Survivability Profiles[4]

The following victim/location count was compiled by *USA Today* using available records.

Fatalities:

North Tower:	1,434
South Tower:	599
Firefighters, police, EMS, and other public safety employees:	479
Hijacked jets:	157
Location unknown:	147
Bystanders outside buildings:	10

The following are some interesting facts and statistics about the WTC attacks:

- American Airlines Flight 11, a 767 loaded with 15,000 gals of jet fuel, struck the North Tower at 500 mph, striking the 93rd through 98th floors with a force equal to 480,000 lbs of TNT.
- United Flight 175 struck the South Tower from the 78th to 84th floors. The higher wing cut into the offices of Euro Brokers, and the fuselage went into Fuji Bank offices on the 79th through 82nd floors.
- There were no survivors on the 92nd floor of the North Tower; however, everyone on the 91st floor survived.

- In the South Tower, 99% of the building's occupants below the impact zone survived.
- Just 1 hr, 42 min, and 5 sec elapsed between the first impact and the collapse of the North Tower.
- Nearly everyone who could escape did survive.
- The WTC was half-empty when the attacks occurred. Between 5,000 and 7,000 people were in each tower, according to estimates.
- The WTC includes more than 200 acres of floor space in the two towers.
- Columbia University scientists measured the impacts on seismographs. The North Tower was struck at 8:46 A.M. with an impact that registered 0.9 on the Richter scale (about the size of a small earthquake). The South Tower was hit at 9:02 A.M.

Stairways. One stairway in the South Tower remained serviceable above the impact zone, but few people escaped in it. Stairway A, one of three stairwells in each tower, was open from bottom to top. A few people escaped down these stairs from the 78th floor; many others were trapped by flames and debris, or too badly injured to escape. One person escaped from the 81st floor, two from the 84th floor, and one from the 91st floor. Some people went up the stairwell in search of helicopter rescue from the roof; none were rescued.

The stairwells above the 92nd floor were wrecked, and 1,360 people above that level died with no survivors. Below the impact zone, 72 people died and more than 4,000 survived.

Stairways A and C, on opposite sides of the core of each building, were 44 in. wide. The center stairway (B) was 56 in. wide. In a 44-in. stairway, a person must turn sideways to let another pass in the opposite direction (e.g., firefighters heading up the stairwell). In a 56-in. stairway, two people can pass without turning.

Elevators. The 83-person elevator crew of ACE Elevator was directed to assemble at a command post in the South Tower lobby after the first attack. When the South Tower was hit, the entire crew withdrew from the building, leaving no elevator technicians to assist firefighters with the rescue of dozens of people trapped in elevators that were stuck. There were 99 elevators in each tower and no fewer that 10 were stuck in the North Tower alone. One mechanic of another elevator company responded from another building to assist with the rescue of victims from elevators. He was killed when the towers collapsed. Many were never rescued from stuck elevators.

In the 1993 terrorist bombing of the WTC, ACE Elevator mechanics (some of whom were helicoptered to the rooftops) helped firefighters rescue dozens of victims trapped in stuck elevators, working in thick smoke and other dangerous conditions.

Elevator mechanics helped Oklahoma City firefighters rescue people from Murrah Building elevators after Timothy McVeigh bombed that building in 1995.

"Nobody knows the insides of a high-rise like an elevator mechanic," claims Robert Caporale, editor of the trade magazine *Elevator World*. "They act as guides for firefighters, in addition to working on elevators."

Coming to Grips with New Forms of Terrorism

As of this writing, the civilized world is still coming to grips with the consequences of the 9-11 attacks on New York and Washington D.C. Why is this an important issue for firefighters and rescuers everywhere in the civilized world? The answer is simple: the methods of attack being used by modern terrorists are often intended to inflict massive damage and casualties, sometimes bringing down large iconic structures full of people. Figure 1–3 shows remains of the collapsed section of the Pentagon after firefighters and Federal Emergency Management Agency (FEMA) USAR Task Forces "delayered" the collapse pile and recovered all the victims, a job that required two weeks of round-the-clock operations.

Fig. 1–3 Partial Collapse of the Pentagon, Shown 12 Days Later, after the Search & Rescue Operations Have Concluded

Rescue is an important feature of consequence management because the most deadly terrorist attacks have resulted in USAR disasters that required days, weeks, and even months of intensive and dangerous collapse SAR operations. To be sure, the response to terrorist attacks may also require multi-casualty treatment and transportation systems, hazardous materials (hazmat) response units, firefighting resources, incident management teams, and other fire department-based capabilities. It's true that terrorist groups are capable of using chemical, radiological, and biological materials in attacks on civilians. But it's also true that more than 80% of terrorist attacks around the world involve the use of explosives, which translates to potential structure collapse that can trap victims and require the use of USAR resources.

Today we know that some terrorist groups are intent on committing even larger and more deadly atrocities, including multiple attacks on large structures and public assemblies. Some terrorist groups are considered capable and willing to use small nuclear devices and other methods to create *urban canyons* that would make the WTC collapses look small by comparison.

The nature of modern terrorism makes it clear that rescuers and firefighters will be confronted with even more horrific events in the years to come. Now is the time to prepare for situations that were inconceivable to most first responders just a decade ago. There is a new paradigm in radicalized religious terrorism, particularly Islamic terrorism. But this rabid new brand of radicalism is not limited to Islam. Radicals like Timothy McVeigh, his co-conspirator(s), and others have also committed heinous acts of domestic terrorism. And we can find examples of terrorist acts by both domestic and international perpetrators who profess allegiance to other religions and causes conducting terror campaigns in hot spots around the world.

Despite some assumptions to the contrary, terrorism is not new to the shores of the United States. The use of bombs and other tools of terror to achieve political or economic means or to solve grievances both personal and otherwise has deep roots in our country. As one example of the long history of terrorism in the United States, the Los Angeles Times Building was bombed in 1910 by anarchists, killing 20 people and injuring scores of others. Ten years later, a Wall Street bank owned by J.P. Morgan was bombed, an attack that killed 33 people and sent dozens to hospitals. Mail bombs were sent by terrorists in the United States as early as the 1920s, arriving at the homes of prominent people such as John D. Rockefeller and Oliver Wendell Holmes.[5] The Ku Klux Klan (KKK), the Weathermen Underground, the Symbionese Liberation Army (SLA), and Ted Kaczynski are other examples of domestic terrorists who have used explosive devices in the United States.

The 9-11 attacks and other recent terrorist acts have established new paradigms of cruelty. Modern terrorism is evolving to a state in which the level of barbarism knows no boundaries. Firefighters and rescuers are well advised to take heed and maintain proper vigilance because it's not likely to end any time soon. In fact, all indications point to an escalation of cruelty and cowardice among terrorists and terrorist groups in the coming years before any decline in their activity, and this means that firefighters and rescuers are likely to become targets of secondary or even tertiary attacks. Figure 1–4 shows examples of commonly available books describing how to build and use explosive devices and other tools of the trade for terrorists.

Fig. 1–4 How-to Books

Like it or not, the fire service is today confronted with new paradigms in terrorism that could scarcely have been conceived just a decade ago, beginning with the emergence of domestic and international terrorists with the means to mount attacks with the potential for incredible lethality. They are guided by a *terrorist doctrine* that makes it acceptable (and perhaps even inviting) to inflict huge civilian and rescuer casualties. Some foreign and domestic groups have adopted the *leaderless cell* and *sleeper cell* modes of operation, making it all the more difficult for law enforcement agencies to locate and disarm them

In the new paradigm of terrorism and enhanced response to deal with its consequences, the concept of national and even international response is likely to be become commonplace. In Figure 1–5, a rescue company captain from the Los Angeles County Fire Department is functioning as a FEMA Incident Support Team Safety Officer to help coordinate operations conducted by FEMA USAR Forces, which in turn are in New York to support the operations of FDNY firefighters, with the FDNY chief officers remaining in command of the overall incident and operations. The concept of West Coast firefighters and rescuers operating on the scene of a disaster on the East Coast (and vice-versa) was considered to be beyond the realm of possibility just two decades ago. Now this level of national response is common for major disasters in the United States and some other nations.

Fig. 1–5 Coordinated Efforts

Inevitably, firefighters and rescuers will be confronted with attacks that have evermore lethal results for citizens and rescuers. It didn't take the 9-11 disasters to alert observant fire/rescue professionals to this new fact of life. The signs have been there for all to see since events like the 1993 WTC bombing, the Oklahoma City bombing, and the 1998 bombings of U.S. embassies in east Africa. In 1995, in a published article on USAR operations at the Oklahoma City bombing, this author wrote:

> *The Oklahoma City bombing, the 1993 WTC bombing, and other recent terrorist incidents dramatically demonstrated that urban terrorism has arrived on the shores of the United States. Practically any city or town may be the target of terrorists with a wide variety of agendas. These developments should be of grave concern to the fire service, whose members are generally first on scene and, therefore, extremely vulnerable to secondary attack (a common terrorist tactic more commonly seen in other parts of the world), secondary collapse due to structural instability, and other life hazards. The potential for terrorism raises the specter of future incidents in which massive damage is inflicted upon large, multi-story buildings.*
>
> *This is not the first time United States governmental facilities have been targeted by terrorists. The Oklahoma disaster was preceded in the 1980's by equally horrendous bombings of the Marine compound and the U.S. Embassy in Beirut. Unfortunately, there were no FEMA USAR task forces available to help during those incidents. Today, United States embassies and other government buildings in other nations are under constant threat of terrorist bombings. The possibility of further domestic terrorism must be taken seriously by the Fire Service.*[6]

For firefighters and rescuers, the drive toward evermore spectacular and lethal terrorist attacks means that yet another new paradigm has been reached. Future attacks by some groups will be designed to leave more innocent victims trapped in collapsed buildings or stranded by uncontrollable fires that leave firefighters no choice but to make a desperate last attempt. Our response to some future attacks will be complicated by multi-hazard devices like dirty bombs that spread radioactive contamination after being dispersed by conventional explosives powerful enough to collapse large buildings. Some attacks will trap people in other predicaments that will require extensive rescue operations to save lives and recover the dead in a respectful manner.

The Importance of Effective Consequence Management

Because terrorist attacks of the new paradigm may require massive, multi-agency USAR response to locate and remove multitudes of missing and trapped victims, local fire/rescue resources may be overwhelmed. Large segments of the nation's state and federal USAR assets may be required to deal with the consequences of one large attack or multiple, simultaneous attacks.

The 9-11 attacks caused a nationwide shutdown of commercial air traffic and a slowdown in the military-based aerial transportation of some FEMA USAR task forces traveling west to east (another new paradigm).[7] This situation demonstrated the need to consider alternate plans, for example, staging international USAR teams within the United States to augment remaining USAR resources during the aftermath of a major attack. This would be a first for a nation accustomed to managing disasters of every size and shape without the need for outside assistance. For the first time, a terrorist attack (or a series of them) could cause the deployment of international USAR teams—and possibly other emergency resources—within the United States (yet another new paradigm).

Because both domestic and international terrorists have demonstrated their increasing willingness to target first responders as well as citizens, incident commanders (ICs) responding to a possible attack must consider tactics and strategies borne of decades-long experience with such attacks in places like Ireland, Scotland, England, Israel, Spain, and elsewhere. Included in the incident action plan (IAP) for a possible terrorist attack must be some contingency for rapid intervention operations to locate and rescue firefighters and other first responders who become lost, trapped, or injured by the detonation of a secondary device.

Effective management of rescue-related consequences of modern terrorism is particularly important for reasons that may not be immediately evident to the first responder. We all know that the primary mission of fire departments responding to a terrorist attack is to save lives. This is accomplished by taking the necessary steps to ensure a reasonable level of training, adequate personal protective equipment (PPE), and a well thought-out plan of action. These prerequisites help ensure firefighters can locate and rescue victims without delay, that they can extinguish fires that threaten trapped victims, and they can treat and remove the injured. This is, in its most basic form, consequence management through effective rescue in the aftermath of a terrorist attack.

But there is another sometimes little understood reason why effective fire/rescue response is so important. The reason is that the *goal* of terrorists is to create havoc and a sense of demoralization and insecurity that changes the political reality by destabilizing the government. This seemingly chaotic condition can lead the populace to question the government's ability to protect public safety. Consequently, if the fire/rescue response to a terrorist attack is (or appears to be) haphazard and ineffective, the effect on public morale may be devastating and lead to even darker consequences. A chaotic and unreasoned emergency response would essentially play into the hands of the terrorists and give them a sort of double victory that we must be prepared to deny them.

Conversely, if the fire/rescue response is massive, rapid, and orderly, characterized by an unbowing and resolute spirit of determination and cooperation among firefighters and other emergency responders, and if it results in timely control of the incident, the terrorists are denied a victory. True, they may have inflicted damage, but they are denied the relative and temporary advantage of proving they can cause panic, social chaos, and demoralization. This is one of the most powerful lessons learned from the response to terrorist acts like the 1993 WTC bombing, the Oklahoma City bombing, and the 9-11 attacks of 2001. When the spirit and dedication to mission of the fire department and other public safety agencies remain unbroken, and when the firefighters and rescuers continue to demonstrate a spirit of will to do their jobs no matter what comes their way, it sets the stage for the public to do the same. In Figure 1–6, This assemblage of civilian heavy equipment and operators in support of the local fire department and FEMA USAR Task Forces is emblematic of the new paradigm in rescue and response to terrorism.

Fig. 1–6 A Heavy Equipment Staging Area near the Pentagon Collapse

The best way to ensure that firefighters and rescuers can meet this goal of denying victory to terrorists is to be aware of potential hazards and be prepared to manage them. With modern terrorism taking its inevitable course toward increasingly spectacular and deadly attacks perpetrated by both foreign and domestic groups and individuals, fire/rescue professionals must consider the scope of their mission with a far wider perspective than ever before in history.

We are confronted with a new paradigm for the foreseeable future: the emergence of domestic and international terrorism that will continue to raise its head from time to time, perhaps with more lethal results for citizens and rescuers. This is a new fact of life that has been obvious to many fire/rescue professionals since events like the 1993 WTC bombing, the Oklahoma City bombing, and the 1998 bombings of U.S. embassies in east Africa.[8] As the primary responders to terrorist attacks, firefighters, rescue team members, and EMS personnel must look far beyond the obvious in planning for (and responding to) acts of foreign and domestic terror. Today, *thinking outside the box* is a mere starting point for progressive fire/rescue organizations.

The 9-11 attacks have emphasized the need for fire service managers, public policy makers, USAR team members, and tailboard firefighters who can think outside the box. These members must consider how various forces within the fire service and within their own agencies can be brought to bear against the consequences of modern terrorism.

Many of the necessary forces have already been developed to address other threats such as earthquakes, hazmat emergencies, urban conflagrations, and mass casualty disasters. Fortunately, we have the ability to adopt these response systems to create an integrated terrorism response system. But it takes time, funding, and a commitment to that vision. We don't have much time, the funding is still lagging, and we need to make sure the commitment to preparing for terrorism does not waver.

The historic progression of terrorism tells us that we as firefighters must be ready for lethal and spectacular attacks that threaten to radically change the way we live and work. Aside from preventing attacks (the first choice and primary role of law enforcement), the best that we as firefighters can do to defeat terrorism is to prevent the chaos and disruption that can follow. We can ensure an effective, rapid, and massive response to terrorist events that will help the government maintain control of the scene and thereby ensure that trapped patients will be located, extracted, and treated, and that loss of first responders will be prevented.

It's necessary to ensure that the lawmakers and those who control funding understand the importance of supporting firefighters and other first responders to effectively manage the consequences of terrorism. To do this, we must be much more proactive in planning, training, and preparing for terrorism.

Today, it's important to learn about terrorism response from those who have faced it much longer than we, including the fire services of Israel, Great Britain, Turkey, Greece, South Africa, and South America. Unfortunately, this nation has experienced events that no other fire service on earth has seen firsthand. The 9-11 attacks resulted in the largest and most devastating structure collapse in the history of mankind. Consequently, we too have a responsibility to share what we've learned with firefighters in our own country and those from other nations (Fig. 1–7).

Fig. 1–7 FEMA USAR Task Forces Conducting SAR Operations at the Pentagon Collapse

A higher index of suspicion: the new face of terrorism

From the events of 9-11 and beyond, it becomes clear that the fire/rescue services need to start viewing airline crashes and other unusual emergencies with a higher index of suspicion. These types of incidents may be acts of terrorism and may include the potential for secondary devices or secondary attacks. We must give serious consideration to scenarios that might have been dismissed as preposterous just a decade ago.

Following the 9-11 attacks, many people seemed inclined to focus exclusively on the threat of terrorism emanating from other nations and particularly from the Middle East. And for good reason: it's increasingly clear that active international terrorist cells are operating within the borders of the United States and other civilized nations. A recent congressional investigation into the 9-11 attacks found that between 70,000 and 120,000 terrorists were trained by al Qaeda in "the skills and arts of terrorism." A significant number of al Qaeda operatives are in the United States, reported Senator Rob Graham of the Senate Select Committee on Intelligence, which conducted the investigation.[9]

While this is a valid concern, it would be a colossal mistake (and a sign of serious naïveté) for fire/rescue personnel to assume that the threat of domestic terrorism has somehow been eliminated. Lest anyone be misguided, there are plenty of domestic terrorist groups consisting of people born and raised in North America operating within the borders of Western nations. Many of these groups are fully intent on bringing down the governments of Western nations.

In the United States, this includes racist groups attempting to establish an *all-white* nation in the Northwest, far-right and far-left groups intent on replacing the current system of governance with something more to their liking, and religious groups whose goal is to replace secular government with authority based on religious commandments and protocols. Members of some of these groups have made it clear they are perfectly willing to commit atrocities to achieve their goals.

Based on the current progression of terrorism, it's obvious that the future will bring even more new paradigms in terrorism in North America. These might include suicide bombings or the use of dirty bombs and other nuclear devices. Secondary/tertiary explosive devices might also be used to wipe out fire/rescue professionals. These attacks would likely be aimed at provoking the government to respond in such a way as to cause a backlash leading to martial law and a possible civil war between citizens and the government.

It's even possible that we may see unholy (and improbable) alliances of far-right Christian extremists collaborating in various nations with radical Jewish or Islamic terrorists to attack governments and targeted groups for whom they share lethal animosity. Such examples of cooperation between individuals and groups from opposite ends of the religious and political spectrum may have seemed preposterous just a few years ago, but one need only consider events listed in the daily newspapers to realize that the most improbable situations are sometimes possible when it comes to terrorism.

Consider, for example, a situation unfolding in Germany at the time of this book's writing. In 2002, German officials indicated they had uncovered evidence that certain German Nazi extremist groups were showing increasing support for the activities of radical Muslim terrorist groups and their collaborators and fund-raisers. It's general knowledge that the 9-11 attacks were cheered equally by many neo-Nazis and Islamic extremists who believe that the world is controlled by Jewish groups whose will is enforced by the U.S. military.

As reported in the *Los Angeles Times* on January 3, 2003, Udo Voigt, chairman of Germany's far-right National Democratic Party, attended speeches by leaders of Hizbut-Tahrir, an Islamic organization identified as dangerous by German authorities. Voigt and his cronies also support Ahmed Huber, a Swiss far-right extremist who is calling for closer ties with Muslim radicals and is suspected to be linked to Al Qaeda.

In a 2002 interview on German television, Voigt told a university crowd, "I think I speak in the name of all German nationalists when I say, if it comes to a great clash (between civilizations), we will not stand at the side of America."[10]

"The common ground they (far-right Christian groups and Islamic radicals) share is deep on two issues," said one Western diplomat in an interview with the *Los Angeles Times*. "They cannot tolerate the existence of Israel, and they share a conspiracy theory that the United States wants to control the Middle East and the world's energy supply. It's a very paranoid world view, but they share it deeply." Based on these and other examples, only the very naïve among us would discount the potential for dangerous sects of the Nazis and radical Islamists to find common ground in their animosity toward the United States and other Western powers that promote freedom and democracy. No one should be surprised if they find it pragmatic to consolidate their efforts to attack Western targets to accomplish their sometimes-divergent goals.

Then back in the United States there is the case of the Montana Militia, whose leaders were arrested in February 2002 and charged with planning to kill as many judges, prosecutors, police officers, and firefighters as possible to provoke the state to activate the National Guard. The group, billing itself as Project Seven, amassed 30,000 rounds of ammunition and hundreds of weapons, as well as, intelligence on the personal lives and comings and goings of police officers, firefighters, prosecutors, and judges (and their families), who were on hit lists found in the leaders' homes.

According to investigators, Project Seven members planned to attack and kill national guardsmen and other law enforcement officials who would be dispatched to Montana in response to the murders. Documents seized during raids indicated that the group ultimately intended to ignite a domestic war that would topple the federal and state governments.

Groups like Project Seven and the individuals who form their membership can no longer be considered isolated cases or anomalies unlikely to replicate. The fact is, domestic terrorist groups have sprouted up across many parts of the United States, and they are fully prepared to take many innocent lives. The Oklahoma City bombing proved that beyond any doubt. Some of the domestic terrorist groups may lay low for a while after events like 9-11. But inevitably, some of these groups will surface to strike out against the government when things quiet down. Or, they may establish yet another new paradigm by conducting terrorism when the government appears most vulnerable, such as in the immediate aftermath of major attacks by foreign groups.

The problem with small bombs

Sometimes it's not a group but an individual that presents the most danger. One example is Ted Kaczynski, the so-called Unabomber. Kaczynski favored the use of small explosive devices that he could mail to targeted individuals, ignoring of course the fact that others might be killed opening the packages or that bystanders might also be killed. There are many other makers of small bombs out there, and they represent a tremendous danger to their intended victims and to fire/rescue personnel (Fig. 1–8).

Fig. 1–8 Day 14 at WTC: Into the Night, Rescuers Worked Round the Clock for Weeks Attempting to Locate All the Victims

According to the U.S. State Department and Justice Department, bombings are by far the most common method used in terrorist attacks. Bombs are being used with greater frequency in the violent settling of personal conflicts in the home and at work. In the United States, the most frequently used explosives are pipe bombs and other small explosive devices. In California alone, an average of more than 400 pipe bomb incidents are reported each year. Since 1990, the incidence of pipe bombings has increased by more than 50% in California.

Pipe bombs and other *simple* explosive devices are of special concern because they can be built by people who do not have a great deal of technical expertise. The materials required to produce these bombs are readily available to the public. Even crudely made explosive devices can be used with deadly accuracy. Powerful bombs may be hidden in mailboxes, small packages, and (as we will see) normal household and recreational items for which suspicion would normally be low (Fig. 1–9).

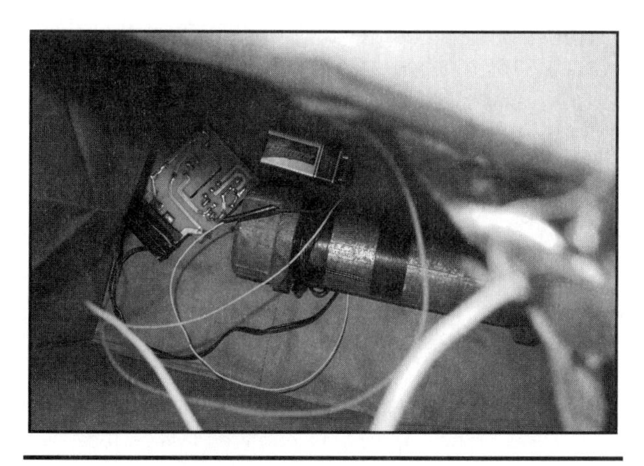

Fig. 1–9 An Example of a Small Explosive Device

A case in point occurred on April 29, 1997. On that day, County of Los Angeles Fire Department (LACoFD) personnel were dispatched to rescue an unconscious woman from the beach at the base of a rocky coastal cliff. They found themselves face to face with a powerful bomb in disguise. This author was the duty captain of the USAR company assigned to the incident. Afterwards, I wrote an article in *Fire Engineering*[11] describing how the firefighters discovered a homemade bomb concealed in an Igloo® cooler next to a waterlogged, unconscious victim they were attempting to rescue (Case Study 1 and Fig. 1–10).

Fig. 1–10 Print from the Article "A Close Call" in *Fire Engineering*

The bomb was discovered only after unsuspecting L.A. County firefighters picked up the cooler to make room to package and treat the victim, who was also about to become a suspect/subject. As they were preparing to move the victim away from the waves, one firefighter noticed wires protruding from beneath the lid; he alerted the others.

Adding to the mystery, the firefighters also discovered a loaded gun in the back pocket of the victim's seawater-soaked jeans, as well as a cache of black powder and a flare gun stashed in the nearby rocks. She eventually slipped into cardiac arrest and was resuscitated by firefighter/paramedics in the back of an LACoFD helicopter. The helicopter was en route to the trauma center at Harbor General Hospital, the actual site of the fictional Rampart Base Station of the 1970s television show *Emergency* with Randolph Mantooth and Kevin Tighe as LACoFD firefighter/paramedics.

It was later determined that the victim was not really a woman at all but rather a male former military bomb expert in the process of having a sex change. The victim had apparently built a substantial bomb in the cooler for purposes that have yet to be divined. This patient later was charged with threatening to bomb the very hospital that helped save her life.

Although it more closely resembles a movie plot than an actual emergency, this technical rescue operation-cum-mystery is archetypal of a seemingly benign situation whose actual hazards are quite different from its first appearances. It was a seemingly *innocent* rescue that very nearly turned deadly for unsuspecting firefighters and emergency room (ER) staff.

Perhaps the key word here is *unsuspecting*. Until the moment they noticed multi-colored wires dangling beneath the closed lid of the cooler, the firefighters (including this author) assumed that the principal hazards of the incident were the high cliffs and pounding surf that hemmed us in. We assumed that the primary problems would be treating and packaging a patient slipping into full arrest after apparently being washed up on the beach and conducting a rope rescue operation or a helicopter extraction and transportation to a trauma center. None of us—least of all the firefighter who moved the cooler to get better access to the victim—suspected we might be standing next to a powerful bomb. Hours later, this bomb tossed rocks into the air when it was detonated under controlled conditions by sheriff's bomb squad members (Fig. 1–11).

Fig. 1–11 An Example of Small Hidden Explosives

The lesson here is that firefighters and other rescuers would be well served by employing a higher index of suspicion when confronted with situations that even hint at the unusual. And we should certainly develop a more acute awareness of our physical, social, political, and emergency surroundings, remaining aware of conditions that are ripe with the possibility of multiple forms of terrorist attacks from both domestic and international sources. If we fail to do this, we will surely lose more firefighters and rescuers in the coming years. As this author surmised in *Fire Engineering* about the so-called beach bomb incident:

> *A nagging concern for modern firefighters is the increasing frequency with which they are confronted by explosive devices and other dangerous weapons during incidents that (at first) appear routine. Today's well informed firefighter knows that legitimate concern for the presence of bombs at otherwise benign emergency scenes is no longer limited to places like Beirut, London, and the Middle East. Bombings and bomb threats now constitute a major threat to everyday life in many parts of the United States.*
>
> *Firefighters and other public safety personnel across the nation can expect to encounter more bombs, booby-traps, and other dangerous weapons for which there may be no logical warning signs. Recent examples of this abound, including New York City firefighters finding booby-trapped apartments during fireground operations, Los Angeles firefighters who encountered a large fuel/ammonium nitrate bomb in the back of a burning truck they had just extinguished, Kansas City firefighters who were killed when an intentionally set fire caused the detonation of explosives at a construction site, and Bakersfield (California) firefighters who narrowly escaped death when a bomb exploded inside a burning passenger van. In short, there is no end to the combination of otherwise benign circumstances in which firefighters may be confronted by explosive devices.*

Five years after the beach bomb incident and five and one-half years after the Oklahoma City bombing, uncomfortable supposition about firefighters becoming victims of future terrorist acts became tragic reality when international terrorists struck on September 11, 2001. As predicted, firefighters (including some who had participated in SAR operations in Oklahoma City and the 1993 bombing of the WTC) were first on the scene. Consequently, they were among the most vulnerable to secondary attack, as well as the citizens who occupied the towers at the time of the attacks.

Case Study 2: Close Call with an Igloo® Bomb

At 8:35 A.M. on April 29, 1997, the LACoFD command-and-control facility received a 911 call reporting a woman over the side of a 300-ft coastal cliff near the westernmost point of the Palos Verdes Peninsula, a prominent landmark in southern California. The peninsula is actually a diamond-shaped mountain that juts out to sea from the L.A. coastline, abruptly cutting off the flat and sandy shores of nearby Redondo Beach. The peninsula was created by a large earthquake fault. Where it meets the ocean, it is outlined by cliffs ranging in height from 150 to 300 ft. Its extremely rugged and boulder-strewn beaches are pounded by treacherous surf that has caused many harrowing rescues over the years.

Companies from the LACoFD and the department's lifeguard division are called upon to perform many technical rescue operations each year along this stretch of coastline. It is not uncommon for the department to handle a dozen or more technical rescues in a month's time there, including cliff hangers, boat wrecks, falls from cliffs, scuba accidents, hang glider accidents, etc. On the morning in question, it was not immediately clear whether the victim was trapped somewhere on the face of the cliff (a common scenario in this section of coastline) or had fallen all the way to the rocky beach below.

The rough terrain in the area of the incident frequently necessitates the use of technical rope rescue systems, helicopters, or lifeguard boats, so it has become standard LACoFD practice to dispatch a *coastal rescue* assignment to these incidents. The first alarm assignment for the woman over-the-side included engine 53, engine 2, Paramedic Squad 2, USAR truck 106 (a USAR-trained ladder company), and USAR-1 (the LACoFD's central USAR company before USAR Task Force 103, which superseded it in 2000). Also called out was copter 18, a Bell 412 fire/rescue helicopter staffed by a pilot and two firefighter/paramedics and equipped with a hoist rescue system. Dive 1, a dive rescue team from the LACoFD's lifeguard division and Baywatch Redondo were also assigned, as were a lifeguard division rescue boat from nearby Redondo Beach Harbor and Battalion 14.

At the time of the incident, copter 18 and USAR-1 were in quarters at the LACoFD Special Operations Division headquarters, a large heliport and office facility that housed eight fire/rescue helicopters, the Technical Operations Section, the USAR offices, and the department's FEMA USAR task force. Because USAR-1 and the air squads were often dispatched to the same incidents, it was not uncommon for the crew of USAR-1 to fly on the helicopters to provide timely technical support and backup for the air squad crew during helicopter hoist rescues.

As the on-duty captain of USAR-1 that day, this author employed that strategy for the incident in question. It so happened that a local ER physician was also flying on copter 18 as part of a program to better acquaint local ER personnel with the unique conditions encountered by air squad paramedics. As we lifted off and headed across the sprawling L.A. Basin toward the coast, we listened for the size-up report from engine 53.

Routine rescue starts to go sour

Upon arrival, the captain of engine 53 established command and radioed his size-up report. He gave the name Coast Incident to the emergency and reported that the crew of engine 53 was about to descend the cliff to assess the situation below it. He noted that a female victim had been spotted in the rocks near the base of the cliff, near the surf line. At first, it occurred to me that this might be a fairly routine technical rescue operation for this area of the county, something we had seen and handled over and over during the past years.

Upon arrival, Battalion 14 assumed command and assigned the captain of engine 53 head of operations. Captain Gene Horner of truck 106 was designated the rescue group leader, and Squad 2 (staffed by two firefighter-paramedics) was designated the medical group. As they climbed down the cliff, engine 53's crew angled toward the woman lying on the rocks directly above the surf line. After hiking across the rock-strewn beach, they found her unconscious and unresponsive and therefore unable to provide information about the circumstances of her predicament.

As they began their medical exam, the firefighters noticed that the victim was wet, cold, and had severe abrasions across her body. She appeared to have been submerged in the water and dashed on the rocks by the waves. This led the crew of engine 53 to believe that she may have washed up on the beach, perhaps the victim of a capsized boat or some other type of marine accident.

Concerned that there might be additional victims somewhere off the coast, engine 53's captain radioed his follow-up report to the coast IC and requested the assignment of Baywatch and Dive 1 to begin a search for additional victims. He also asked that the coast IC request the response of the U.S. Coast Guard to assist with the ocean search.

By this time, pilot Gary Lineberry was flying copter 18 over the ocean, approximately one-quarter mile from shore, paralleling the beach just north of Venice and below the takeoff pattern of L.A. International Airport. Sitting in the cabin of the copter, we listened to the radio traffic from the on-scene units and discussed (via the copter's intercom system) various contingencies for rescuing any victims in the water, as well as, how best to extract the first victim from the beach. We decided that if we encountered any victims in the open ocean en route to the incident, we would use the department's marine disaster helicopter rescue swimmer insertion evolution to deploy two rescue swimmers into the ocean. This is the same operation we employ to conduct rescues in the event of a plane down or a capsized boat in the open sea. Any victims would be hoisted aboard using the copter's rescue hoist with one rescue swimmer accompanying each victim.

Back on the ground, the crew of engine 53—now joined by firefighter/paramedics from Squad 2—performed a head-to-toe primary survey and began treating the 190-lb victim for hypothermia, near-drowning, and injuries. She was wearing jeans, but her top had been partially ripped off. The rescuers checked her legs and pelvis for signs of fracture, but left her jeans on to provide protection from the rocks and to help conserve her body temperature.

The firefighters noticed several items scattered in the rocks near the victim, including a 5-gal Igloo® cooler and some articles of clothing. Initially, there was no indication that these items had any connection to the victim; it appeared that they might have been left by other beach goers the previous day or washed ashore from some other source. And since there was no outward indication of danger, they proceeded to stabilize the patient as per normal procedures in preparation for a helicopter extraction.

One firefighter-paramedic from Squad 2 picked up the cooler and moved it out of the way to clear enough space to place the patient on a backboard. Moments later, copter 18 approached the scene and requested an assignment from the IC. Copter 18 and USAR-1 were directed to report to the rescue group leader.

Something amiss

Meanwhile, Squad 2 paramedics were becoming concerned about the patient's deteriorating condition. Her pulse was extremely weak, and they had difficulty determining a blood pressure. The base station ordered large-bore IVs, which they proceeded to start. The patient remained deeply unconscious and barely responded to painful stimulus. Her condition appeared consistent with someone who may have been washed off the rocks and suffered a near drowning, or someone who washed up on the beach after a boating accident. Both are common incidents there, and both scenarios seemed equally plausible. The paramedics worked quickly, trying to stabilize her before the copter landed.

As the firefighters maneuvered around her, they took the opportunity to take a closer look at the other items scattered around the patient. What they discovered quickly changed the course of the incident. To their surprise, they found a loaded pistol wedged between two rocks and a flare gun lying on a nearby rock. Then one of the firefighters found several opened cans of black powder. They immediately recognized that this might not be a simple near-drowning case anymore but perhaps something altogether different. However, with the woman in a state of unconsciousness and no other witnesses, there was no way to determine exactly what led her to be found on the rocks. There was no way to ascertain whether there were additional victims or if there were other unknown dangers. Within moments, they would discover that they had been working in direct proximity to lethal danger since the time of their arrival on the beach.

Copter 18 was directed by the rescue group leader to evaluate the scene to determine whether a hoist operation or a landing on the rocks would be most appropriate to remove the victim from the beach. Pilot Lineberry determined that he could safely land on the rocks. He carefully hovered into position and set the copter down on a set of boulders that provided a solid platform for the skids to support its weight.

We've got a bomb here

We climbed out onto the rocks with the evacuation stretcher and medical gear. Firefighter/paramedic Fred Findlay and I reached the location of the victim just moments after the other firefighters found the guns and black powder. USAR-1 engineer Brian LeFave and Firefighter/paramedic Dana Rickman followed, carrying the stretcher. Captain Horner began briefing me on the situation and needs. He was interrupted by Findlay, who tapped on my shoulder and pointed at the cooler on a rock at our feet. I glanced down and saw the focus of his concern; several wires resembling primer cord, or fuses, protruded from beneath the lid of the cooler.

I looked at Findlay and we both mouthed the same word simultaneously: "Bomb!" Two feet to my right was the firefighter who had moments earlier moved the cooler away from the patient. The look on his face was not unlike that of someone who had narrowly missed being struck by a falling rock. It wasn't quite shock but the realization that he had dodged a bullet, and if circumstances had been different, he and some of his colleagues might have been splattered across the rocks.

Findlay strongly suggested that we expedite packaging the patient on a Miller board and move her to a safe location away from the cooler. Horner directed his personnel to work with Findlay to get the woman packaged and moved. Findlay was concerned (with good reason) that one of us might accidentally kick the cooler if we tried to carry the patient anywhere near it, so Horner picked out a route that would lead us away from the cooler.

Naturally, we realized that it might not be a bomb after all. For starters, an Igloo® cooler wasn't exactly the sort of container we expected. Perhaps there was another reasonable explanation for the wires protruding; making that determination was the job of the bomb squad. Later it occurred to me that an innocent child walking along the beach that morning might have been inclined out of curiosity to open or move the cooler.

There was no question in our minds about doing anything with the cooler except leaving it and denying entry into the area. Obviously someone had been motivated enough to create something that at least resembled a homemade bomb. Anyone who would go to that trouble might also have the inclination and know-how to build a real one. We agreed that if the cooler was in fact loaded with explosive material and an effective detonating device, it might be a powerful bomb. The first question was: What distance from the cooler would provide sufficient shielding from such a blast? In reality, that question was academic because we were on a narrow strip of beach, hemmed in by cliffs and the ocean.

The next questions that arose in my mind were: Who would put a bomb on a beach next to an unconscious woman? What was the connection with the pistol and flare gun? Was our victim the perpetrator, perhaps as part of a suicide attempt? Was it a drug deal gone bad with the flare gun used to signal a boat coming into the cove? Or, was the woman the victim of some type of murder plot? Was there a boat foundering in the ocean with other victims aboard? Could someone have blown up a boat on the water, not suspecting that this victim might wash up on the beach? At first, these scenarios began to seem a bit far-fetched to me. But then I remembered that I was in L.A., where far-fetched is a decidedly relative term.

It occurred to us that there might be some sort of shock-sensitive detonator planted in the cooler. But if that was the case, why hadn't the bomb exploded when the firefighter from Squad 2 picked it up? It was possible that the device might be on some type of timer. But what type? And if so, was it set to go off at a particular hour and minute, or in a certain interval of time after it was moved (i.e., by an unsuspecting citizen or firefighter)?

Then, one more problem presented itself. The cooler was perched on a rock that was quite close to the waterline, and the tide was now coming in. The waves were steadily pounding closer and closer on nearby rocks. Findlay was right; it was time to move away from the cooler and get everyone else off the beach.

By this time, two sheriff's deputies had climbed down and were watching the scene from the base of the cliff, perplexed by our sudden attention to the cooler and our hasty retreat. When I called over to them, "We've got a bomb here," they seemed to think I was kidding. When they realized I was serious, they carefully approached the cooler at a safe distance to look for themselves. They came back a little faster this time and one of them said, "You're right, it looks like a bomb." With that, they radioed their dispatcher to respond the L.A. County Sheriff's Department's bomb squad.

Meanwhile, the IC was informed of the presence of a possible bomb on the beach, and the sheriff's deputies hurriedly waved off several citizen bystanders in the vicinity and established a wide exclusion zone.

Situation becomes critical for victim

The woman was quickly packaged and carried away toward the waiting helicopter. Pilot Lineberry had kept the blades turning in anticipation of a quick turnaround time. As the firefighters struggled to carry her over the boulders, she suddenly vomited copious amounts of seawater. We found a safe place to set her down to ensure a patent airway. The woman's rate of respirations quickly decreased. The paramedics, assisted by the physician, scrambled to suction and intubate her.

Efforts to intubate the patient included several attempts by the physician, lying on the rocks, to get a visual on her vocal cords with the laryngoscope. Then, the oscilloscope went flat and there was no pulse. Cardiopulmonary resuscitation (CPR) was begun and the paramedics followed full-arrest protocol as the woman was quickly moved to the copter. The woman's weight created some difficulty with moving her across the boulders, and at one point, a firefighter slipped on the rocks and went down beneath the stretcher. The others quickly lifted her off his leg and slid the stretcher into copter 18.

As copter 18 flew toward Harbor General Hospital, the closest trauma center, aggressive resuscitation efforts continued unabated. Findlay and Rickman were busy administering IV drugs in an effort to regain the woman's cardiac rhythm. Within 10 minutes, Lineberry set the Bell 412 down on the helipad, and the patient was transferred to a waiting ambulance for the trip across the parking lot to the ER.

As Harbor General's ER team took over treatment, copter 18 paramedics informed them of the unusual factors at the scene. They emphasized that because of the sequence of events, they had not had a chance to remove her jeans to check for possible injuries from gunshot wounds, etc. The attending physician assured us that they would thoroughly check the patient's back.

Later we learned that when they removed the woman's jeans to check for hidden wounds, ER personnel discovered a loaded Derringer pistol in the cocked position in her back pocket. Then they discovered that she was not exactly a woman; she was a man undergoing a prolonged sex change procedure to become a woman.

As a result of aggressive treatment in the field and in the ER, personnel at Harbor General Hospital were able to resuscitate the subject, who lived. The subject was committed for a time to the hospital where she apparently made terrorist threats against the facility. As of this writing, the subject has reportedly refused to discuss the details leading up to being found on the beach that morning.

Bomb blast

The sheriff's department's bomb squad responded to the scene and conducted a detailed survey of the cooler and the rest of the scene. They determined that indeed the cooler appeared to be some sort of bomb or an attempt to make the cooler look like a bomb. Late in the afternoon, the cooler was detonated on the beach. The explosion was powerful and would have caused grave injuries or death to firefighters and other bystanders.

Incident conclusions

The ocean search conducted by Baywatch, Dive 1, and the U.S. Coast Guard proved negative for additional victims or vessels. Sheriff's department officials provided the following information about the incident:

- The bomb was produced and placed in the rocks by the subject.
- The subject was apparently psychotic.
- Cans of black powder were jammed into the cooler with a fuse system providing detonation capability.
- Upon detonation, the cooler itself would have acted to contain the initial blast for an instant. This in turn would have caused the cooler to rupture into shrapnel, which would have caused grievous injuries to anyone standing nearby.
- The subject was reported to have a military background associated with the use or disposal of explosives.
- The subject survived the ordeal, but the hospital will not release additional medical data, as is common in such cases.
- Hospital personnel reported that the subject made threats to blow up a section of the hospital because she was held there against her will during treatment.
- It is not known whether the subject had any association with any terrorist organization.

Lessons

First responders should carefully reevaluate the manner in which they approach seemingly innocent situations where there may be hidden danger of explosive devices or other conditions/substances immediately detrimental to life and health. In this case, there was little forewarning of the danger actually present on the scene. The first-responding firefighters approached the incident with a reasonable assumption that they were responding to some form of technical rescue. Initially, they had no reason to believe that a bomb or weapons might be present.

Upon discovery of the possible presence of a bomb at an emergency scene, standard precautions should be implemented to mitigate the possibility of a detonation. In this case, the incident had progressed for nearly 30 min before there was any indication that a bomb might be present. Once the bomb was discovered, personnel reacted appropriately and changed gears to protect the patient, firefighters, police officers, and the public from the newly identified hazard.

During our critiques of the incident, we agreed that we should have refrained from using fire department radios in the vicinity of the cooler once the wires were observed protruding from the cooler. Radio silence is standard procedure for bomb threat incidents to reduce the threat of activating a radio-controlled detonator. However, the need for radio silence simply didn't occur to us immediately because of the initial confusion about what we were dealing with and because the sequence of events did not follow the typical profile of an incident in which one would suspect the presence of a bomb.

Once weapons (especially loaded ones) or other suspicious items are discovered at the scene of an emergency like this, the index of suspicion for other unseen dangers should be immediately raised. In this case, it is fortunate that the cocked gun (in the patient's back pocket) did not fire during efforts to load, treat, and transport the patient. There is no telling where the bullet from such a shot might have gone; however, it might have proved catastrophic had it occurred during the helicopter flight to the trauma center.

Helicopter flight and landing should be restricted in the vicinity of a suspected bomb. The rotor wash of a helicopter, with its attending atmospheric pressure changes and violent wind turbulence, might be sufficient to set off some bombs. As was proven by the helicopter crash that killed actor Vic Morrow and two child actors during the filming of the movie *Twilight Zone*, even minor bomb explosions in the vicinity of a helicopter in flight may cause it to crash with catastrophic results. In this case, the rotor wash from the Bell 412 helicopter might have been sufficient to blow the cooler across the rocks if air squad crewmembers had not had the presence of mind to keep copter 18 at a distance once the protruding wires were discovered.

There is a need to reevaluate the general manner in which the fire service approaches incidents involving bomb threats, explosions of unknown origin, cases where victims are adversely affected by unknown substances, situations where victims are found under suspicious conditions, and other incidents where the index of suspicion for explosives or other lethal hazards is high. It has been said that no emergency is truly routine. The phrase might be overused in some cases, but this incident was proof that some seemingly *routine* emergencies can quickly turn deadly for firefighters and other public safety personnel who let down their guards.

Thinking the Unthinkable

Francis Brannigan recently wrote in *Fire Engineering*,[12] "We must learn to think the unthinkable; we must go beyond EXPERIENCE to competent risk analysis." We can conclude that it was past time for wake-up calls, even before the 9-11 attacks. And now in the aftermath of the attacks, it's time for the fire/rescue services to get serious about terrorism. It's time for those who hold the purse strings to allot sufficient funds to ensure that first responders can do their job of protecting the public with some reasonable expectation of protection, including appropriate PPE to deal with nuclear/biological/chemical (NBC) and other weapons of mass destruction (WMD) threats. It's time for a sea change in the way fire departments train and prepare for the potential for terrorism to take new and previously unfathomable forms. It's time for fire departments to become more flexible and adaptable to fast-changing conditions and new paradigms (Fig. 1–12).

Fig. 1–12 Device Rigged to Kill and Maim Unsuspecting Responders

Yet, in all too many cases, it remains *business as usual* in many fire departments, especially at the level of the first responders who will face imminent danger from future terrorist attacks. Since the 9-11 attacks and as of this writing, less than one year later, a paltry share of U.S. anti-terrorism funding has been designated to properly equip firefighters, police officers, EMS workers, and other first responders to deal with the hazards posed by the new terrorist threats. And much of the money earmarked for them will be siphoned off to other places as it follows its inevitable course through a Byzantine funding system.

Likely scenarios require significant rescue response

Consequence management programs must be prepared for potential biological and chemical attacks. They must also be prepared for the ever-present dangers related to the use of conventional nuclear devices, including so-called dirty bombs that use conventional explosives to contaminate victims, firefighters, EMS personnel, police officers, civilian rescuers, and property with radioactive materials.

Ironically, many fire departments and other public safety agencies began turning in their civil defense radiological monitoring devices after the fall of the Berlin Wall and collapse of the Soviet Union (and the supposed end of the threat of nuclear attack). The potential for terrorists to obtain and use nuclear materials for attacks on U.S. targets apparently was not considered valid by emergency and civil defense planners. Consequently, instead of being carried on fire and rescue apparatus to provide immediate accessibility, many of the nation's civil defense radiological monitoring kits can be found stashed in storerooms and fire station attics, where they have remained without calibration or servicing or serviceable batteries since the mid-1990s.

Radiological emergency training, which just a decade ago was a standard part of municipal fire department training academies, has been curtailed in many cases. Yet one of the threats of most concern as of this writing is dirty bombs. Without radiological monitoring equipment, first responders cannot determine the presence of a nuclear component to a terrorist attack, they cannot designate proper exclusion zones, nor can they determine how long it is safe to operate in a contaminated area.

Without dosimeters, first responders have no way of documenting their total exposure doses for a given emergency operation. In short, in an age when NBC attacks have become the primary focus of many terrorism consequence management programs, we have failed to address the most basic threats, including the N in NBC.

Preparing and Protecting First Responders

In March 1998, Battalion Chief Ray Downey of the FDNY Special Operations Command gave prescient testimony before Congress about the potential for terrorist attacks to endanger the lives of firefighters and other first responders (See Appendix 1). His goal was to convince a Congressional panel of the need for more funding and support for programs that would put WMD consequence management equipment and training in the hands of the nation's firefighters and other first responders (Fig. 1–13).

Fig. 1–13 First Responders and Security Forces Attempt to Gain Control at the Pentagon after the 9-11 Attack

"It is the first responder who will be facing the challenges that WMD presents," Downey testified. "Unless you have been there, you cannot fully appreciate what firefighters face during an incident of WMD terrorism."

Firefighters and rescuers responding to terrorist attacks are ill-served by shortsightedness and lack of imagination. There is a saying that goes: To counter terrorism, we must think like terrorists. We must be able to anticipate their next moves and to develop a purposeful and effective response. Today, it's painfully evident that this is not happening. Many decision-makers remain trapped *inside the box*, either unable or unwilling to strategize outside it. The money necessary to address the true needs of first responders simply is not being directed to them. Instead, much of the nation's anti-terrorism funding is diverted to programs that will not provide timely consequence management.

The fire service faces a range of dilemmas related to terrorism. Take for example the issue of PPE. In October 2001, the National Institute of Occupational Safety and Health (NIOSH) issued preliminary recommendations for firefighters operating at potential bioterrorism attacks. As of this writing, the recommendations are being evaluated by fire departments nationwide, and chief officers across the country are wondering how they will meet these requirements with current budgets.[13]

Among the NIOSH recommendations[14] is one related to PPE for firefighters: "NIOSH recommends against wearing standard firefighter turnout gear in potentially contaminated areas when responding to reports involving biological agents." Instead, NIOSH recommends an approved self-contained breathing apparatus (SCBA) and a Level A protective suit. In Figure 1–14, FEMA USAR Task Force rescue squads conduct shoring operations to stabilize upper floors of the Pentagon in order to allow SAR operations to continue without causing secondary collapse. These stabilization operations continued night and day for nearly two weeks, while other rescuers delayered the collapse pile searching for live victims and the deceased.

Fig. 1–14 Shoring Operations at the Pentagon

Some fire departments have concluded that they simply cannot meet the standard set by NIOSH, and others have decided that the NIOSH recommendations are overreaching and cost-prohibitive. Some major departments have already implemented protocols that allow firefighters in SCBA and full turnouts to respond to and investigate the scene of potential biological attacks like anthrax and to upgrade to Level B protection if evidence points to potential contamination of the scene by anthrax.

Other departments have taken the stance that they will treat potential bioterrorism attacks like any other hazmat emergency. This includes isolating the area, denying entry, creating an exclusion and contamination reduction zone, and establishing a decon area and support zone. Others respond hazmat units and require Level A protection until personnel can determine conclusively that a lesser level of protection is appropriate.

Unless decision-makers and those who control funding address the true lessons learned from the 9-11 attacks, the future of terrorism will include waves of first responders confronting attacks that they are neither equipped nor trained to manage in a reasonably safe manner. Many of them will suffer the same fate as the 479 firefighters, police officers, and other first responders who perished in the WTC attack. A terrorist attack involving chemical and explosive agents in L.A. County is simulated in Figure 1–15.

Fig. 1–15 Decontamination Operations at a Simulated Terrorist Attack

Prior to 9-11, the only terrorist attack in American history comparable to the WTC and Pentagon attacks was the bombing of the Alfred P. Murrah federal building in Oklahoma City on April 19, 1995. All that changed on 9-11 when four airliners were hijacked and crashed into buildings or struck the ground, shocking the nation and forcing fire/rescue services to recognize the growing dimensions of modern terrorism.

Two months later, on November 12, yet another airliner crashed in New York City. Just 3 min after an American Airlines A300 airbus took off from John F. Kennedy International Airport, it plunged into the Belle Harbor neighborhood in the Queens borough. Killed in the crash of Flight 587 were at least 265 people. A series of new precautions were taken within minutes of the crash. Fighter jets were scrambled over New York, air space was closed, and four local airports were immediately shut down. From the east coast to L.A., public safety agencies immediately implemented new plans to deal with the potential for additional airliner *events*.

It's an indication of the paradigm shift in terrorism on U.S. soil that the first question many people asked after the Belle Harbor crash was, "Will any more airliners come down before the day is through?" Additional crashes would indicate yet another terrorist attack in progress.

In May 2001, President George W. Bush announced the formation of the FEMA Office of National Preparedness (ONP) to help prepare the nation's response agencies to manage the consequences of terrorist attacks. The ONP's formation, which prior to 9-11 was hardly noted by the media, suddenly took on much greater importance after the attacks. In the aftermath of the attacks, President Bush created the White House Office of Homeland Security to oversee multiple federal agency efforts to confront, prevent, and respond to the consequences of terrorism on U.S. soil. The ONP, which links vertically with the Office of Homeland Security, is (as of this writing) the lead agency for consequence management related to terrorism. It will be part of a multi-prong approach to preparing the nation's first responders and specialized units to deal with the potential effects of terrorist attacks. For the first time, it will place a high degree of emphasis on providing equipment and training for the first-arriving firefighting, police, and EMS personnel at the site of a terrorism event.

Looking beyond the obvious

As the primary responders to terrorist attacks, firefighters must look far beyond the obvious in the process of planning for (and responding to) acts of terror. Today, being extra vigilant to the winds of change and adopting a doctrine of thinking outside the box are mere starting points for progressive fire/rescue departments. Hazmat technicians work with L.A. Sheriff Department explosives experts in simulated terrorist attack in Figure 1–16.

Fig. 1–16 HazMat Firefighters and Explosives Experts in Simulation of Bomb Planted in Tanker Car

After the events of 9-11, it's quite clear that firefighters and rescuers must start thinking very *far* outside the box. Those who find themselves responding to airline crashes and other unusual emergencies must approach these events with a higher index of suspicion for terrorism. This includes the potential for secondary devices or attacks. We must give serious consideration to scenarios that might have been dismissed as preposterous just a decade ago.

Rescue and Terrorism from Domestic and International Sources

Following the 9-11 attacks, many people seemed inclined to focus exclusively on the threat of terrorism emanating from other lands and particularly the Middle East. This is a deadly mistake. While foreign terrorists are a huge concern, it is a colossal error and a sign of serious naïveté for firefighters to assume that the threat of domestic terrorism has somehow been eliminated.

Today there are many domestic terrorist groups and individuals operating within the borders of the United States. Many of these groups are fully intent on bringing down the government of the United States, some to establish an all-white nation in the Northwest and others to bring in another form of government more to their liking. The members of some groups have already demonstrated their willingness to commit all manner of atrocities to achieve their goals. Why are the distinctions and similarities between international terrorism and domestic terrorism of any concern to rescuers and firefighters? Because understanding from where terrorism emanates and recognizing the patterns of its different iterations helps rescuers and firefighters prepare for the most likely scenarios.

For example, domestic groups may eventually adopt the practices of international terrorists deemed most successful. If, in the coming years, international terrorists find that dirty bombs and small nuclear devices are effective, we may very well see domestic terrorists attempt similar attacks in the United States. For these and other reasons, it makes sense for firefighters and rescuers to be aware of what domestic and international terrorists are doing and what they are likely to try next.

The problem with homicide/suicide bombers

Through the ages, there have been occasional instances where suicidal or combined suicidal/homicidal attacks and retreats have been used as potent weapons against enemies of various peoples and states. In certain situations, it has become acceptable or even venerated to perpetrate acts of self-destruction, melded with the slaughter of innocents and potential combatants. Such is the case of Palestinian murder/suicide bombers active in various parts of the world (Fig. 1–17).

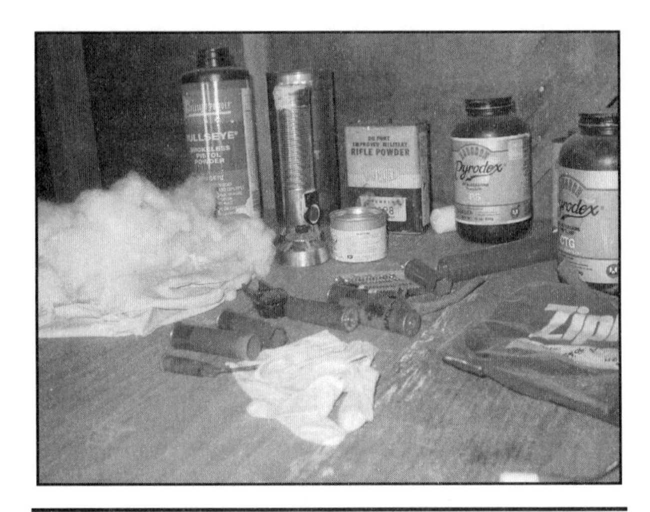

Fig. 1–17 Potential Homicide/Suicide Bomber's Tools?

The concept of committing suicidal acts in the name of a larger cause is not at all new.[15] Despite the basic human imperative of self-preservation, we can trace such acts to at least the 1st century, when extremists among the Jewish Zealots established the fortress of Masada, high on a cliff-shrouded plateau. After a long siege by the Roman army, who surrounded the plateau and built a colossal earthen ramp to breach the ramparts, the Sicaril (most radical group) threw themselves and their families off the cliffs rather than be enslaved. To this day, this singular act is remembered for the powerful effect it had on the Romans, who for all their ruthlessness had yet to witness such *zealousness*.

One thousand years later in northern Persia, the fearsome Shiite Muslim sect known as the Hashshashin conducted a long string of crude suicide attacks. These attacks were directed at the Christian Crusaders who moved through the region on their way to attempt to retake the Holy Land from the Muslim conquerors.

Nearly a thousand years after the Hashshashin struck fear in the hearts of Crusaders for their willingness to sacrifice themselves for what they considered a larger cause, Japanese Kamikaze (meaning *divine wind*) pilots took a similar tact. In the WWII Pacific theater, Kamikaze pilots purposely crashed their airplanes into American ships and other targets. No less than 34 ships were sunk by the Kamikazes, and hundreds more were damaged.

In 1945, about 1,000 Japanese Kamikazes killed more than 5,000 American soldiers at Okinawa. This ratio of Kamikazes to enemy soldiers (5:1) was considered an acceptable level of loss-for-gain by Japanese commanders. In fact, they considered it a huge victory. Nearly 60 years later, Palestinian homicide/suicide bombers and their handlers attempt to achieve similar loss-for-gain ratios in their murderous attacks against Israeli civilians and military personnel alike. Only the most naïve members of the fire/rescue services delude themselves to believe these tactics will not spread to other Western nations like the United States.

In April 1983, Americans began experiencing the consequences of modern homicide/suicide attacks at the U.S. Embassy in Beirut. The embassy was destroyed by a 400-lb truck-bomb driven directly into the compound by a member of the terrorist group Islamic Jihad. The Jihad were trying to dislodge the U.S. military from its peacekeeping role in Lebanon. Among 63 people killed in the attack was the CIA's Middle East director. At first, the United States resisted the pressure to leave Lebanon and Beirut. However, leaders of Islamic Jihad were convinced that if Americans were dealt enough pain and suffering, they would abandon the peacekeeping operation, which the Jihad felt favored Israel. Six months later, their theory would prove correct and serve as a lesson on the virulence of the radical Islamists.

On October 23, Islamic Jihad struck again in Beirut. Two suicide truck-bombers drove into French and American military compounds, causing huge explosions that collapsed multi-story reinforced concrete barracks. Fifty-eight French soldiers were killed in the explosion and collapse. The American compound suffered devastation that resulted in the deaths of 241 U.S. service personnel and the entrapment of many more. Unfortunately for the soldiers trapped alive in the debris of the collapse, this was before the days of highly trained/experienced and internationally deployable USAR teams. Consequently, soldiers were forced to search for and extricate their comrades without many of the USAR resources considered standard in modern collapse rescue disasters.

Two years later, the radical Islamic group Abu Nidal struck a dubious blow for Palestinian freedom when it dispatched homicide/suicide squads to international airports in Rome and Vienna. Tossing grenades and spraying the crowds with machine guns before they themselves were killed, the Abu Nidal killed 20 innocent people and forced many nations to position armored personnel carriers, tanks, and other military assets at their international airports.

Throughout the 1980s/90s and continuing into the new millennium, the Tamil Tiger rebels of Sri Lanka have conducted more than 250 homicide/suicide missions in their war for an independent state.

Two U.S. embassies and adjacent buildings in East Africa were ripped apart by homicide/suicide bombers who struck within minutes of one another in a coordinated attack linked to the Al Qaeda. The resulting structural collapses killed hundreds and required the response of international USAR teams from the United States, Israel, France, and other nations.

On October 12, 2000, the U.S. Destroyer Cole was struck by homicide/suicide bombers. The attackers maneuvered a skiff right up to the side of the ship in the Port of Aden (Yemen) and detonated a powerful bomb that killed 17 U.S. service personnel and injured 39 others.

For decades, Israelis have been the targets of homicide/suicide bombings that continue to this day and show no signs of ending soon.

To firefighters, rescuers, and officers who took the time to understand the evolution of terrorism, it probably came as little surprise when foreign terrorists exported suicide attacks to U.S. soil on September 11, 2001. They would have recognized and tracked the modern trend toward evermore deadly homicide/suicide attacks. While they may not have predicted the exact mode of attack, well-informed firefighters/rescuers anticipated the time when terrorist groups would begin committing suicide attacks within the borders of the United States.

To many of us who have watched modern terrorism evolve, this type of attack was practically inevitable, in part because terrorist groups had found homicide/suicide bombings so successful in causing havoc in other nations. Only the very naïve among us will persist in believing that the 9-11 attacks were an anomaly. All of the evidence (for those who choose to study it) indicates that 9-11 was just the leading edge of what will become a wave of homicide/suicide attacks on U.S. soil in the coming years.

If the evolution of international terrorism is any indication, firefighters and rescuers in the United States will increasingly become the target of secondary devices. These devices are aimed at causing thorough chaos and panic at the scene of terrorist attacks. We have already seen at least one domestic terrorist act target rescuers through the use of a secondary device placed at an abortion clinic in Atlanta. Soon, other domestic and international terrorists will find reason to employ the same tactics that (in their eyes) have worked well in places like Northern Ireland, Israel, and Sri Lanka. The integration of Hazmat and USAR resources is increasingly important to address the explosive, collapse, and contamination hazards that may be found at the scene of terrorist attacks (Fig. 1–18).

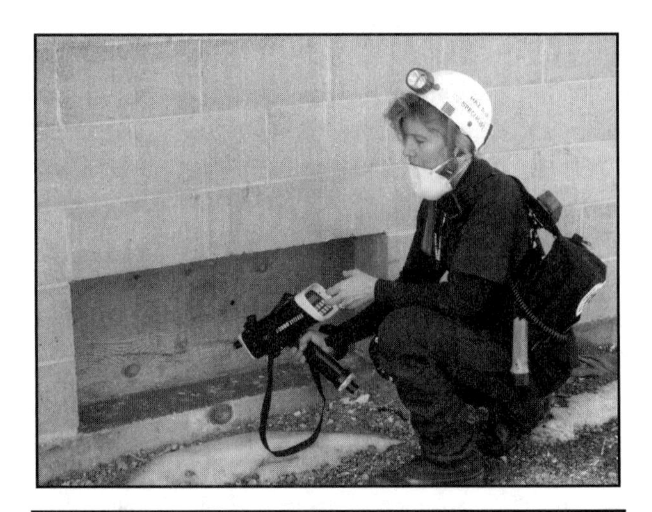

Fig. 1–18 Integrating Hazmat and USAR Resources

Based on a review of patterns and trends, it is not at all inconceivable that domestic terrorists will at some point begin adopting the radical Islamic strategy of conducting homicide/suicide attacks in the United States. Some observers may dismiss this possibility because they reason that groups and individuals of Christian faith would never adopt suicide tactics. But how many of these observers would have expected supposedly fundamentalist Christian terrorists to blow up a federal building full of innocent children, women, and men in Oklahoma City? The fact is, when individuals or groups become convinced that systematic destruction of innocent lives presents a path for achieving political, social, religious, or cultural goals, few possibilities can be eliminated. The perceived *success* of radical Islamic terrorists in accessing targets and causing devastation will almost certainly convince some domestic Christians to employ similar tactics.

For firefighters and rescuers (and their supervisors), the trend toward increasing use of homicide/suicide tactics and secondary attacks creates a dilemma. The new paradigm is already here. There is no need to wait until the next group of terrorists decides to take their own lives in the course of carrying out a murderous attack. There is no need to await the next secondary attack aimed at fire/rescue personnel. The time to recognize this hazard and make appropriate adjustments is *now*. Fire/rescue agencies in the United States should adopt policies and protocols that take into account the likelihood of homicide/suicide attacks and the use of secondary devices.

New Paradigms

It has been suggested that some radical domestic Christian groups might at some point find common ground with foreign terrorists. Some radical groups in the United States vehemently oppose certain U.S. government policies and agencies or certain other groups. They might find it advantageous to collaborate with Middle Eastern Islamic terrorist cells to attack the government and other groups for whom they share lethal animosity. These suggestions might have seemed preposterous a few years ago, but one need only consider events listed in daily newspapers for examples of the extremism to which some domestic terrorist groups aspire.

Consider, for example, the recent case of the Montana Militia, whose leaders were arrested in February 2002. The group was charged with planning to kill as many judges, prosecutors, police officers, and firefighters as possible to provoke the state to activate the national guard. Calling itself Project Seven, the group had amassed 30,000 rounds of ammunition, hundreds of weapons, and intelligence on the personal lives of police officers, firefighters, prosecutors, and judges (and their families). These individuals' names were on hit lists found in the leaders' homes. According to investigators, Project Seven members planned to attack and kill national guardsmen and other law enforcement officials who would be dispatched to Montana in response to the murders. According to documents seized during raids, the group ultimately intended to ignite a domestic war that would topple the federal and state governments.

Groups like Project Seven can no longer be considered isolated cases or anomalies unlikely to replicate. The fact is, domestic terrorist groups have sprouted up across many parts of the United States, and they are fully prepared to take many innocent lives. The Oklahoma City bombing proved that beyond any doubt. Some of the domestic terrorist groups may lay low for a while after events like 9-11. But inevitably, some of these groups will surface to strike out against the government when things quiet down. Or, they may establish yet another new paradigm by conducting terrorism when the government appears most vulnerable, such as in the immediate aftermath of major attacks by foreign groups.

WMD and Rescue Operations

Many anti-terrorism programs narrowly focus on potential biological and chemical attacks to the exclusion of conventional explosives and nuclear attacks. Likely conventional attacks include the use of so-called dirty bombs that contaminate victims, firefighters, EMS personnel, police officers, civilian rescuers, and property with radioactive materials Firefighters and other first responders who encounter scenes as shown in Figure 1–19. During the course of other operations like structural firefighting and EMS response should strongly consider a careful operational retreat looking for booby traps and immediately notify law enforcement to investigate.

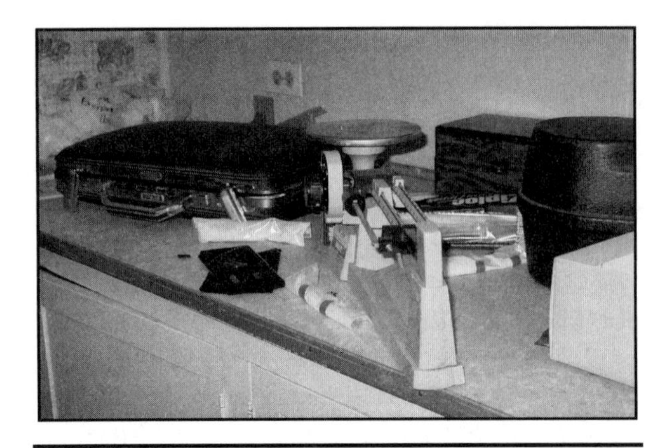

Fig. 1–19 Possible Components of a Homemade Bomb

Many fire departments increased their readiness to manage the consequences of structural collapse just as they began turning in their civil defense radiological monitoring devices. This occurred after the fall of the Berlin Wall and collapse of the Soviet Union, which many people apparently believed represented an end to the threat of nuclear attack. The potential for terrorists to obtain and use nuclear materials (perhaps with dirty bombs) to attack U.S. targets presumably was not considered valid by emergency planners. Consequently, many of the nation's first responders are left without reliable radiological monitoring capabilities. This is a huge tactical and strategic error that almost certainly will come back to haunt us.

The use of dirty bombs is an emerging terrorist threat that could cause massive structural collapses *and* radiological contamination at the same sites. Without radiological monitoring equipment, first responders have no way to determine whether there is a nuclear component to a terrorist bombing. They are incapable of designating proper exclusion zones, and they are incapable of determining how long they can safely operate in a contaminated area. Without dosimeters, first responders have no way of documenting their total exposure doses for a given emergency operation.

Urban Canyons

The history of the United States is rife with attacks by various domestic groups and individuals who have employed explosives, firebombs, sabotage, mail bombs, and even chemical and biological agents. Despite the variety of weapons used, the use of explosives has always constituted the major threat. Terrorists often see explosions or other spectacular events such as airplanes flying into buildings as dramatic statements of their power to disrupt society when they are aggrieved. Moreover, explosive attacks create *moments in time* and *instants of terror* that few witnesses will forget. For some terrorists and terrorist organizations, this sort of unforgettable event is part of a statement, or message, they are sending. They are saying in essence that (1) they will not be ignored, (2) they are capable of disrupting the lives of their enemies (and supporters of their enemies), and (3) they are deadly serious about achieving their goals (Fig. 1–20).

Fig. 1–20 Collapse Operations at the WTC, 13 Days after the 9-11 Attacks

In terms of potential for structural collapse and other consequences that would necessitate massive rescue operations, bombings appear to be the most potent and prevalent threat.

As already discussed, bombs have been and remain by far the most common weapons used in terrorist attacks in the United States. This is true of many nations that experience terrorism within their borders. The fact is, explosive devices continue to be favored by both domestic and international terrorists and groups intent on bringing down monuments, government or civilian buildings, bridges, and dams, etc., or simply blowing up people.

Beyond their banter about being freedom fighters or martyrs, there is something unmistakably egotistical about people who do the planning behind many of the worst terrorist attacks. This is especially the case for those who send others to do the dirty work while they slink back into the shadows and observe from a place of safety. We see it in the egocentricity that allows them to target entire populations of nonbelievers, including civilians, as *free game* for attack.

Unfortunately, many of these egos are drawn to the spectacular. It is perhaps a form of narcissism mixed with other traits. This is not to say that they're not dedicated to their particular causes; it would be a mistake to discount the willingness of terrorists to do almost anything to achieve their goals. And that—combined with ego—is why we must be prepared for evermore spectacular forms of terrorist attacks. Today, the most destructive threat is probably a nuclear device, one of the few true WMDs. One would be naïve, perhaps even negligent given the course of recent events, to discount the potential for one or more nuclear devices to be detonated in the West in the foreseeable future.

What are the greatest terrorism-related threats? Unfortunately, we have already witnessed the successful use of airplanes, tanker trucks, and other common forms of transportation. These vehicles have been turned into portable weapons, capable of wreaking havoc that requires massive USAR operations that last days or weeks to deal with destruction and trapped victims. The potential for additional terrorist attacks (foreign and domestic) using these modes remains high because terrorists see these attacks as successful. As of this writing, we have yet to see trains, ships, or shipping containers used as bombs, but it would be naïve to rule out the potential for terrorists to graduate to that next level of destruction.

Geoff Williams, fire master of the Central Scotland Fire Brigade, has for years warned about the need for fire/rescue agencies in various nations to prepare for *urban canyons*. Urban canyons are created by catastrophic events like earthquakes or huge detonations by terrorist groups that leave entire sections of a city slashed to bedrock, with surrounding buildings forming canyon walls.

For years, Williams' warning was a sort of voice in the wilderness because many fire/rescue officials found it difficult to imagine an event with sufficient force to create such an urban canyon. But now we have all seen an urban canyon in the form of the 9-11 attacks. It's a certainty that some terrorist or terrorist organization is plotting, discussing, or dreaming about an event that will make the WTC attacks look like a foreshock preceding something even bigger. Something bigger could perhaps occur simultaneously in one or more different cities in the United States or within an ally nation. And certainly there is the ever-present potential for a catastrophic earthquake or tsunami to cause similar—if not worse—destruction in some cities. In major collapse emergencies, the services of civilian structural engineers, construction workers, and other specialists may be required. Progressive fire/rescue agencies recognize this and have contingencies in place to request and use these specialists in a timely manner when conditions warrant (Fig. 1–21).

Fig. 1–21 Calling on Specialists in Major Collapse Emergencies (Courtesy Gene Romano)

The question is, how does the fire service plan and prepare for such an event, and how will fire departments manage the consequences when it happens? Fortunately (if the word fortunate can be used here), there is an answer. We employ the same basic strategies and resources that we have used to manage the consequences of other terrorist attacks and earthquake disasters. Only we do it quicker, on a much larger scale, with many more resources, with much better coordination, and for much longer periods of time.

The downside is that we may suffer many more casualties among first responders and perhaps also among secondary responders. But the point is that we handle the consequences and do not allow the evil of men or the capriciousness of nature to overcome us. This means that fire chiefs, company officers, firefighters, and rescue/hazmat/EMS specialists must be prepared to quickly expand the scope of their operation in the aftermath of such an event. Personnel must be prepared to expand even when they are employing the same basic strategies and tactics to handle the individual and composite problems that confront them.

Rescuers will still do rescue; they will just do much more of it for a much longer period of time. USAR teams will still conduct USAR operations using the five stages of structure collapse SAR; they will just do it on a much larger scale. FEMA USAR task forces (trained and equipped to operate in WMD environments) will still be deployed, but they will probably be reinforced by USAR teams from other nations.

Rescuers will still employ the concepts of the incident command system (ICS) or standard emergency management systems (SEMS). They will still use the lookout, communications, escape routes, safe zones (LCES) system, structural triage, medical triage, risk-vs.-gain decision-making, personnel accountability, operational retreat, and rapid intervention. They will just do all these things on a scale not imagined a decade ago. If the urban canyon and surrounding areas are contaminated by radiation or other poisons, there may be a need for wholesale evacuation. But we will determine the perimeter, contain the area, do what we can to reduce casualties, and handle the problem. If command-level leaders are lost, there will be a succession plan based on ICS principles. In short, the problems associated with an urban canyon event are ultimately manageable (Fig. 1–22).

Fig. 1–22 A FEMA USAR Safety Officer Confers with Structural Engineers at the Pentagon

Before the 9-11 attacks, the only terrorist attack on American soil comparable to the WTC and Pentagon attacks was the Oklahoma City bombing in April 1995. Today we have been thrust into an entirely new paradigm of terrorism. The 9-11 attacks challenged the world and the fire/rescue services to recognize the growing dimensions of modern terrorism. In the United States, which for so long has felt somewhat immune to terrorism, there is a deeper understanding that fire/rescue services must change the way they do business. The U.S. fire/rescue services must start looking with greater interest at the lessons already learned by firefighters and rescuers in places like Israel, Ireland, Scotland, England, South Africa, etc.

Case Study 3: Report from California Task Force 2 at Oklahoma City Bombing

One example of the scope of operations required to manage an urban canyon disaster can be seen in the experiences of FEMA USAR task forces dispatched during the 16-day SAR operation in Oklahoma City. Deployment of the LACoFD FEMA USAR task force provides one perspective on how the rescue-related consequences of large terrorist attacks can be managed. It also gives first responders and chief officers a glimpse into the FEMA national USAR response system, the Federal Response Plan, and the operations of USAR task forces. Finally, it is an example of how FEMA USAR task forces can be put to work to help the local IC manage large, complex, unusual, and highly dangerous USAR-related disasters.

It should be emphasized that the USAR response to Oklahoma City was a unified effort that included 10 other FEMA USAR task forces, the entire Oklahoma City Fire Department (OCFD), and dozens of other organizations. Hundreds of USAR-trained firefighters and technical experts assigned to these task forces—many of whom were strangers before the blast—deployed from all points of the nation to work side by side in the rubble of the Murrah Building for 16 dark days. The spirit of cooperation and professionalism was unprecedented at the time. Of course, we have since seen a similar level of cooperation and unity following the terrorist attacks of 9-11. Some would say that the Oklahoma City response established a blueprint for the manner in which many diverse agencies that respond to such events must cooperate and operate under the unified command concept.

One example of the bittersweet quality of disaster work was demonstrated during an operational planning meeting in the Murrah Building on day 15 of the operation. Personnel from FEMA, the OCFD, several FEMA USAR task forces, and other agencies sat wearily describing the progress of the search effort. The news was not good; there had been no live rescues since day 1. Many victims were still buried in the debris, and the danger to rescuers still lurked at every turn. The only positive news was that no rescuers had been killed or maimed since the first dark day when Timothy McVeigh slinked away after lighting the fuse that would take so many innocent lives. Under the circumstances, this was a victory of sorts, amidst a sea of tragedy. The Oklahoma City bombing left a bittersweet legacy indeed.

Activation

The LACoFD USAR task force (California Task Force 2, or CATF-2) was activated and dispatched to Oklahoma to reinforce Oklahoma City firefighters and FEMA USAR task forces already deployed from Sacramento and Phoenix. They were joined by other USAR task forces from Menlo Park, Orange County, and San Diego (CA); New York City (NY); Miami-Dade County (FL); Fairfax County and Virginia Beach (VA); and Seattle (WA). They would conduct the most lengthy, complex USAR operation in American history to that point.

Oklahoma was the third mobilization for CATF-2 since 1992. Previous missions had included Hawaii's hurricane Iniki and the Northridge Meadows collapse during the 1994 L.A. earthquake. Previous experiences with building collapse disasters from earthquakes included Sylmar in Los Angeles (1971), Mexico City (1985), Whittier (1987), Loma Prieta (1989), Northridge (1994), and Kobe, Japan (1995). These experiences, in combination with local collapse rescue operations and technical rescues endemic to L.A. County, proved to be an advantage in operating in the collapse environment of the Murrah building. Thus, early predictions about the advantages of fire service participation in the FEMA USAR task force program were validated.

Following FEMA guidelines for all USAR task forces, the CATF-2 roster included three personnel assigned to each USAR task force position.[16] This redundancy is built into the FEMA USAR system to ensure the ability of every USAR task force to staff and respond within mandated times.

All uniformed members assigned to CATF-2 had worked together in training sessions and disasters since 1991; some had worked together for more than 20 years. This familiarity and cohesiveness has been found to markedly enhance teamwork. It is an important factor during the lengthy and dangerous deployments to disasters like the one that occurred in Oklahoma.

Mobilization

Following procedures outlined in the LACoFD's FEMA USAR task force mobilization plan, CATF-2 members were immediately contacted and instructed to report to the special operations division for activation. Check-in and mobilization procedures for the 62-person task force and the 31-ton equipment cache were completed within 3 hrs. Task force leaders were Assistant Chief Mike Idol and Battalion Chief Mike Sandeman of the LACoFD's special operations bureau. A convoy of two 40-ft tractor-trailers, two buses, and several support vehicles was given an escort by the California Highway Patrol for the 2-hr response to March Air Force Base in Riverside. Upon their arrival, personnel from March AFB immediately loaded the equipment cache onto a C-141 transport plane.[17]

Meanwhile back in Oklahoma City

On the afternoon of April 19, the OCFD met with the FEMA USAR incident support team (IST) and the leaders of the first FEMA USAR task forces to arrive from Sacramento and Phoenix. It was clear that it would take perhaps two weeks to find and rescue (or recover) all the victims missing in the Murrah building. The OCFD had already cleared all known live victims from the building, a remarkable feat considering the level of destruction, number of victims, and difficulty encountered trying to reach them. Still, many rescuers felt it was possible that live victims might be trapped in protected void spaces somewhere under the mountains of debris that covered the lower three levels of the building.

The OCFD, under the direction of Fire Chief Gary Marrs, was going at full speed. The Oklahoma firefighters showed no signs of slowing down. Under the direction of Battalion Chief Mike Shannon (battalion chief of OCFD special operations) and other officers, they were vigorously working to remove debris from *the pit* and other areas where (it was thought) live victims might still be found.

The FEMA USAR IST and the task forces from Phoenix and Sacramento worked with OCFD commanders to set the stage for effective long-term SAR operations. The safety of rescuers and the possibility of finding any live victims hinged on their ability to work hard and smart, and the ability of the commanders to make wise decisions throughout the incident. There were many monumental tasks to be tackled, some of them simultaneously. Those in charge were faced with tough decisions about priorities as the incident progressed.

During the first planning meeting, the foundation of the IAP for operations of the FEMA USAR task forces was established. Goals and tasks were identified for the firefighters and USAR task force members who would be working in the Murrah building for the next 12-hr shift. A variety of short-term and long-term strategies was established, and each task force was given a list of assignments for the day. The OCFD, still in charge of the overall incident, provided first responder firefighters and those from its specialized units to assist the task forces and complete other special tasks (including body recovery).

Point of arrival operations

After the C-141 with CATF-2 aboard set down at Tinker AFB in the evening, there was a flurry of activity. Personnel from Tinker were extremely helpful and efficient. Within 15 min, CATF-2 personnel were rolling through the streets of Oklahoma City in a military bus with police escort, their equipment cache close behind.

The outline of the monolithic Murrah building was lit by intense floodlights from the ongoing search efforts. The sight was at once incredible and surrealistic. The incredible force of the blast practically tore the building in half, with the largest mass heaped in three-story piles at its base. CATF-2 Rescue Team Manager Wayne Ibers later recounted, "Trying to describe the (Murrah) building is like describing the Grand Canyon to someone who's never seen it: there is simply no frame of reference for the mind to compare it to."

Another CATF-2 member said that the Murrah building looked "like something from (the movie) *Aliens*, like the scene when they first stepped foot onto the derelict ship and everything was torn up and silent." Geoff Williams, the Central Scotland assistant fire master, might have described the Murrah building as the first urban canyon caused by a terrorist attack on U.S. soil. Unfortunately, six years later, this scene would be dwarfed by the magnitude of the destruction of 9-11.

Base camp

By the time Tinker AFB personnel delivered the CATF-2 equipment cache to the Myriad Convention Center, the task force leaders were attending a briefing with the FEMA ESF-9 cell. Also in the meeting were the FEMA USAR IST, which had been deployed to coordinate the operations of the USAR task forces, and the OCFD. FEMA USAR task force leaders from Phoenix, Sacramento, and Virginia Beach also attended. A 12-hr rotation of USAR task forces was developed to ensure that SAR operations proceeded at full pace, without letup for 24 hrs a day. This was the routine until every victim was located and rescued or extracted and until the structure was made sufficiently safe for law enforcement teams to conduct investigations.

A Base of Operation (BoO) was established at the convention center to organize deployment of USAR equipment to the collapse site.

The BoO also tracked the location and movement of personnel and equipment and developed USAR task force operational plans. The base was also set up to establish a forward BoO and a rehab area. Finally, all the supplies and equipment needed the next day had to be readied.

Daily routine

The daily routine was soon established. CATF-2 would rotate every 12 hrs with their counterparts from the New York City USAR Task Force (NYTF-1). New York would work the night shift and CATF-2 would work the day shift. CATF-2 was awakened at 5:00 A.M. every morning for breakfast. After feeding, they loaded their equipment onto a large flatbed truck. Included were two portable shelters to protect equipment and personnel from the elements during rehabilitation periods. The shelters would prove to be essential for protecting the equipment from wind, hail, and rain that buffeted the city for days.

Prior to the beginning of each operational period, all USAR task force leaders attended an operational briefing with officials from OCFD, the FEMA IST, NYTF-1, other task force leaders, and the FBI. The purpose was to review the activities and progress of the past 12 hrs and set the plan for the following 12-hr operational period.

After the operational briefing, CATF-2 leaders convened for a task force briefing. The day's IAP was handed out and assignments were reviewed. Significant safety issues were also discussed. Then, everyone went to work for the next 12 hrs. At the completion of the operational period, the task force was transported to the convention center for feeding and rehab. Afterwards, CATF-2 leaders conducted a final briefing before the force turned in for the night.

Operations

For the next 16 days, firefighters of the OCFD and members of the USAR task forces encountered almost every conceivable situation to be found in a disaster of this nature, including:

- Multiple secondary bomb scares.
- An extremely unstable building, subject to potential collapse at any time from a variety of causes.
- Winds and rain that challenged the physical capabilities of the members and threatened to collapse parts of the building and dislodge overhead hazards.
- Hundreds of large and potentially lethal overhead hazards including four-ton concrete slabs, bowling-ball-sized concrete chunks, and office furniture. Occasionally, these items turned into projectiles that barely missed the rescuers below. At least two firefighters were struck by debris during the 16-day operation that caused moderate injuries.
- Extreme carnage.
- Difficult operating conditions, i.e., confined spaces, dust, multiple entrapment hazards, cold weather, nighttime operations, falling hazards, high-angle work, constant maneuvering of heavy equipment in and around the building, and the constant threat of secondary collapse, etc.
- Sheer hard labor of logistics and moving heavy equipment onto upper floors and other collapse zones. Also, removing hundreds of tons of debris by hand or with buckets and shovels.
- Routine difficulties, such as communications, variations of the ICS system, etc. that are expected whenever hundreds of people from dozens of agencies across the country converge on a single incident site for 16 days.

- Sometimes-noisy sleeping arrangements at the BoO at the Myriad Convention Center.
- Constant safety hazards caused by multiple, simultaneous SAR operations involving cutting, breaching, lifting, and removing major structural elements throughout the building.

CATF-2 operations

During the Oklahoma City mission, CATF-2 members used practically every piece of equipment in the equipment cache. The wide variety of work assignments provided a substantial test for the equipment and the logistics system.

The search dogs, coordinated by LACoFD Captain Rory Rehbeck (canine search manager), were used extensively throughout the operation and proved to be extremely accurate in pinpointing the location of victims buried in the rubble. This helped firefighters concentrate on critical areas for rescue/recovery and reduced danger to rescuers by limiting their exposure to victimless areas.

Interaction with local heavy equipment and crane operators by LACoFD Battalion Chief Don Hull and heavy equipment/rigging specialists Al Fortune and Bob Reill expedited the removal of hundreds of tons of rubble in a safe, effective manner. Structural specialists Keith Martin and Rod Spears made critical decisions about the removal, breaching, shoring, and cutting of large debris throughout the operations.

Physicians Bruce Cummings and Steve Chin (medical team managers) provided valuable safety support and medical treatment services to the task force. This included constant mandates for all task force members to wear eye and respiratory protection and advice about personnel rehab issues. Several minor injuries required their attention, as did a more serious foreign-body-type eye injury that required treatment by an Oklahoma City eye specialist.

Communications specialists Mike Steel and Joe Silva dealt with vexing radio problems caused by the massive influx of media that crowded critical frequencies with interference. This caused havoc with SAR operations during the first few days.

Under the direction of CATF-2 rescue team managers Wayne Ibers and Mike Minore and four rescue squad officers, CATF-2 members conducted the following types of operations:

- Multiple searches of the entire Murrah building and several additional structures by canine specialists and technical search specialists.
- Massive debris removal by hand, with jackhammers, and with other tools. This was often combined with crane work. It was estimated that more than 600 tons were removed from the Murrah building during the operation, including more than 100 tons in one 24-hr period.
- Assistance to the OCFD, the FEMA IST, and the FBI with continuous updates of the victim list.
- Mitigation of hundreds of overhead hazards. This intense multi-day operation required members of various task forces to conduct extensive high-angle rope work to rig heavy debris for removal from the face of the building. The task also required heavy cutting, heavy lifting, and close coordination with crane operators.
- Assisting the FEMA IST with support of various command, logistics, documentation, and planning functions.

- Assisting the FBI by performing *surgical* demolition of dozens of vehicles in front of the Murrah building for evidence collection.
- Massive shoring operations in conjunction with other task forces and Oklahoma City firefighters to help save the building from collapse so SAR operations could continue.
- Assisting the FBI with evidence collection and documentation inside the Murrah building.
- Handling of heavy cutting/breaching/lifting equipment to break up and lift large slabs (assisted by cranes).
- CATF-2 members with special skills were called upon by FEMA, California OES, the FBI, and other task forces to assist with logistics, communications, and other special needs.

CATF-2 canine search operations

Due to the overwhelming importance of a rapid search effort to locate potential survivors (and the anticipation of a long-term incident), the California OES dispatched six canine search specialists and six search dogs to Oklahoma City with CATF-2. To make the most effective use of these search resources, CATF-2 leaders decided at the last minute to staff a special canine coordinator position for the mission. LACoFD Captain Rory Rehbeck, normally assigned as a search team manager on CATF-2, was chosen for this position because of his extensive background with USAR search dogs.

This proved to be an invaluable asset to CATF-2 and to the overall search operations at the multiple buildings damaged by the bomb. With years of canine search experience, Rehbeck was extremely effective as a liaison between the dog handlers and the task force. He effectively coordinated all canine search operations for CATF-2, including the following:

- Establishing a canine search plan for each day of the mission.
- Evaluating and responding to the requests of the CATF-2 four rescue squads for canines to search portions of the building as operations progressed.
- Coordinating all CATF-2 canine search efforts with other task forces and the OCFD.
- Generally keeping search dogs operating at optimum performance throughout the mission, including established rest periods and other essential tasks.

One of Rehbeck's main priorities was to ensure maximum safety for canine specialists and dogs, who often found themselves working in very precarious positions. Rehbeck's vast knowledge of structural issues, USAR task force operations and needs, search methodology, and other aspects of the operation was an important factor. For example, on day two, he halted CATF-2 canine search operations when a storm pounded the building and began dropping debris onto the dogs and their handlers. He kept the dogs and handlers out of the danger zone until reasonable safety could be reestablished.

Rehbeck's attention to safety aspects of the canine search function allowed the dogs and handlers to concentrate more fully on the actual search, knowing that their welfare was being attended to. He later stated:

> *It was evident after a week or so that the probability of finding live victims was relatively low. This did not deter*

> *the canine search mission because we felt we could locate deceased victims and help the families with closure. The dogs were able to direct rescue squads to almost the exact point where victims were located. This allowed the squads to focus their efforts and helped maximize the available manpower.*

Some of the greatest lessons of the incident were related to the actual search function. The following passage is from Rehbeck's after-action report on the Oklahoma City bombing:

> *As areas were exposed, canine teams would search the areas. Approximately 80% of the time a victim would be found within 2 to 3 ft of where the dogs were showing interest. The dogs would also show interest toward body parts and fluids. The number of victims found by CATF-2 canine teams during our operations was 19.*

The dogs and handlers were well trained. Everyone worked well during this incident. One reason the dogs worked well is because we brought their portable kennels to the staging area of the incident. This provided the following:

- A comfort zone for the dogs.
- An area to rest the dogs and keep them warm.
- It freed handlers to assist with other task force duties when not actively involved in searches.

The dogs were rested and were only removed from the kennels to exercise and to search. This helped maintain their search drive. In my opinion, the use of the portable kennels led to the success that the canine teams achieved. The dogs were kept well rested, and when the time came to search, they were like racehorses at the starting gate.

The dogs were used on a rotational basis. I paired an experienced dog with a less experienced dog. This allowed us to keep two teams on standby in staging while the others searched. When there was downtime from the search, I found other work assignments for the handlers assisting the task force. This kept them busy and functioning as task force members.

At several points, it became necessary to hoist dogs aloft to place them on upper floors for search operations. This was especially true on the east end of the building, where all nine floors were isolated without stairways or elevators. Here it was necessary to hoist them with cranes and rope systems. There were complications because not all dogs were equipped with harnesses built for vertical lifting. After the Oklahoma disaster, Rehbeck recommended vertical lifting harnesses be included as standard equipment for all FEMA USAR search canines.

Heavy equipment and rigging operations

LACoFD Captain Al Fortune was one of two CATF-2 heavy equipment specialists assigned to the Oklahoma City bombing. He was assisted by LACoFD heavy-equipment operator Bob Reill. Fortune was supervisor of the LACoFD heavy-equipment section, whose fleet included six Type I heavy (Caterpillar D-8) bulldozers, two D-6 dozers, eight heavy-equipment transports, several cranes, backhoes, graders, front-end loaders, and various other types of equipment. The D-8 bulldozers were strategically located at fire stations around L.A. County for immediate response to wildfires, floods (rescue and road clearing), earthquakes (road clearing), and other emergencies. The other heavy equipment was used for a variety of emergency and routine operations, including building and maintaining fire roads, overhauling major structure fires, and various fire department construction projects. As a heavy-equipment and rigging specialist attached to a FEMA USAR task force, Fortune was responsible for the following tasks:

- Interfacing with local heavy-equipment and crane operators.
- Educating local heavy-equipment operators about their role in the SAR process at the scene of a disaster. This included informing them about their place in the ICS, the capabilities of the FEMA USAR task forces, communications issues, and tactics/strategy used in an incident.
- Coordinating emergency activities of heavy equipment at the rescue site.
- Making recommendations about heavy-equipment issues to the local IC, the FEMA USAR IST, FEMA USAR task force leaders, and various task force components as necessary.
- Assisting in the planning process.
- Determining tactics and strategy for heavy-equipment use.
- Dealing with the myriad heavy-equipment issues and problems that may arise during SAR operations.

The Oklahoma City SAR operation was heavily dependent on the capabilities of local crane operators and other heavy-equipment resources. Throughout the incident, heavy-equipment and rigging specialists from the FEMA USAR task forces played a vital role in maintaining safe and effective operations. The following passages are from the after-action report written by Fortune and Reill. They offer a glimpse at the complexity of heavy-equipment operations at the Murrah building and provide valuable considerations for future disasters.

As photo documentation of the incident indicates, less than 10% of the heavy equipment on-scene was utilized on day two. By day 10, 100% of the heavy equipment was being fully utilized. The main stumbling blocks to initial utilization appeared to stem from the following:

- Lack of communication between the heavy-equipment operators and rescue teams
- Lack of direction for the heavy-equipment operators
- Lack of understanding by rescuers and firefighters of heavy-equipment capabilities and limitations

As heavy-equipment and rigging specialists, our first order of business was to establish a communications avenue and a resource inventory with the civilian operators. One of the significant problems identified prior to our interfacing with them was the inability of the civilian operators to dovetail their expertise and equipment into the rescue evolutions performed by the task forces and firefighters. As can be imagined, this was extremely frustrating and demoralizing to these individuals. At that point, it became the duty of the heavy-equipment and rigging specialists to perform the following:

- Determine and communicate clear direction to civilian heavy-equipment operators in understandable terms regarding the IAP and individual objectives assigned to cranes and other heavy equipment.

- Communicate to rescue personnel the capabilities and limitations of various types of heavy equipment.
- Communicate to rescue personnel the procedures, protocols, and techniques (in understandable terms) related to heavy-lifting applications and other heavy-equipment operations.
- Develop and maintain a comfortable level of assurance for the firefighters and other rescuers regarding the competency and abilities of local heavy-equipment operators to perform specific tasks throughout the incident. For example, "Yes, that lattice crane with a jib is capable of lifting that piece of debris..."
- Develop and maintain a comfortable level of assurance for the heavy-equipment operators regarding the strategy, tactics, and equipment being employed by the firefighters and rescuers. For example, "The reason two firefighters are rigged to the side of the building with ropes is to provide backup support for one another..." or "Yes, that little thing (shot cutter, rebar cutting device) will cut through that rebar..."

We were able to successfully maintain a mutual comfort level by assigning one heavy-equipment/rigging specialist on the ground to work directly with the heavy-equipment operators and the other heavy-equipment/ rigging specialist in the building at the actual worksite. Quite often, we had several *picks* (a crane sling-loads a large piece of debris from the building) in progress simultaneously with up to four rescue squads (24 firefighters) assisting. This resulted in the frequent question from the crane operators, "Just how many of you 'yellow-suiters' are there?"

Due to the physical properties of the work environment and the precarious terrain of the site, creativity was the rule of the day. Some examples of situations that required creativity include:

- Rigging a 20-yd dump box from one of the cranes and using it as a platform for firefighters to work from while removing hanging debris from the exterior of the building eight or nine floors up.
- Rigging smaller dump boxes from hydraulic cranes and positioning them in hard-to-reach areas for firefighters to work from.
- Rigging or slinging overhead hazards (*widow-makers*) in place by securing them to the building rather than wasting valuable time trying to remove them from the exterior of the building.
- Using manlifts as observation platforms or as a means to transport rescue equipment and personnel to upper floors.

The enthusiasm and adaptability of the heavy-equipment operators and the rescuers was tremendous. When we first arrived on the scene, most of the firefighters didn't know the difference between a sling and a choker, but by day 10, they were setting chokers, and rigging sling loads two to three picks ahead of the cranes. The L.A. County guys were beginning to talk with Oklahoma accents and dip snuff like the crane operators!

In other words, they became a team. They knew and trusted each other. This enabled the heavy-equipment riggers to give each pick (rigged by the firefighters) the once-over and then...*pick it.* Productivity was greatly enhanced, and the search operation moved faster. This may make a great difference next time the L.A. County USAR task force is deployed; personnel will be ready to start working with the heavy-equipment operators right away. Without a doubt, there are many Oklahoma firefighters and members of other USAR task forces who came away from this incident with a better understanding of how to use heavy equipment to the best advantage, and with some valuable rescue skills.

The heavy-equipment and rigging specialists provided the medium for firefighters, other rescuers, and civilian heavy-equipment operators to work together safely, efficiently, and productively, thereby meeting critical goals and objectives during our operational periods.

Other Lessons for Collapse-Related Terrorist Disasters

The following is a list of lessons learned by CATF-2 members at the Oklahoma City bombing, some of which would be reinforced on 9-11:

- The FEMA USAR task force system (including the FEMA USAR IST and 11 USAR task forces) was extremely effective in dealing with most of the situations that occurred during the incident. The big problem, according to local officials, was the time it took to deploy these rescue teams from across the nation. It's critical to get the USAR IST members on the scene as soon as possible to provide critical technical advice to the IC and his staff during the initial hours of a disaster, while the USAR task forces are often still in transit.
- The organization, training, and previous disaster/collapse rescue experience of CATF-2 members proved effective in managing the numerous tasks assigned during its 10-day mission in Oklahoma City.
- LACoFD was able to staff the entire 62-person USAR task force within 1 hr and 20 mins of activation by FEMA and mobilize within 3 hrs. This is the result of development and utilization of an elaborate mobilization plan that relies on a web of actions by the fire chief, the department's dispatch center, the special operations bureau, the LACoFD Forestry Division (which handles much of the logistics), the heavy-equipment/transportation section, the California OES, and other components of the system.
- It was imperative that all members assigned to USAR task forces be properly trained and fully cognizant of the demands of their position(s). All task force members should be familiar with the responsibilities and capabilities of all other positions on the team. The interplay between the different task force positions was a dynamic and constantly changing one. Everyone needs to know his own job intimately, and everyone needs to be familiar with what others are doing.
- All members of CATF-2 (except certain positions such as canine search and structures specialists) were cross-trained to operate in more than one position on the USAR task force. This proved extremely effective in Oklahoma City. Members could *fill in* for others who became injured or needed rehab, and safety and effectiveness was improved because of the enhanced familiarity with what other task force members were doing.
- FEMA's field operations guide (FOG) for USAR task force operations proved extremely helpful. All CATF-2 members were required to review the guide during their flight to Tinker AFB, and they

were required to keep it on them during the mission. This provided a quick reference when questions about policy and procedure arose.

- The Oklahoma City SAR operations placed high demands on cutting and concrete-breaking tools. It is necessary to consider additional equipment and resources to cut, lift, and break reinforced concrete in future collapse operations at concrete buildings.[18]
- Many new techniques and tools were used or developed to deal with the demands and conditions found at the Murrah building. These lessons will be incorporated into future rescue operations and training.
- Position checklists developed by FEMA and California OES proved helpful during the mission.
- Operations involving heavy equipment and cranes were a critical part of the Oklahoma incident. The familiarity of CATF-2 members with heavy-equipment and crane techniques proved extremely important to the success of the operations. In the future, training and exercises should be considered to improve this aspect of the USAR system.
- CATF-2's equipment cache was preloaded in boxes on military-spec pallets with a computer and hard copy inventory system and was kept response-ready on two 40-ft tractor/trailers. All the paperwork required by the U.S. Department of Defense for air transportation was completed so that the entire cache could be loaded directly onto military aircraft with no delay. These preparations proved crucial to the rapid mobilization of CATF-2.
- CATF-2 has procedures for repacking the equipment cache back into its original configuration before returning from missions. This ensures that CATF-2 is immediately ready to respond to additional disasters following any mission.
- In Oklahoma City, the BoO was secured by local law enforcement officials. There was little need to worry about the base camp while CATF-2 was working in the Murrah building. However, if a task force is deployed to less-secure areas in future disasters (including those on foreign shores in potentially *unfriendly* regions), security will be needed to prevent the possibility of theft and/or sabotage.
- The BoO at the Myriad Convention Center was noisy due to the high level of activity there 24 hrs a day. This is a consideration in future disasters when determining the location to set up base camps to provide maximum rehab of rescue teams.
- The feeding and other support provided at the Myriad Convention Center by the Red Cross, the City of Oklahoma City, and local citizens was tremendous. It is important to remember that such support will not always be available to rescue teams during long-term disasters. For this reason, all FEMA USAR task forces are equipped to be self-sufficient for up to 10 days, including food and water.
- The canine coordinator position was an important addition to CATF-2 during the Oklahoma mission. The canine search capabilities were enhanced by his ability to coordinate the efforts of search dogs and to interface with other search/rescue components of the FEMA USAR task force system.
- Radio communications proved to be a daunting problem. Since Oklahoma, mobile repeaters have become a priority to ensure constant communication between all task force components.
- There was extreme noise in various command locations due to generators and other mechanical equipment. FEMA USAR task forces now have the capabilities for operating essential generators at a distance from command positions.

- It was critical for the medical team component of each USAR task force to immediately determine the local incident medical plan to ensure immediate treatment and transportation of injured task force members. This information should be communicated to all task force members verbally and as part of the written task force operational/action plan.
- At the end of each operational period, CATF-2 leaders conducted task force briefings to review the day's progress, review safety issues, deal with critical incident stress debriefing (CISD) issues, and prepare for the next operational period. This proved extremely helpful and important to all members of CATF-2 and improved the safety and effectiveness of SAR operations.
- Lifting harnesses were required to move search dogs to upper floors and to allow the dogs to work in precarious positions supported by rope systems. They have since been authorized for purchase by USAR task forces.
- Decontamination of dogs with warm water and hair dryers was important to the effectiveness of canine search capabilities throughout the mission.
- When they were not actively working, search dogs were kept resting in low-traffic areas and out of adverse weather. This helped assure the readiness of dogs to work long hours in the rubble pile.
- Each rescuer working in a large collapse site like the Murrah building should have a personal signaling/alarm device (i.e., whistles, boat air horns, personal alert devices [PALS], etc.) at all times. This enables them to warn of impending adverse events such as collapses and to call for help if they become trapped.
- If possible, each rescuer should be issued a hand-held radio.
- Tin snips and wire cutters were in high demand by rescuers trying to make their way through jungles of wire left hanging from drop ceilings and tangles of phone wires.
- Each rescue squad should have boxes of hand tools on-site.
- Each rescuer should carry electrical side cutter pliers.
- Extension cable for come-along tools proved extremely helpful.
- All rescuers working on upper floors should wear personal rope rescue harnesses to operate in high-hazard positions with support of rope systems.
- Scraping tools (i.e., McLeod tools and heavy rakes) proved essential for clearing debris from the edges of upper floors to prevent objects from falling on rescuers below.
- Squad leaders should meet after each operational period to determine which tools will be required on the worksite during the next period.
- All rescuers should be educated to recognize the signs/symptoms of physical and emotional stress in other team members.
- More precise methods are sometimes needed to remove victims and body parts during extrication operations. In one case, a USAR task force used a chain saw to section the body of a victim for removal. Task force medical team managers should be involved with these decisions.[19]

CATF-2 returned home to Los Angeles on May 2, 1995. Several LACoFD members remained behind as part of the FEMA USAR IST. With the exception of a nurse killed by falling debris in the initial moments after the bombing, the 16-day search, rescue, and recovery operation at the Murrah building was completed without serious injury or fatalities to rescuers. CATF-2 members suffered several minor injuries and one moderate eye injury. The lack of serious injuries and fatalities among rescuers is remarkable considering the constant and numerous life-threatening hazards that existed throughout the operation.

The FEMA USAR system performed as designed, ensuring the greatest chance of rescue for any victims trapped alive. FEMA and other agencies in the USAR task force system deserve the support of government and the public/private sectors for sufficient funding that enhanced capabilities through additional training, equipment, and other needs.

It is a sad fact that the potential for terrorist acts in this nation and at U.S. government facilities on foreign shores is ever-present. The Oklahoma City incident remains a sobering reminder of the need for the fire service to continue improving its capabilities to manage structural collapse disasters resulting from natural, as well as, man-made causes.

Managing Rescue-Related Terrorism Incidents

Terrorist attacks, unlike other forms of disaster or emergency, require a multifaceted, multi-agency, multidisciplined response. They require that ICs exercise rapid but cooperative decision-making, exhibit a good working knowledge of applicable laws, implement terrorism-response protocols, and practice diplomacy. The aerial view of the WTC collapse shown in Figure 1–23 is reminiscent of the "urban canyon" terrorist attack scenario postulated by Geoff Williams, Fire Master of the Central Scotland Fire Brigades, long before the 9-11 attacks brought it to reality.

Fig. 1–23 The Collapse of the WTC Towers Created a Prototypical "Urban Canyon"

In SAR operations, the mission of the fire/rescue agencies is to locate and rescue live victims and identify and extract deceased victims; this process is known as *consequence management*. The mission of law enforcement on the other hand is to prevent attacks, investigate the causes when they occur, and pursue the perpetrators, a process known as *crisis management*. In SAR operations, there is a necessary choreography between the two missions. In short, the scene of any potential terrorist attack is both an SAR operation *and* a crime scene. The scene must be treated as such to reduce loss-of-life and suffering, capture/prosecute those responsible for the attack, and hopefully prevent future attacks.

As one might expect, there may be a natural and inevitable tension between the priorities of the fire/rescue mission and the law enforcement mission. In years past—in the case of both threatened and actual terrorist attacks—this natural conflict has materialized in the form of miscommunication, reticence to share information, delayed decision-making, and a lack of coordination. These problems have created safety hazards for firefighters/rescuers and law enforcement agents alike, not to mention to the public, for whom we all work. In Figure 1–24, military, civilian, FEMA, and local fire department personnel operate in concert at the Pentagon collapse operations after the 9-11 attacks. This model of collaborative operation is seen more frequently in the modern era of disaster response and the management of the consequences of terrorism.

Fig. 1–24 Command Operations at the WTC

Fig. 1–25 Example of Unified Command in Action

Fortunately, many of these conflicts have been resolved through cooperative agreements for planning, preparedness, and response to threatened and actual terrorist attacks in the United States and its territories. Today it can be accurately stated that the U.S. systems of crisis management and consequence management have become models of effectiveness and cooperation. At the same time, we can say that it's unfortunate that this level of efficiency and cooperation has resulted—in part—from our experiences dealing with some of the most deadly terrorist attacks in history.

In the aftermath of the 1995 Oklahoma City bombing, a presidential directive identified the FBI as the lead agency for terrorism crisis management in the United States, and FEMA as the lead agency for consequence management on U.S. soil. These agencies operate as partners at the scene of terrorist events involving SAR operations and other federal consequence management responses. Adoption of the ICS and SEMS and development of the Federal Response Plan have helped clarify the manner in which these agencies (and other involved federal agencies) cooperate to ensure the best possible outcomes. In Figure 1–25, the IC from the Arlington Fire Department discusses SAR strategy with FEMA USAR IST Assistant Leader Carlos Castillo and Operations Chief Ruben Almaguer. This is an example of the unified command concept in action at the site of the 9-11 attack on the Pentagon.

For local ICs, this means that if a major terrorist attack[20] occurs, it is essential to establish unified command between fire/rescue/EMS and law enforcement and to incorporate the FBI and FEMA into the IAP. In the event of a major terrorist attack (or possible event) that involves missing or trapped victims, ICs should expect the FBI, FEMA, and a plethora of other agencies to immediately respond and become integrally involved in decision-making.

For firefighters, rescue team members, and officers, it means working hand-in-hand with FBI agents and other law enforcement authorities[21] in and around actual collapse zones. It means helping to secure the area for their investigation through shoring and other operations. Terrorist events also require that firefighters, rescue teams, and officers assist in the identification, location, and recovery of evidence such as airliner black boxes. They should be prepared to dissect vehicles and buildings to help investigators locate and recover evidence and allow investigators to accompany them into collapse rescue areas to document the rescue and recovery of victims. This evidence is required for later use in investigations and potential prosecution of suspected perpetrators. Often, law enforcement officers also need assistance in identifying and collecting sensitive or secret documents and information in collapsed buildings.[22]

This was the case at the WTC and Pentagon in the aftermath of 9-11. At that time, this author wrote:

> *Based on past collapse disasters and applicable protocol for this incident, the following approach was applied to the recovery of Pentagon Incident fatalities: When a victim was located, work in the area was halted to protect the body, personal belongings, and evidence. An FBI evidence team (one of several on constant standby in front of the collapse) was called in to photograph and gather victim-related evidence. If physical extrication was required, a Rescue Squad from the assigned USAR task force freed the victim to the point where he or she could be easily removed. The next step in the process was a Military Honor guard who collected the victim, handling the victims with both dignity and respect.*[23]

During the unified command meetings, the USAR IST leader provided input from the IST and task force operations leaders that identified operational goals during each 12-hr operational period. Once agreement was achieved by command, the IST conveyed these goals into tactical objectives included in the USAR action plan for each operational period. The IST then distributed the action plan to all USAR resources, conducted briefings with task force leaders, the FBI, and other affected parties before each operational period, and provided continuous coordination and supervision of USAR operations and support.

In turn, each USAR task force developed and distributed a written IAP specific to that task force's operations during every operational period. This is a standard FEMA USAR approach to ensure that task forces operate in a manner consistent with the needs of the local IC and that all resources at a disaster are on the same page. This is particularly important at large-scale and/or complex disasters where different resources can find themselves employing opposing tactics because the action plan isn't transparent or is ineffectively communicated.

> *There were many different stakeholders in this operation," said John Huff, daytime leader of the FEMA USAR IST. "This was not only a major disaster with early reports of nearly 800 people missing, but it was also a crime scene on federal property, in the nation's military command center." Huff went on to say, "Therefore, the stakeholders included the local fire departments, the military, local law enforcement, the FBI, the FEMA teams, and many volunteer agencies. Each had a legitimate purpose for being there and had needs and wants to be met.*

During the SAR phase of the Pentagon operations, FBI special agents and supervisors worked hand-in-hand with local fire/rescue agencies, the IST, and task forces to document and collect evidence as victims were being located and recovered. The agents were well trained and organized, attired with proper PPE for the environment, and ready to help rescuers wherever and whenever possible In Figure 1–26, FEMA USAR Task Force members cut shoring materials outside the Pentagon collapse while local firefighters and USAR team members continue stabilization operations to prevent damaged sections of the building from suffering secondary collapse.

Fig. 1–26 Shoring and Stabilizing Collapsed and Unstable Areas of the Pentagon

FBI supervisors assigned to the Pentagon operations were integrally involved in the decision-making process on the fire/rescue side. They attended all essential planning and operational meetings, provided essential assistance and information, and generally ensured that everyone was on the same proverbial page. Cooperation between firefighters, rescuers, FBI, military, and other agencies was optimal; cooperation was one reason there were no casualties among the rescuers and investigators during the 12-day operation. Much the same can be said for the SAR operations at the collapse of the WTC.

The template for this cooperative interagency approach to the consequences of terrorist attacks was established in part from the experiences of law enforcement and fire/rescue agencies at the Oklahoma City bombing. There, similar SAR and investigatory/security/prosecutorial challenges were encountered and overcome in real time. In Oklahoma City, a federal building was bombed in the middle of a major city, within a large county, in a midwestern state. This situation presented a number of jurisdictional issues and challenges never before encountered. The FBI and other agencies recognized the necessity for fire/rescue agencies (including FEMA) to have complete and total access to the Murrah building to locate, rescue, and recover victims. At the same time, FBI and other federal agents needed to fulfill their responsibility to detect, deter, prevent, and respond to terrorist actions that threaten the United States. The assigned FBI members and all the other agencies simply *made it work* during the middle of the disaster (Fig. 1–27).

Fig. 1–27 LACoFD USAR Operations at Train Derailment in a Residential Neighborhood

Before USAR task forces departed from the Pentagon, one of their missions was to ensure that the structure was sufficiently stable. They had to enable the FBI and other investigatory/law enforcement agencies to conduct crime scene investigations with a reasonable degree of safety from secondary collapse and other lethal hazards. After the last victim was recovered, the task forces continued to work with the local fire/rescue departments to shore up the Pentagon and remove overhead hazards. These hazards could have endangered the lives of the FBI agents, the National Transportation and Safety Bureau (NTSB), the military, and others involved in the process of investigating the attack.

Homeland Security

In the United States, the 9-11 attacks clearly changed the way that crisis management and consequence management are viewed. More than any other single event, the 9-11 attacks made it clear to the public that international terrorism has indeed landed on the shores of North America. The attacks demonstrated that international terrorists can infiltrate the strata of American society through a years-long development of terrorist cells and the insidious presence of individuals whose common goal is to attack so-called infidels.[24] These individuals seek to destroy a way of life that has become the heritage and cornerstone of free societies. They would forcibly impose their own cultural beliefs on others or destroy those whose belief in a supreme being differs from their own.

In addition, a host of domestic terrorist groups and individuals have demonstrated a willingness to harm their own countrymen to impose their own versions of government, order, or religion. These combined developments make it clear that terrorism has achieved a new paradigm. This new paradigm requires an equally revolutionary response to deal effectively with its threats and its consequences.

These events have created an emerging realization that terrorism is a threat likely to confront us in new and challenging ways for many years to come.

This realization has prompted the U.S. government to revamp its entire approach to the issue of terrorism. Part of that new approach has been the creation of the Department of Homeland Security, which represents the most drastic reorganization of the U.S. government since World War II (Fig. 1–28).

Fig. 1–28 FEMA USAR Task Force from Virginia Conducts Memorial for the Victims of 9-11 Attacks

Conclusion

The topic of terrorism consequence management could fill an entire volume, and there will certainly be more attention paid to this subject in the years to come. In the meantime, local ICs are advised to study the Federal Response Plan, the FEMA USAR response system, the ICS FOG, and case studies of terrorist events, especially 9-11. They are also advised to develop or enhance close relationships with local law enforcement officials and local FBI representatives, including national joint terrorism task force members (Figs. 1–29 and 1–30).

Fig. 1–29 Near Where the Airliner Hit the Pentagon

Fig. 1–30 Searching for Victims at the Pentagon

Endnotes

1 John F. Murphy, Jr., *Sword of Islam: Muslim Extremists from the Arab Conquests to the Attack on America* (59 John Glenn Drive, Amherst, NY: Prometheus Books, 2002): 235.

2 "A Nation Challenged: The Doomed Flights," *New York Times* (October 16, 2001); Larry Collins, "Report From CATF-2," *NFPA Journal* 3,4 (1995); Larry Collins, "A Close Call," *Fire Engineering* (August 1997); Francis Brannigan, "The Old Professor: Laminated Beams and Arches," *Fire Engineering* (November 2001).

3 Megan Garvey and Richard Marosi, "Those Who Seized the Moment," *Los Angeles Times* (November 22, 2001).

4 Dennis Cauchon, "For Many on Sept. 11, Survival Was No Accident," *USA Today* (December 19, 2001).

5 Cicilia Rasmussen, "Muckraker's Own Life as Compelling as His Writing," *Los Angeles Times* (May 11, 2003).

6 Larry Collins, "Report From CATF-2" *NFPA Journal* 3,4 (1995).

7 The shutdown of U.S. airspace also threatened to delay an east-to-west return of USAR task forces and other emergency resources if it had become necessary to move some of them to the west coast in the event of a major earthquake or a secondary attack on some western state—yet another new paradigm.

8 The east Africa bombings required a massive international response that included U.S. international USAR task forces from Fairfax (VA) and Miami-Dade (FL) to locate and remove more than 200 victims from the collapsed buildings. This raises an interesting point: what will be the USAR response to future terrorist attacks that require more than two USAR task forces?

9 "Al Qaeda Trained at Least 70,000 in Terrorist Camps, Senator Says," *Los Angeles Times* (July 14, 2003).

10 Fleishman, Jeffrey, "East, West Radicals Find Unsettling Bond," *Los Angeles Times,* January 3, 2003.

11 Larry Collins, "A Close Call," *Fire Engineering* (August 1997).

12 Francis Brannigan, "The Old Professor: Laminated Beams and Arches," *Fire Engineering* (November 2001).

13 Tim Elliott, "Tools, News, Techniques: NIOSH Controversy," *Fire Rescue Magazine* (February 2002).

14 NIOSH Interim Recommendations for Bioterrorism PPE, www.cdc.gov/niosh/unp-intrecppe.htm.

15 Benedict Carey, "Method Without Madness?" *Los Angeles Times* (July 20, 2002).

16 LACoFD required each firefighter and officer assigned to CATF-2 to be certified to operate in at least two task force positions to ensure that there would be no loss of efficiency if a member became injured or ill during the lengthy deployments. Some CATF-2 members were qualified to operate in three or more different task force positions. They have experience in various types of disaster response, including earthquake-spawned structure collapse, floods, mud and debris flows, landslides, hurricanes, terrorist attacks, transportation accidents, wildland interface disasters, and even riots. Each task force leader has commanded major emergency operations and USAR incidents. Task force members are listed on a special roster used by the LACoFD to document their qualifications, training, and other data. Efforts are made to assign a mix of the most experienced personnel with those who have less experience in disaster rescue. This helps ensure that the task force has sufficient depth of experience on its missions.

17 As we know, the importance of logistics cannot be overemphasized in campaign disaster operations. CATF-2 technical team/logistics managers Mike Layhee and Don Roy worked closely with March AFB and members of the Riverside USAR task force (CATF-6) for years to develop an innovative method of packaging the 50-ton USAR task force equipment cache for immediate response. This system is a model for some other FEMA USAR task forces.

18 Five and one-half years later, the 9-11 attacks further proved that all FEMA USAR task forces should be prepared for major metal-burning operations. This was proven by the total collapse of both WTC towers, which left huge mountains of metal and very little reinforced concrete.

19 In the Pentagon collapse operations, it became necessary to section the body of one deceased victim to remove her from the collapse without potentially killing rescuers because of precarious conditions. A USAR task force physician was assigned to perform a surgical

amputation of the victim's shoulder using a bone saw. This proved far more effective than a chain saw (as used in Oklahoma City) and has become the standard for dealing with victims' bodies that must be sectioned.

[20] This also applies to certain *threatened* terrorist attacks and also to more *local* events that may not be sufficiently large to require a federal USAR response.

[21] Depending on the nature and location of the attack (e.g., plane crash, nuclear device, etc.), there may be authorities and investigators from a number of other agencies including the NTSB, Alcohol, Tobacco, and Firearms (ATF), the Nuclear Regulatory Commission, the U.S. Department of Defense, etc.

[22] During USAR operations at the Pentagon collapse, rescuers assisted with the location and recovery of items such as computers, manuals, and other sensitive documents and information, in cooperation with teams of FBI agents assigned to work with the FEMA USAR task forces and local firefighters. On the second day of the operation, members of the Virginia Beach (VA) FEMA USAR Task Force located the black box from Flight 77. This occurred after the FBI, working with the Arlington County Fire Department IC, the FEMA USAR IST and task force leaders, distributed photographs of the black boxes for easier identification. This is an example of the effectiveness of cooperative efforts between law enforcement and fire/rescue agencies at the scene of terrorism incidents.

[23] Author's article on SAR ops at Pentagon in *Fire Engineering*, December 2001.

[24] *Infidel* is an unfortunate term coined by some Islamists who display their own egocentricity and profound lack of attention to truth by characterizing as nonbelievers anyone who does not subscribe to the belief that Mohammed was the last prophet and whose pronouncements were the final word on the existence of God. This ignores, of course, the fact that Christians, Jews, and those of many other faiths also believe in a single supreme being. Naturally, Christianity has not been entirely free of similar biases and misrepresentations over the course of its history. How much different might be the world today if the whole of Islam would at least acknowledge the undeniable fact that people of other faiths—whom some classify as infidels—are in fact equal believers in a supreme being. Perhaps if this degree of accuracy were demanded within every faith, the demonization of those whose beliefs differ might be curtailed, with a corresponding reduction in the level of violence from terrorism that hangs like the Sword of Damocles over the head of modern civilization. Rescuers could then concentrate on the consequences of natural disasters and man-made mishaps, instead of those caused purposely by the hands of men.

[25] U.S. Fire Administration Web Site (www.usfa.fema.gov/dhtml/fire-service/cipc.cfm).

[26] Paul M. Maniscalco, Hank T. Christen, and Gerald Dickens, "Understanding the Response to Acts of Terrorism," *Weapons of Mass Destruction Journal, Volume I* (Scott Health and Safety, 2002).

2

Structure Collapse SAR Operations

Well-wrought this wall: Fates broke it.

The stronghold burst...

Snapped rooftrees, tower fallen,

the work of Giants, the stonesmiths,

mouldereth...

And the wielders and wrights?

Earthgrip holds them—gone, long gone

fast in gravesgrip while fifty fathers

and sons have passed...

Bright were the buildings, halls where springs ran,

high, horngabled, much throng-noise;

These many meadhalls men filled

with loud cheerfulness: Fates changed that...

(Unknown, The Ruin*)*[1]

An entire book (and apparently the occasional poem) can be written about the effects of structural collapse, both ancient and modern. The past two decades have produced a number of well-researched books and instructional manuals on the broad-ranging topic of SAR operations in structure collapse emergencies. Some of these include the FEMA USAR *Operational System Description*, the FEMA USAR *Rescue Specialists Training Manual*, the manuals used to teach Rescue Systems I and II, and numerous others. The intent of this chapter is to highlight key points and strategies that have proven helpful in preparing for and managing structure collapse operations in the real world. Combined with formal collapse rescue training and augmented by experience, research and development, and experimentation by rescuers and their commanders, this information is intended to improve the safety and effectiveness of structure collapse SAR operations

In a sign of the times and an indication of the need for national and even international scope of emergency response to major terrorist attacks, this LACoFD USAR captain (Fig. 2–1) was assigned as a safety officer on the FEMA IST. The team was helping to coordinate the operations of FEMA USAR Task Forces responding from 19 states to assist the FDNY after the WTC 9-11 collapse.

Fig. 2–1 Safety Officer from a FEMA USAR Incident Support Team at WTC

Creativity, common sense, and advice from structural experts (structural engineers, firefighters, and rescuers with construction, engineering, and collapse rescue experience) are essential components of successful collapse SAR operations. Coordination of USAR operations by experienced rescuers and highly competent ICs is also important. We already know that many of the most successful rescuers make it a practice of thinking *outside the box*. Collapse rescue operations are a case in point because the myriad potential complications require flexibility, rapid recognition of changing conditions, and timely adjustments.

We Are Mining for People

The goal of firefighters and other rescuers responding to a structure collapse is to save lives by quickly assessing the scene and understanding what has happened and what is likely to happen. Rescuers must locate trapped people wherever they may be and treat them during the extrication process. They must also remove victims without causing further harm to themselves, to other rescuers, or to other trapped victims. All of this must be done without causing the rest of the building to come down. In entering collapsed buildings to conduct SAR, rescuers place themselves in a hostile and alien environment not meant for human habitation. It's a dangerous proposition but also a necessary one, which brings up a true story that helps illustrate the essence of collapse SAR operations.

Following the 9-11 attacks, the administration of the LACoFD convened a group of its firefighters who had responded to the Pentagon and WTC collapses. The meeting was part of the FEMA national USAR response system, which conducted a *post-incident analysis* to capture lessons learned during the response to these events.

One USAR company captain assigned to the FEMA USAR IST dispatched to New York presented a case study of the USAR operations at the WTC. He discussed the operational issues and the many personal safety hazards that the FDNY and other rescuers from across the nation encountered there. Another captain assigned to a FEMA IST at the Pentagon discussed the intensive shoring, cutting, breaching, and other operations conducted in that massive building. He showed slides of local firefighters and FEMA USAR task force members searching for victims inside the burned-out skeleton. Then the captain compared the Pentagon collapse operations to those that might be required if a major earthquake or terrorist attack were to cause similar damage to large buildings in the L.A. area.

Hearing this and seeing photos of rescuers tunneling through collapse debris, one chief officer stood up and made the observation, "We need miners to do this kind of work!" The chief's statement brought some subdued grins around the room. It's true that other nations have used miners to help search for victims in collapsed buildings, for example in the Philippines and Mexico. Modern USAR methods however have adopted principles from several disciplines. Mining techniques have been combined with construction and demolition equipment/ methods and the expertise of structural engineers to improve the manner by which we locate and rescue victims from collapsed buildings. Besides, where would we find miners in the middle of Los Angeles?

Still, the chief's comment was based on a kernel of truth and on his intuition about the essence of collapse SAR operations as he had just observed them. As firefighters and rescuers at the scene of a collapsed structure, it is our job to *mine* the *mountains* of collapse debris to locate and extract live humans. It may require us to sink shafts into large piles of debris, shoring them with timbers just as one would place shoring in a mineshaft. It may then require us to dig our way through. We are in fact miners, and the *gold* we seek is human beings, trapped inside and waiting for assistance.

It's important to understand how and why the structure failed, but it's equally important to understand what additional collapses could occur and how to get the job done without losing more people. The strategies and tactics of collapse SAR will necessarily vary according to the type and cause of the collapse(s); the level of training and experience among firefighters, rescuers, supervisors, and ICs; the available resources; the number of trapped or missing victims and their predicaments; the presence of uncontrolled fires, floods, and other external hazards; the anticipated length of operations; and other factors.

For example, when determining the most appropriate strategies and tactics for a technical rescue, Captain Wayne Ibers of LACoFD USAR Task Force 103 always asks the following two questions: (1) Is it (relatively) safe? and (2) Will it work?

If both of these questions can be answered in the affirmative, it's a good indication that the tactics being considered are appropriate to handle the emergency at hand. Asking these two questions seems like a simplistic way to evaluate the tactics and strategies being considered for the job. However, it's a method that's proven to be very effective over many years and hundreds of complex and dangerous rescues. Plus, it doesn't require memorization of any mnemonics or formulas. In a highly dangerous, highly fluid situation, these two questions address the essential dilemmas facing firefighters and other rescuers and their supervisors.

Understanding Why Structures Fail

Structural collapse can result from many causes, including earthquakes, terrorist attacks, explosions, tornadoes, hurricanes, landslides, tsunamis, flooding, dam failure, transportation accidents, and other *impact loading* events In Figure 2–2, canine search teams and rescue squads from a FEMA USAR Task Force comb debris at the collapsed WTC, searching for some sign of survivors or deceased victims.

Fig. 2–2 Searching for Survivors (Courtesy Amanda Bricknell)

In the most simple terms, we can say that structure collapse is most commonly related to a loss of the stability designed into the building. We can also say that the shape of the building has been changed by forces emanating from one or more of the aforementioned events. Furthermore, the new shape of the structure was not designed to carry loads or resist forces, and the building will continue to change shape until stability is reached. In its journey to stability, the building may suffer partial or total failure, depending on a number of factors.

We can also say that gravity is constantly attempting to *pull* all structures to the earth. A significant loss of stability, caused by horizontal offset or insufficient vertical support, creates a moment of opportunity for gravity to overcome the structure's resistance. When complicated by lateral loads from earthquakes or winds and vertical/lateral loads from explosions or impacts, down comes the building, in part or in total.

The big question for firefighters and rescuers at that point is: Did the structure reach its final state, or is it likely to collapse further as gravity continues to pull it? This is where well-trained and highly experienced rescue companies, USAR units/task forces, and other rescue assets really earn their keep. It's their job to help answer these questions and to guide the IC about what should be done to protect the lives of those trapped, those in potential collapse zones, and those rescuers who will be sent in to get them (Fig. 2–3).

Fig. 2–3 Partial Collapse of a Multi-story Medical Office Building after the Northridge Earthquake Struck Los Angeles

For firefighters and rescue teams assigned to search a building (Fig. 2–3) in this condition (and to rescue trapped victims), it's important to anticipate the possibility of additional collapse. The potential (or likelihood) of additional collapse can come from traffic and other vibrations, rescuers moving through the structure, heavy equipment operations, spontaneous failure of structural members, or aftershocks.

Structural failure is a fact of life in older cities like New York, where collapses of some sort occur practically every day.[2] Collapses occur as well in younger cities in the West occasionally damaged or destroyed by earthquakes and other collapse-related disasters.

Cities even older than New York suffer their share of *spontaneous* collapses. Consider Jerusalem, where in May 2001, a three-story banquet hall suddenly collapsed during a wedding celebration, killing 25 people and injuring more than 300.[3] And who can forget the devastating effects of 9-11 when the WTC collapsed and the Pentagon suffered a major collapse?

Even before the 9-11 attacks, a tremendous number of resources were devoted to structure collapse operations in New York. In comparison, consider fire/rescue agencies in regions that don't experience collapse operations on a daily basis, but which still are vulnerable to collapse-related disasters from earthquakes, landslides, floods, terrorism, and other causes. Their firefighters, rescue teams, and ICs may be quite inexperienced in the strategies and tactics required to manage structure collapse SAR operations. Nevertheless, in the event of an earthquake or other disaster, they may face hundreds, thousands, or even tens of thousands of collapsed buildings. Many individual fire/rescue agencies (and entire cities, states, or even some nations) with fewer USAR resources than the city of New York would quickly become overwhelmed by the simultaneous collapse of large buildings with dozens, hundreds, or even thousands of victims trapped within.

The comparison also highlights the importance of effective planning, training, and equipping of fire/rescue units in places susceptible to major disasters such as earthquakes. Without it, they cannot expect to provide a reasonably effective response to these infrequent events. The comparison also highlights the importance of quickly requesting additional assistance once it becomes clear that local fire/rescue agencies are overwhelmed by the magnitude of the event.

Of course, many cities in Europe, the Middle East, Asia, Asia Minor, and North Africa have buildings much older than those found in the United States. Some have experienced structural collapses on a far grander scale, not just from earthquakes and other natural disasters, but also from man-made disasters like war. London, Dresden, Beirut, Sarajevo, and parts of Northern Ireland are only a few examples of places where structure collapse from man-made causes has been a common threat. Consequently, fire/rescue brigades in these locales have become highly experienced with the realities of structure collapse operations. Their time-tested approaches to collapse SAR have also influenced this chapter (Fig. 2–4).

Fig. 2–4 Collapse of Large Apartment Complexes in the 1999 Taiwan Earthquake Caused Significant Challenges

Even in relatively young cities, structures fail without the prompting of earthquakes, explosions, floods, or other disasters. The east coast experiences spontaneous and impact-induced structure collapses far more frequently than the Midwest and the West. This is due to its older buildings, economic downturns, and harsh weather, etc. The West/Midwest nevertheless still contains aging structures whose time may be short-lived.

Earthquakes

Earthquakes tend to cause problems such as loss of shear strength, failure of beam/column joint strength, tension/compression failure, wall-roof interconnection failure, local column failure, and other problems that can cause partial or total collapse. Earthquakes are such a huge factor in structure collapse that they have earned their own chapter in this series of books (see Volume III), where their effects are covered in further detail.

Suffice to say that the ground motion related to earthquakes may cause damage to structural components that hold buildings together. This damage in turn allows gravity and the lateral forces of ground motion to overcome the structural systems, resulting in partial or total structure collapse and entrapment of occupants. The inevitable aftershocks tend to further weaken already-damaged structures, making them vulnerable to secondary collapse in the hours, days, and weeks following any major earthquake.

Wind

Severe wind conditions damage structures by generating static and dynamic pressures on them. This situation can be complicated by the effect of objects flying in the air and striking the building, which is already under attack from the wind. Roofs can be ripped off buildings, reducing building stability and potentially becoming missiles that create impact loads on other structures. Walls may also be blown in, causing instability leading to collapse of the roof. Metal-clad buildings, mobile homes, and other light metal structures are particularly vulnerable to wind damage.

Floods and mud/debris flows

Floods come in many forms, including slow-rise floods, flash floods, and walls of water from dam failure or tsunamis. Moving water is notorious for causing damage ranging from collapse to buildings actually floating away to be demolished or deposited somewhere downstream. Water surging against structures can collapse the most stable of them, although lighter structures are most vulnerable. Moving water can rip buildings from their foundations, causing them to overturn, slide, or disintegrate. Water can undermine foundations and cause impact damage by carrying large objects such as automobiles and other structures.

The failure of a dam, or the occurrence of flash flooding or tsunamis, may leave a path of total destruction in the main impact area. These disasters can cause varying levels of damage and leave victims trapped in the outlying areas affected by lower water levels with lower velocities and less debris to impact structures.

By their nature, floods complicate collapse SAR operations because firefighters and other rescuers may be forced back by the initial wall of water. They will be forced to adopt swift-water/flood rescue methods to make access, move equipment and personnel, and conduct extrication and evacuation operations.

Mud and debris flows can carry rocks, boulders, trees, structures, people, and the remains of entire hillsides. The impact effect is obvious. Mudslides can leave structures filled to the roof in mud, with people trapped somewhere within.

Explosions

Explosions can result from many sources, including natural gas, propane, or gasoline leaks that find an ignition source. They can occur during mining, tunneling, and construction accidents. Explosions can also be caused by dust, back drafts, arson, terrorism, and innumerable other causes. Most structures are not designed and built for the tremendous internal and external pressures, or initial uplift and overpressure, created by explosions. Even when their duration can be measured only in milliseconds, partial or total structure collapse is possible.

Explosion blast waves consist mainly of a strong main pressure wave followed by a slightly longer but less intense reversed pressure. The shock waves travel rapidly from one point of a building to another. Blast forces can directly produce damage to non-structural components, although failure of structural elements can further compound this damage. When large structures are exposed to explosive forces, they can move in measurable amounts in reaction to the passing of the initial shock wave. The vector of the net force will be determined by a number of factors, including the configurations of the structure and its surrounding buildings, the terrain, the origin of the blast, and other factors (Fig. 2–5).

Fig. 2–5 What is the Potential for Secondary Collapse?

In Figure 2–5, Firefighters and FEMA USAR Task Force members assess structural stability at the bombing of the Alfred P. Murrah Building in Oklahoma City in April 1995. Note the debris suspended from concrete slabs by reinforcing bars, as well as the way debris can hide damage to the column/slab joints. When large explosions impact reinforced concrete buildings of this type, firefighters may be faced with collapse potential that is difficult to predict.

The initial blast wave can instantly destroy key structural components such as columns, transfer beams, and other main supports. Floor-wall connections, roof-wall connections, steel I-beam welds, etc. can also be destroyed. In the next instant, gravity takes over. When gravity effects take over again, the building will quickly *realize* that one or more support systems are missing and will either redistribute the load—or collapse.

Terrorists often place bombs in places that maximize the destruction of columns, each of which supports a significant portion of the structure. Very few structures are capable of redistributing the load when one or more columns suddenly disappear in an explosion. Some new federal facilities require the structural system to accommodate the loss of building columns from explosions or sudden impact. Even if heavier columns remain intact, entire floors that provide loading and lateral bracing to keep the columns in place may be obliterated. Because of their surface area, floors are more vulnerable to the blast wave.

Steel frames, beams, and columns may survive the initial blast wave, but the bracing and connections that hold them in place may be damaged or destroyed. Thus, they will fall apart when gravity takes over. Wall and floor panes are particularly vulnerable to being blasted away, which further erodes the stability of structures and may lead to partial or total collapse.

The effects of bombing events on structures are primarily due to three factors: (1) the quantity of explosives (e.g., the size of explosion), (2) the separation distance of the explosion from the structure, and (3) the type of construction and the design criteria of the structure.[4]

It can be said that blast events are the result of either interior or exterior explosions. Extreme structural damage is likely from interior explosions, particularly for truck/car bombs having 500 or more pounds of high explosive placed (driven) inside the structure. Shock loading from interior blasts can fracture brittle connections of slabs to beams and columns and even disintegrate nearby support columns within the first few milliseconds. Expanding gas pressures will then load the now-damaged structure with loading pressures up to or exceeding 40 times the static design loadings for durations of time (depending upon venting characteristics of the particular building). Gravity will then act on the fractured structure, causing it to collapse (partially or totally) about one second after the explosion.

Interior blasts are more likely to cause damage to a particularly important part of the building, the slab-column connections. An intact slab-column connection enables a column to carry significantly larger loads. A building column 12 ft high might have nearly three times the load-carrying capacity of the same size column 24 ft high. Destruction of the connection can result in the column having a much larger unsupported length and therefore significantly reduced load-carrying capacity. Practically speaking, it is difficult for engineers and designers to mitigate the risks from internal explosive effects. The best solution is to prevent the attack from reaching the interior of the structure.

Interior explosion effects may even result from an *external* bomb placed outside the building but located very close to it (or under an overhang). One example of this effect was seen in the bombing of the Murrah building.

A truck bomb was placed on the street at the entrance, a place that allowed the blast wave to be directed into the front of the glass-skinned structure. The resulting damage was similar to what would be expected if the bomb had been detonated in the lobby (Fig. 2–6).

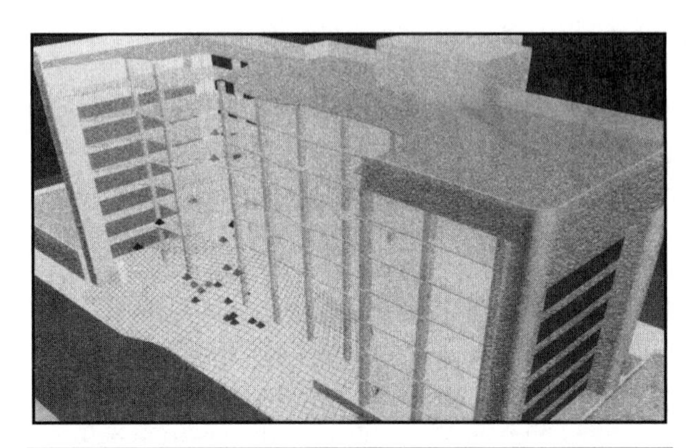

Fig. 2–6 A Computer-Generated Rendering of the Damage to Structural Components of the Alfred P. Murrah Building in the Oklahoma City Bombing (Courtesy California Office of Emergency Services GIS Unit)

Exterior explosions may be more survivable for a building and its occupants, depending in part upon the size of the blast, the distance of the explosive device from the building, and the construction of the building. Currently, intensive work and testing is in progress to provide design engineers more reliable criteria with which to design structures that can resist the effects of external blasts. In the coming years, this may prove critical in the construction of government buildings (including police and fire stations), public assembly structures, and other potential terrorist targets.

Explosion effects on earthquake-resistant buildings. In many regions prone to earthquakes, building codes require a higher level of redundancy to resist ground movement. It's natural to assume that these buildings are more resistant to the effects of explosions, but the issue is more complex, and certain factors should be considered when assessing the level of hazard from explosion-related collapse. It's an issue for firefighters and rescuers who may be called upon to respond to collapse emergencies in earthquake-resistant buildings (Fig. 2–7).

Fig. 2–7 Are There Any Survivors?

In the center of this photo shown in Figure 2–7, Los Angeles City and County firefighters are seen working on the surface, supporting USAR company members in a 9-hour operation to rescue the driver of a street sweeper truck trapped when this three-story reinforced concrete parking structure collapsed during the Northridge earthquake in January 1994. Note the potential for secondary collapse during aftershocks, and the vulnerability of rescuers operating within the footprint of the damaged buildings. How long can any victims survive?

We can say that earthquake explosions push the structural and non-structural elements of a building to their ultimate, near-collapse limit. Therefore, the design tools and strategies to enhance building performance in blast or earthquake events are somewhat similar. But significant differences exist between the effects of earthquakes and blasts on any particular building or structural component.[5] For example, earthquakes generally originate beneath the surface of the ground, and energy from the rupture of earthquake faults is transferred to buildings through ground shaking. These vibrations affect the entire structural system, and they can last up to a minute or more. Damage to non-structural components can also occur as a consequence of excessive structural behavior during earthquake events.

In contrast to earthquakes, terrorist bombs are typically detonated above ground and outside the target building. In some instances, terrorists have placed bombs inside the target structures and occasionally below grade. For example, the 1993 WTC bombing originated in a van in a below-grade parking structure. In both cases, the effects of explosions are localized in comparison to an earthquake. And whereas earthquakes commonly last up to one minute, the blast forces from explosions last just a few milliseconds.

So it can be said that earthquake-resistant design alone does not inherently confer a sizeable measure of blast-resistant design, nor does blast-resistant design automatically provide earthquake resistance. However, it is conceivable that more blast-resistant structural systems could be developed to take into account both threats. The result could produce greater effectiveness than conventional approaches that consider each hazard independently and sequentially. Research within the engineering community is ongoing to help identify how this can be accomplished.

Concrete column sections (particularly those with closely spaced ties) have survived blast-testing effects reasonably well. Minimum rebar requirements in later concrete-related codes have resulted in redundancy capacities that help accommodate the loss of a building column due to a blast. Steel building structures appear particularly vulnerable to the severe twisting caused by uneven initial shock loadings on structural members.

Firefighters and other rescuers should understand that the first explosion may not be the last. Secondary explosions are common after dust explosions and those involving stored explosive materials set off by smaller detonations. Secondary terrorist attacks are (unfortunately but predictably) becoming the norm in places where such tactics were previously unheard of.

Collapse of burning buildings

Firefighters face the threat of structural collapse on the fireground every day. It's a constant fact of everyday life and part of our natural environment. Every year firefighters die or suffer injury when structures fail. Failures can occur while firefighters are attacking fires within buildings or attempting to locate and rescue victims before the buildings fall or become fully engulfed in flame and smoke. The collapse potential is so much a part of the backdrop of our environment that some of us scarcely seem to take note of it, especially when the fire has been knocked down and it's time for overhaul (Fig. 2–8).

Fig. 2–8 The Most Deadly Structural Collapse in History Begins

The two largest and most deadly structure collapses in the history were ultimately the result of fire eating away at the structural support of the WTC towers (further complicated by structural damage caused by the impact of the jetliners that struck them). Both collapses occurred about an hour after the fires began, contradicting the long-held belief that firefighters will have several hours to rescue victims and suppress fire in high-rise buildings without worry of catastrophic structure collapse. The collapse of WTC 7 (a 47-story high structure) on the afternoon of 9-11 further highlighted the danger of total structural failure in high-rise buildings affected by out-of-control fire.

It's a temptation to take for granted that fire-damaged buildings will remain standing once the bulk of fire has been knocked down. Some officers let their guards down and assume that walls will remain standing, roofs will remain on top of walls, and columns will continue to support weight. Occasionally we are reminded that structures can fall, even after the greater part of a fire has been extinguished, especially when high winds, rain, snow, or earthquakes are present.

After the 9-11 attacks, no one should need to be reminded of the insidious effect that fire can have on structural components and how unpredictable fire can be when it attacks buildings in force. Who would have anticipated that the WTC towers could fall in so short a time? Who would have predicted that both towers could collapse with such totality and ferocity from the effects of fire?

Today there is newfound recognition of fire as one of the most significant causes of structural failure. The new awareness is causing fire/rescue services to rethink old assumptions and look at the potential for collapse in structure fires in a new light. This is especially true for fires caused by explosions, impacts, terrorist attacks, etc.

Signs of Impending Structural Failure

Impact loading can result from partial collapse, contact with adjacent buildings, falling objects during construction accidents, or impact from cars, trucks, trains, airplanes, and other vehicles (especially if heavy fire is involved). Other causes of impact loading are crowd movement (e.g., stadium seating collapses during sports events) or collapses of suspended walkways (e.g., those in a hotel), etc. (Fig. 2–9).

Fig. 2–9 Beginning of Partial Collapse at the Pentagon (Courtesy FEMA)

In this photo (Fig. 2–9), firefighters race to escape partial collapse of the Pentagon about 30 minutes after it was struck by a commercial airliner on 9-11. Although the jetliner caused serious damage to structural supports as it penetrated the first floor, it was the ensuing jet-fuel-fed fire that ultimately precipitated the collapse during active firefighting and rescue operations.

Weakening of load-bearing structural elements can be caused by earthquakes, fires, floods, wind, explosions, building renovations, undermining, etc. Weakening can also be caused by overloading from excess storage, water flow during firefighting operations, or crowds exceeding design capacity, etc. Some signs that may indicate structural weakening include:

- Floors or roof lines bowing downward, indicating weakened columns, load-bearing walls, and other supports.
- Distortion of door frames and window frames or glass windowpanes breaking, indicating wall or floor movement or torsional twisting of the structure.

- Walls that lean more than one-third the width of their base indicate the structure has been exposed to loads or forces exceeding the design strength (as well as actual strength). Any wall that appears out of plumb should be carefully examined for signs that connections are damaged or that the integrity of the wall itself has been compromised.
- Stepped cracks along masonry are indications that something is amiss. The cause could be deflection of the foundation, floors, or other supports.
- Smoke or water issuing from cracks in walls or floors of burning buildings may be indications of internal heat and pressure that can split masonry walls and lead to collapse. It could also indicate damaged connections or slabs. Beware of water coming *in* through cracks in the walls. This author has been on the scene of fires where serious compromise of structural integrity was first noticed when water from aerial-directed water curtains and other exterior (defensive) firefighting operations began pouring *into* exposure buildings whose walls had been damaged by intense radiated heated for prolonged periods.
- V-shaped cracks are indicative of excessive pressure being applied to structural components, which can lead to collapse. A V-shaped crack may indicate upward pressure that can cause a roof or floor to fail if the support end is in line with it. An inverted V-shaped crack may be a sign that extreme downward pressure is being applied, as in the case of floors that have fallen away (pulling down the wall with them) or the failure of support columns.
- Failure of roof components indicate that the connections may be damaged, that an excessive load may have been imposed, or that the restraining measures to keep walls in place may have been damaged.
- Sliding plaster, plaster dust, bricks, etc. indicate a moving wall, floor, or structure.
- Cracks in pad stone supports and rusting of external wall brackets should be closely evaluated because they indicate excessive forces have been applied, or that long-term damage may be leading to collapse conditions. Damaged iron wall brackets of the type used to support fire escapes should be treated with caution because the extra weight of rescuers and the movement of people and equipment can cause failure.

Typical patterns of structural collapse

Structural collapse can be broadly classified according to which part of the structure has failed, how it fell, the configuration of the collapse, the characteristic void spaces that are left, and the hazards they present to firefighters and other rescuers. One school of thought says that structure collapses can be characterized as either *internal* or *external*. With interior collapses, walls or floors have failed, but the exterior walls and sometimes the roof remain intact. In external collapses, outside features such as fire escapes, scaffolding, roofing materials, chimneys, and outside walls have fallen. *Total* collapse occurs when the frame, walls, floors, and other components have failed, or in some cases, combinations thereof.

Structural failures leave typical patterns of hidden void spaces that may be potentially survivable void spaces. If we understand how collapses occur and can recognize potential survivable void spaces, we can exploit the strengths and weaknesses of collapsed buildings to find better and safer ways to locate, reach, and extract trapped victims. This knowledge can lead to round-the-clock, nonstop USAR efforts that provide trapped victims the best chance of rescue. This is because the knowledge helps ensure that firefighters/ rescuers will continue searching until all potentially survivable void spaces have been searched by physical and/or technical means.

Pancake collapse

Pancaking of floors can result from several causes, including an imposed load caused by a crane dropping a heavy beam or slab on the top floor during construction or a helicopter crashing on the roof. Floor or roof truss failure during the course of a major fire can also precipitate a pancake-type collapse that can involve one or more floors. Earthquakes and other events can cause the failure of load-bearing walls or other supports on one or more floors, which can sometimes result in the pancaking of floors in the middle levels of a building (mid-story collapse) or all the floors down to ground level. Once a single floor falls onto the next floor, there is a potential for that next floor to fail and for the remaining floors to fall in succession. This has occurred in earthquakes, explosions, construction accidents, and the 9-11 attacks on both the Pentagon and the WTC (Fig. 2–10).

Fig. 2–10 Pancake Collapse

The pancake collapse of the three-story reinforced concrete parking structure (Fig. 2–10) challenged firefighters who spent 9 hrs rescuing a man trapped in a crushed vehicle. The victim was driving a sweeper truck on the first floor when the Northridge earthquake struck L.A. in the predawn hours

Pancake collapses tend to leave potentially survivable void spaces beneath strong structural and non-structural objects. Such objects can include desks, beds, cabinets, beams, and anything else that may prevent the floors from coming in direct contact with one another. Only a small space is required for a person to survive, and the smaller void spaces may provide safe harbor for small people or children and infants. In some earthquake-prone areas like Istanbul (Turkey), it's not uncommon for people to stack books and furniture about their sleeping areas to increase the likelihood of survival in case their building fails during a major quake.

The roof and upper floors are usually found collapsed in such a way as to give the impression of a complete structural failure. Subsequently, pancake collapses are sometimes mistakenly referred to as total collapses. To the layman and even some untrained or inexperienced first responders, it may appear hopeless for victims who are trapped inside. To the contrary, pancake collapses are notorious for harboring live victims for many days when people are trapped in survivable void spaces, but hidden from the outside.

Firefighters/rescuers who see a pancake-style collapse should immediately think, "Live victims are likely to be found inside this collapse unless some other factor (like fire) has sealed their fates." Survivable void spaces are more likely than not to be found inside pancake collapses. Therefore, round-the-clock, nonstop USAR efforts should be initiated until all potentially survivable void spaces have been searched by physical and/or technical means.

For SAR, it may be possible to create vertical shafts into the pancaked section, working your way down from the top and searching each floor laterally as you make access. This stage-3 operation is often (but not always) the recommended approach. Each collapse situation will dictate the tactics and approach.

It may also be possible to enter the pancake laterally from the outside. This can be done by placing cribbing and shoring as you work your way into voids. Simultaneously, the team tunnels through the debris sandwiched between the slab, passing the debris outside in bucket-brigade fashion. This stage-3 operation is a high-risk option and one that should be undertaken with the greatest of care with the most experienced rescuers and (when possible) with advice from qualified structural experts.

Another rescue option is to delaminate the pancake collapse, peeling back each layer like an onion and searching beneath each layer for victims and additional void spaces that need to be searched. This would be classified as a combination stage-3/stage-4 operation, consisting of selected debris removal and void-space search. It is an equipment-intensive endeavor requiring close coordination between rescuers, heavy-equipment operators, and (when available) structural experts to be effective and reasonably safe.

In the absence of heavy equipment such as cranes, this same approach can be employed by rescuers with rescue saws, diamond-tipped chain saws, and other equipment that can cut and break apart concrete slabs. Tools for cutting and sectioning wood or metal flooring and roofs can also be used. The use of jackhammers and other pounding devices should be carefully considered because the intense vibrations can be a precursor to secondary collapses. As always, creativity is a benefit.

Yet another option is to approach the collapse from below, breaching your way *up* into each floor and creating one or more vertical shafts moving upward. This approach may be especially effective if begun in a protective basement or underground parking structure beneath the collapsed building.

Mid-story pancake collapse. The effect of a mid-story pancake collapse is similar to that of a full pancake collapse, except that the pancaking does not extend from ground to roof, but rather is restricted to one or more middle floors of the building (Fig. 2–11).

Fig. 2–11 Mid-Story Collapse in Kobe, Japan in 1995

A mid-story pancake collapse in Kobe, Japan resulted from the earthquake that struck there on January 17, 1995 (one year to the day after the deadly Northridge earthquake in Los Angeles).

One problem with mid-story collapses is that the floors above and below are still intact and thus subject to potential collapse from aftershocks, vibrations, cutting and breaching operations, etc. It presents a dilemma for rescuers who must find a way to get into the mid-story collapse to search for victims and extract them without causing the other floors to collapse. One approach is to come in from the floor above the collapse and breach downward through the floor to reach the mid-story collapse zone. This places rescuers in a floor that might be subject to secondary collapse caused by the next aftershock or from the detonation of a secondary device.

Another method is to approach from the floor below the collapse and breach upward through the ceiling slab to conduct a technical and physical search. However, this places rescuers below an existing collapse in a building so badly compromised that secondary collapse is possible. They are working from a floor that may be subject to the same type of failure as the floor above.

A third approach is to use ladders, ropes, or other means to enter the collapsed floor from the outside, tunneling or breaching laterally into the collapse zone looking for victims. This may or may not be feasible, depending on the compactness of the collapsed floor, the materials that must be removed to reach victims, the size of void spaces, and other factors. The advantage (relatively speaking) is that the floor is probably close to reaching its ultimate collapse state, unless another event precipitates a progressive pancake collapse that results in total failure of the building.

Yet another approach is to de-layer the building from the top down. This may be unfeasible and excessively time-consuming and dangerous, depending on conditions. The advice of experienced structure specialists, construction/demolition experts, and experienced rescuers is helpful in decisions of this nature.

Lean-to collapse

In fires, earthquakes, and other collapse events, it's possible for a single bearing wall to collapse or for the end of one or more beams to pull away from an outside wall and fall. In either case, the result is a lean-to collapse. A lean-to collapse occurs when a floor, roof, or beam ends up leaning against a remaining bearing wall, debris, or even another building (Fig. 2–12).

Fig. 2–12 Lean-to Collapse

Nearly two weeks after the 9-11 attacks, rescuers struggled to remove debris and search void spaces in the WTC collapses.

Imagine a V lying on its side. That's what the void space below the floor or roof of a lean-to collapse looks like if you were to draw a cutaway diagram. The void spaces are characteristically found within the two lines that make up the V, which places them below the sloped floor or roof. That's where live victims are likely to be found. Even if the sloped floor is piled high in debris, there may be live victims in the survivable void space below. Victims may also be found in the debris that has piled up on top of the sloped floor, but their survivability is often reduced because of the direct contact with (and crushing by) the furniture and other debris.

V-shaped collapse

Imagine a cutaway view of a floor, a roof, or a floor slab breaking in the middle, with the center falling to the next lower level. You should be able to imagine that the floor, roof, or floor slab forms a V shape, with all the furnishings and victims who were on that level piled in the middle of the V. You should also be able to imagine that below and on either side of the sloping walls of the newly formed V, there will be void spaces (Fig. 2–13).

Fig. 2–13 V-Collapse

Collapses that are *V*-shaped are commonly caused when heavy dead loads or imposed loads cause the floor or roof to break in the middle. It is a common consequence of earthquakes, renovation mishaps, construction accidents, or severe fires that cause steel columns to fail, dropping the floor or roof above.

Victims will typically be found in potentially survivable void spaces beneath the collapsed floor, on either side of the sloping collapse. They may also be found trapped within and beneath the furniture and debris that has slid to the center on top of the collapse.

It may be possible to remove the debris from atop the collapsed area, extracting victims in the process, to gain access to the bare floor. Then rescuers can consider breaching downward through the floor to search the void spaces below.

A more risky approach is to breach the walls that usually remain standing on either side of the V. This approach should always be carefully considered and should always be avoided if the wall is brick or masonry. Removing or breaching even a small section of brick/masonry wall may precipitate a catastrophic collapse of the wall and other parts of the structure. If the wall is wood-framed and you are merely breaching the fiberboard between the studs (especially to conduct a technical search), the results may be less dangerous. A concrete slab floor (as in a tilt-up building) may be breachable without collapse, but each situation has to be judged on its own merits. This is where it's good to have a structural engineer or another building expert on hand to advise on the safest approach and to help mitigate any complications.

It may also be possible to breach upward from the floor below to conduct a technical search and—if victims are found—to make access and extract them. Upward breaching is labor-intensive and places rescuers below the collapse zone, which may be an unwarranted risk in some cases. But there may be some advantages to this approach in certain cases.

Tent or A-frame collapse

Imagine a tent with a sloped roof and a pole in the center to hold it off the ground, with vertical walls surrounding it. Now stack debris and victims on top of the tent and imagine victims trapped beneath the roof and between the walls. This is similar to the type of collapse that occurs when an earthquake, fire, explosion, or other force causes floor beams to collapse near the outer walls, leaving an interior-bearing wall to support the center of the floor. It is essentially the same situation you would find if two V-shaped collapses occurred side by side (Fig. 2–14).

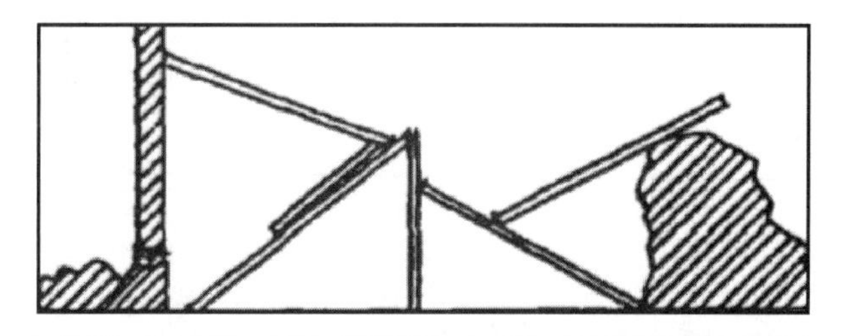

Fig. 2–14 "Tent" or "A-Frame" Collapse

The SAR approach options are essentially the same as for V-shaped collapses. Rescuers must consider the importance of maintaining the integrity of the interior-bearing wall holding the floor and debris off any victims trapped in the voids below the tent. Upward breaching in this situation may be contraindicated because of possible complications to the interior-bearing wall. Breaching the interior-bearing wall to gain access to one void space from the other void space should be very carefully considered.

90° Collapse

A 90° wall collapse is an obvious danger to passersby and rescuers alike. While it's possible for external walls to collapse inward, it's probably more likely that they will fall outward, away from the building. An earthquake, explosion, or some other precipitating event may cause one or more walls to fall outward onto streets and sidewalks. If the wall remains intact and falls over like a tree, it is known as a 90° collapse. The wall may fall its full height away from the building, or it may fall even farther. We should anticipate that walls will fall up to 1½ times their height. Some walls, like those of tilt-up concrete buildings, may create a cushion of air just before impact with the ground, carrying them even farther from the building. Debris thrown from the collapsing wall may travel even farther (Fig. 2–15).

Fig. 2–15 90° Collapse on a Car

In Figure 2–15, as firefighters reconstitute their fire attack and begin search and rescue following the Pentagon collapse, it becomes evident that they have victims missing in a combination lean-to and pancake collapse. The roof structure over the collapse is the size of a football field, and at the time of this photo it was not known exactly how many victims were missing. It took 12 days to locate and extract all of them.

Where will victims be found after a 90° collapse? Naturally, firefighters and rescuers will begin looking beneath fallen walls. They are often preceded by passersby, relatives, and friends who witnessed the collapse.

If the building is of masonry/brick or stone construction, the job may require a great deal of handwork, usually bucket brigades passing material out to the street as rescuers make their way to the victims through the debris. This is fairly non-technical stuff, but responding firefighters and rescuers must evaluate the overall scene and be on the watch for additional hazards such as secondary collapses, rupturing natural gas lines, downed power lines, or aftershocks, etc.

If the building is of wood frame, metal frame, or reinforced concrete construction, wood-, metal-, or concrete-cutting tools may be needed to *section* the wall(s) into manageable sizes. This will better enable rapid lifting and removal of materials by hand to access patients and get weight off them before they succumb. The IC and rescuers may choose to employ air bags and other heavy lifting devices to lift intact wall(s) off victims.

It's not at all uncommon for 90° wall collapses to bury automobiles parked on the street, waiting in traffic, or simply driving by when the event occurs. When this happens, it's sometimes necessary to conduct modified search operations and perform modified vehicle extrication tactics to access victims and remove them from their automobiles beneath the debris.

Curtain fall collapse

Imagine a masonry, brick, or stone wall one or more stories high. Where is the wall going to go if the event destroys the consolidation of the mortar or other supports holding the wall in place? Naturally, the result is often a huge pile of brick, masonry blocks, or stones, beneath which may be buried a number of victims. Even some curtain-wall-constructed buildings can be *defaced* in an explosion or earthquake in such a way that the facing material buries victims and automobiles beneath huge piles of metal, glass, and other materials. A related problem is glass falling from high-rise buildings during explosions and earthquakes. One estimate of potential damage from a major earthquake in L.A. County includes scenarios in which victims are buried on streets and sidewalks beneath 13-ft high mounds of broken glass from high-rise buildings (Fig. 2–16).

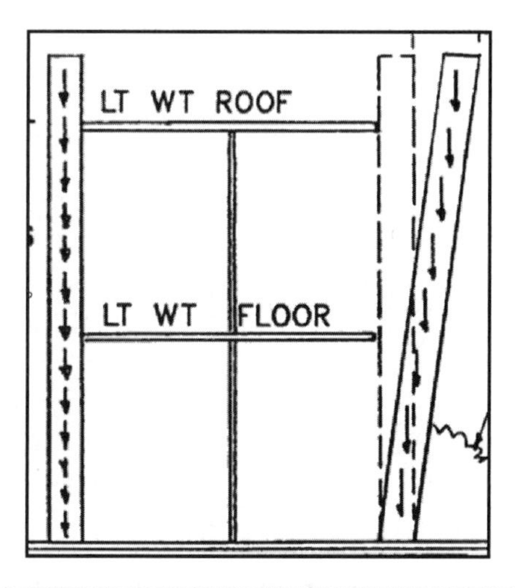

Fig. 2–16 Curtain Wall Collapse

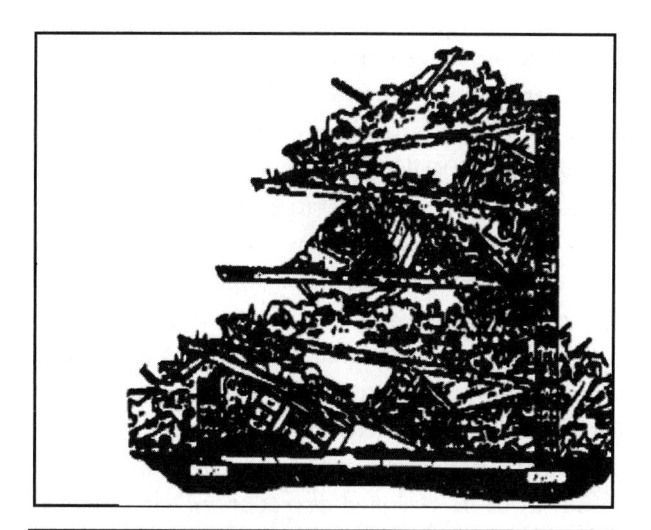

Fig. 2–17 Cantilever Collapse

Cantilever collapse

When the outer wall is destroyed, leaving the roof and/or upper floors dangling in thin air as unsupported members, we have a cantilever collapse. Cantilever collapses have also been described as pancake collapses with floors extending as unsupported planes. These are among the most dangerous and unpredictable collapse situations because a serious amount of weight may be suspended in midair, with overloaded and unsupported floors or roof ready to snap and fall without warning. Debris on the upper floors may also be ready to cascade down if the floors begin to sag (Fig. 2–17).

Cantilever collapses are notoriously difficult to assess and stabilize, and SAR operations below a cantilever collapse involve a great deal of risk for anyone near the potential fall zone. Trapped victims are in imminent danger of being further buried by cascading material or by the entire floor or roof coming down, potentially pulling other parts of the building down as well. Likewise, firefighters and rescuers are placed in harm's way just trying to assess these collapses, search for victims, install shoring, treat trapped victims, and extract victims from the rubble.

It's difficult to define the best way to approach cantilevered collapses, other than to say that good assessment is a priority. Employing multiple *lookouts* and safety officers to constantly look for signs of impending collapse and to sound a warning is also essential. The FDNY makes a practice of placing lookouts on ladder towers and other aerial devices to perform this function. Another approach is to use personnel baskets suspended beneath cranes or position lookouts on adjacent buildings throughout the operation. This is when structural experts with the training and technical equipment such as tiltmeters or theodolites are very important to rescuer safety and operational effectiveness.

A priority is to stabilize the cantilever as much as possible without making the situation worse. This may require sophisticated, engineered stabilization methods beyond those of the typical fire/rescue agency, supervised by qualified structural engineers. It may require that heavy equipment be positioned as shields to protect rescuers from falling debris or perhaps heavy equipment or aerial ladders positioned to prevent the building from moving. Obviously, these decisions should be made by highly experienced personnel, with proper advice from experts. There should be effective coordination and communication among all participants, with a well-understood operational retreat signal, good personnel accountability, and rapid intervention crews (RICs)[6] in place.

There is also the potential for victims to be trapped *on* cantilevered floor(s), sometimes complicated by entrapment beneath non-structural materials. The approach to these victims will vary depending on the conditions at the scene. One consideration includes the use of aerial devices to reach victims. Hopefully victims can be extracted while rescuers are suspended from aerial devices in such a way as to minimize extra weight or vibration on the cantilevered floor. Victims can also be approached from the attached side of the cantilevered floors using rescuers attached to rope systems or other safeguards. Clearly, some of these operations are so dangerous as to be classified discretionary, based on the conditions at the scene, the available resources, etc.

Fig. 2–18 Overturned Apartment Building in the 1995 Kobe Quake

Inward/outward collapse

In some cases, the outside walls fracture horizontally along the middle during a collapse event, and the wall collapses in large chunks. Sometimes the top one-half falls inward while the lower one-half falls outward, or vice versa. In either case, the result is a pile of debris at the base of the wall, sometimes projecting more than one-half the height of the wall away from the base. Because the pieces of collapse debris are larger than those of a curtain-fall collapse, heavy lifting operations may be required to remove the debris from atop victims, or it may be necessary to section the debris into manageable pieces that can be removed by hand.

Overturning

When shear walls or foundations fail during a collapse event like an earthquake, heavy floor buildings may fall over sideways by their full height, sometimes remaining intact but lying on their sides. Vertical routes like elevator shafts and stairwells suddenly become horizontal passageways, and horizontal passages like hallways suddenly become vertical shafts (Fig. 2–18).

Total collapse

As of this writing, the most lethal example of total structural collapse in history was the fire and impact-induced failure of the WTC on September 11, 2001. No single collapse in the history of mankind has been as devastating to the occupants and those attempting to rescue them.

Total collapse is different from a pancake collapse, and each must be approached with slightly different tactics. Remember that in a pancake collapse, there are void spaces between relatively intact slabs stacked atop one another. Within those void spaces are furnishings and people, some of whom are trapped alive. Conversely, in a total collapse, there are no slabs left stacked like hot cakes; the entire essence and form of the building has been lost. It's barely recognizable as what it was before the event. The connectors are gone, the framework is often nonexistent or hugely distorted, and the thing has fallen like a disjointed giant (Fig. 2–19).

Fig. 2–19 Total Collapse of a 4-Story Parking Structure in the Northridge Quake

In some extreme cases, the collapse is so total—and the forces so tremendous—that there are no slabs whatsoever, no furniture, no computers, no filing cabinets, no televisions. This doesn't mean that there is no survivability, just that the likelihood of locating survivable void spaces is often minimal. When survivable void spaces do exist, they are somewhat more difficult to predict and locate because of the disjointed manner in which the building has fallen. Whereas in a pancake collapse an experienced rescuer can mentally (or in writing or on a computer) reconstruct the building and predict with reasonable accuracy where victims might be trapped in hidden void spaces, in a total collapse this process is more difficult.

In a total collapse, most or all of the floors have fallen to the ground floor or into the basement. All walls have collapsed outward or inward onto the floors, and the only thing resembling the original building might be the roof lying atop huge smoking piles of collapse debris. There may be survivable void spaces in disparate places throughout the collapse piles, and people may survive for many days in them.

In a total collapse, it may be somewhat easier to get to people lightly trapped near the surface, and it may be possible to get rescuers into the visible void spaces to conduct physical searches. It may also be somewhat easier to snake fiber-optic scopes, search cameras, and snake-eye cameras down into cracks between pieces of material, looking for void spaces and people within them. Other technical search equipment, like the trapped person locator, may be effective in detecting people tapping from the inside on solid objects, although it may be more difficult to trace the signals back to their source because of disjointed debris piles.

Search canines may be very effective in picking up human scent wafting from cracks in the debris piles of a total collapse, but it may be more challenging to find exact locations. It may be possible to tunnel into the pile like miners (a technique employed and perfected by the London Fire Brigade during World War II), angling toward likely places where people might be trapped alive. But this process may be more dangerous than the tactics used to search pancake collapses, and the prospect of being buried by secondary collapse is often higher.

In general, total collapse situations are more lethal than many pancake collapses. The reason is that collapse debris is smaller and more broken up and consequently may be more tightly packed, offering fewer survivable void spaces. Once again however, it must be emphasized that live victims *can* be found and rescued from total collapse situations. But it's important for firefighters and other rescuers to understand the difference between total collapse and pancake collapse and to take advantage of that knowledge to devise the most appropriate strategy and tactics.

Combination collapse

Due to the vagaries of structural failures and their causes, it's not always possible to place collapses into neat categories. Sometimes a building will fail in a way that leaves multiple styles of collapse, each with its own characteristic pattern of void spaces. The Pentagon collapse following the 9-11 attacks is a case in point (Fig. 2–20).

Fig. 2–20 The Pentagon Structure Failed in a Combination of Lean-to and Pancake Collapse Patterns (Courtesy FEMA)

As noted in Figure 2–20, the collapse was painstakingly dissected by heavy equipment under the close control of the FEMA Incident Support Team and USAR Task Force members (including structural specialists, search team personnel, and heavy equipment and rigging specialists). Characteristic void spaces were uncovered. These are the potentially survivable void spaces that we are looking for in Stage 4 collapse operations; they are the places where trapped people may be found alive after buildings collapse. In the case of the Pentagon, some victims were found in these spaces, and it appeared that there might have been a chance for survival if not for the intense heat and smoke that accompanied the terrorist attack.

Because areas surrounding the point of impact at the Pentagon were so sturdy, the collapse area fell in pancake style, with part of the pancaked layers leaning on uncollapsed portions of the massive structure. Consequently, firefighters and FEMA USAR task forces were faced with victims trapped within a combination lean-to/pancake collapse. Victims were sandwiched in void spaces that one would expect in a pancake collapse; only in this case, the sloped angle of the collapse affected their final resting places.

Concurrently, other victims were found where one would expect them in a lean-to collapse, except that the void spaces were overlaid by multiple layers of pancaked slabs. Tragically, the intense, fuel-fed fires that accompanied the airliner impact also rendered some potentially survivable void spaces as non-survivable tombs.

As previously noted, the WTC terrorist attacks caused a total collapse of the twin towers, but the falling debris struck other buildings, causing a combination of collapse types over the entire disaster site. Not only were firefighters confronted with a variety of collapse types, each of which required a certain combination of approaches, but because these different collapse types were on the same site, it created further complications. To make the situation worse, some of the most experienced collapse SAR experts were themselves buried in the debris.

The Five Stages of Collapse SAR

The collapse SAR strategy employed by the modern fire and rescue service has been boiled down to the following standard five stages of collapse SAR:

1. Response, size-up, and reconnaissance
2. Surface rescue
3. Void-space search
4. Selected debris removal
5. General debris removal

These five stages of collapse rescue are applicable regardless of the size of the collapse or the scope of the overall disaster. This system is employed whether the collapse involves a single-family dwelling, a multistory hotel, or a collapse the size of those that occurred in the 9-11 attacks on the Pentagon and WTC. In fact, the larger the collapse (or collapse disaster), the more important it is that all firefighters and rescuers understand and employ the five stages of collapse rescue.

These five stages and the tactics and strategies that underlie them are based on—but not limited to—the following sources:

- Experiences of the London Fire Brigade during World War II
- Experiences of fire departments on the east coast of the United States, some of whom have been dealing with collapse emergencies for two centuries
- Collapse disasters following west coast earthquakes
- Experiences of FEMA USAR task forces and ISTs, whose members have responded to collapse disasters in disparate places such as L.A. County (Northridge earthquake), San Francisco (Loma Prieta earthquake, train derailment—Figure 2–21); Puerto Rico (gas explosion); Turkey, Japan, Greece, Mexico City, and Taiwan (various earthquakes); and East Africa (U.S. embassy bombings)
- Oklahoma City bombing
- The 9-11 attacks
- Experiences of rescue units and USAR teams from other nations

During Stages 1 through 4 of collapse rescue, rescuers typically encounter the most difficult, complex, dangerous, and time-consuming rescue problems, often resulting in *miracle* rescues. Locating and rescuing deeply entombed victims from collapsed structures is bread-and-butter for modern fire department USAR programs. It's a specialty of FEMA's 28 USAR task forces and a mainstay of internationally deployed USAR teams.[7]

Fig. 2–21 Fire Department Units in a Staging Area during Collapse Rescue Operations at an L.A. Area Train Derailment (Courtesy FEMA)

During Stages 1 through 4 of collapse rescue, rescuers typically encounter the most difficult, complex, dangerous, and time-consuming rescue problems, often resulting in *miracle* rescues. Locating and rescuing deeply entombed victims from collapsed structures is bread-and-butter for modern fire department USAR programs. It's a specialty of FEMA's 28 USAR task forces and a mainstay of internationally deployed USAR teams.[7]

It's important to note that certain strategic objectives and tactics (search, recon, shoring, medical treatment, etc.) may be employed through the course of all five stages of collapse rescue, depending on the conditions encountered. The stages of collapse rescue are not divided by empirical walls. Rescuers must use a certain degree of finesse and judgment based on experience and training to ensure that the most good is being done to locate and rescue (or recover) the most victims. Therefore, the five stages of collapse rescue are simply a framework within which all tasks required to extract the victims proceed systematically.

That was certainly the case at the Pentagon attack. For example, structural stabilization (generally shoring, tie-backs, and other means) is something that may be required from the first arrival of units until the end of stage 4 (or even stage 5). At the Pentagon, some sections of the building on the perimeter of the actual collapse required stabilization to prevent the rest of the building from coming down, even as last bits of collapse debris were being removed.

Determining how and when to mitigate such hazards shown in Figure 2–22—and how and when to commit personnel to SAR operations before overhead hazards can be reduced—requires a combination of training, experience, and an evaluation of the risk-vs-gain equation.

Fig. 2–22 Overhead Hazards Confronting Rescuers at the WTC Collapse

Removal of overhead hazards is another potential multi-stage task. At the Pentagon, new sources of suspended debris were continually encountered as each layer of the collapse was being peeled away. Another example of a multi-stage tactic is combined canine search operations with technical searches. This tactic may be required to search different parts of a collapsed structure during the initial reconnaissance and then repeatedly as the collapse area is de-layered.

Void spaces need to be searched whenever they are encountered. Therefore, even after stage 3 is completed and stage 4 (selected debris removal) begins, additional voids may be uncovered by heavy equipment. Those voids must be searched for additional victims before debris removal continues.

Yet another example of a multi-stage tasking is medical treatment, or rubble pile medicine. Rescuers may find themselves treating one trapped victim for many hours in one section of the collapse, while stages 1 through 4 proceed in other areas of the building. Unfortunately, no live victims were found at the Pentagon after the first hours; many of the victims who might otherwise have been survivors were dealt a deadly blow by the immense fire that erupted after the airliner struck.

Size-up/reconnaissance is another process that can last for days (or even weeks) in large or complex disasters through all five stages. That was the case at the Pentagon incident.

The following is a more detailed explanation (in order) of the five stages of collapse SAR (American system):

Stage 1: response, size-up, and reconnaissance

Response. Many collapse incidents are preceded by events that let everyone know something bad just happened. In some cases, firefighters/rescuers are literally shaken from their chairs or beds by the precipitating event, which might be an earthquake, explosion, or something else that causes buildings to fall. When the WTC and Pentagon were attacked, local firefighters didn't need to be told that something big just happened. Many of us have seen the raw video footage shot as the first airliner flew directly over the heads of FDNY personnel on the scene of an outside natural gas leak. There was clearly no doubt about what had just happened (Fig. 2–23).

Fig. 2–23 "Reconning" the Scene at the WTC Collapse

Likewise, Oklahoma City firefighters didn't wait for the alarm to sound after the Murrah federal building was bombed in 1995. They heard the explosion and felt the concussion. Many firefighters have rolled out on still alarms to gas explosions after their stations have been rocked by concussions. Earthquakes also leave little doubt about what has just happened, triggering automatic responses by agencies with predesignated protocols for just such events.

Earthquakes in particular tend to give immediate clues about their strength and the potential for structure collapse occurring in stricken areas. Firefighters conducting windshield surveys of their first-in districts in the moments following a quake will quickly find indications of how bad the quake was (Fig 2-24).

Consider the other possible causes of collapses, such as fires, natural gas explosions, vehicles into structures, bombs, etc. Each of these causes is often associated with particular hazards. Always consider the potential for explosion(s) from post-collapse gas leaks or secondary bombs during terrorist acts, etc. Consider any indication of a terrorist bombing because it *might* be accompanied by the release of nuclear, chemical, or biological agents. Consider precautions against the above-mentioned hazards and review any pre-attack plans for the incident building(s).

Arrival and size-up of the collapse area. What is the extent of the affected area? Is it a single building or an entire neighborhood or city? Conduct an eight-sided size-up of the involved building(s) and the surrounding area. Check the top, bottom-basement, and four sides of the building. Also check the air space around the building for falling hazards from adjacent structures and other aerial hazards. Finally, conduct a rotary sweep of the ground around the structure, looking for hazards such as ruptured gas mains, broken water mains, railroad tracks, and other potential ground-level problems.

Fig. 2–24 Sizing Up Collapse after the 1994 Northridge Quake

What time of day is it? What day of the week? Is it a holiday? What is the building's occupancy type (e.g., office and commercial vs. residential, school, or hospital, etc.)? Have the utilities (natural gas, water, propane, electricity, etc.) been shut off? If this has not been addressed, assign this task immediately and confirm that it has been completed. Are there uncontrolled fires, gas leaks, flooding, or other hazards that may take the lives of victims before they can be located and rescued? What is the overall situation in the immediate area of the collapse? What is the condition of the area surrounding the actual collapse? What is the construction of building(s)? What is the pattern of collapse?

What is the condition of internal and external masonry? What type of roof is on the building, and does it present a danger in its own right, as it did in the case of the Pentagon collapse? Are there basements or subterranean passageways? Are there heavy water tanks, air-conditioning units, heating plants, communications equipment, fuel tanks, or generators on the roof, upper levels, or in the basement? Is there heavy stock that can absorb water and cause secondary collapse? Are there vertical shafts for elevators and stairways?

Is there a potential for secondary devices or possible terrorist attack? Is there a hazmat problem? Is there a need for additional resources? What kind of resources and how many?

Reconnaissance for likely locations of trapped victims. How many potential victims are possibly missing or trapped? How many victims have been rescued and removed? From which rooms? Is there an inventory of the occupants normally found there? Is there a seating chart for offices and schools? Is there an accounting of which occupants are missing, which ones have been accounted for, and which ones have left the area? Mark the building as the process of size-up, recon, and search proceed.

Apparatus placement. Just as in fireground operations, the placement of fire/rescue apparatus and vehicles is both a strategic and a tactical issue. Parking bulky vehicles in the wrong locations can block access, place personnel in the collapse zone, expose units to radiated heat, prevent placement of hoses, hand lines, and aerial ladders, and block egress/ingress of critical resources. Some agencies have developed standard operating guidelines (SOG) for collapse operations, including the placement of apparatus and personnel. The first-arriving officers must quickly develop a strategy for the incident, and the placement of apparatus should reflect and support those decisions.

Aerial ladders/towers. The FDNY SOG for most structure collapse operations include the assignment of at least one tower ladder (aerial platform) at (or near) the front of the affected building(s). The ladder or platform is raised to an elevation from which a rescue/USAR-trained member and designated officers can observe the entire collapse zone and surrounding areas. This provides a constant lookout over the operational area, with particular emphasis on shifting walls, smoke or water from wall cracks, sagging roofs, and other signs of impending secondary collapse. The member assigned to this lookout position should have the means to immediately notify everyone on the scene when signs of impending collapse or explosion are noted. This may include a hand-held radio, an air horn, a whistle, or some other signaling device (Fig. 2–25).

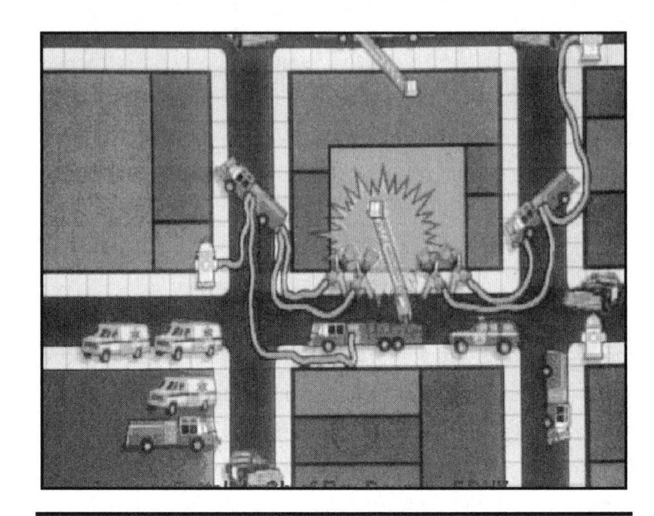

Fig. 2–25 Schematic of Typical FDNY Apparatus Placement at Collapse Emergencies (Courtesy FDNY)

The IC should consider assigning tower ladders (or aerial towers/platforms) to different sides of the collapse zone to maximize the observation capabilities for lookouts. In extreme cases, the FDNY has used tower ladders for emergency stabilization of upper-floor walls to protect rescuers below them while they make immediate rescues. The ladders are not considered shoring per se; in this situation, they're a last-ditch effort to keep the walls from coming in on top of personnel until more reliable shoring can be established.

Obviously the 9-11 attacks on the WTC were anomalous in terms of the conditions that usually confront fire department officers at potential collapse scenes. Therefore it's unfair to use typical guidelines for apparatus spotting to judge the placement of apparatus at the fires, which destroyed many fire/rescue vehicles when the buildings collapsed. But it's an example of how traditional guidelines for establishing collapse zones (i.e., one and one-half the height of the building away from its base) aren't always feasible in the real world, especially when dealing with the tight confines of a crowded urban area full of high-rise buildings.

Engine companies. Because of the potential for fire and explosions during collapse rescue operations, at least two sources of water should be secured for firefighting. Engine companies should be positioned to support effective water supply operations. Because collapse events may cause damage to nearby hydrants and water mains, it may be necessary to conduct relay-pumping operations.

At least two engine companies should be assigned to deploy hose lines at the collapse zone (preferably from different directions) to ensure immediate water on any fires that develop and to provide fire suppression following a secondary explosion.

The IC should consider assigning engine company crews dressed in full PPE as quick-attack fire-suppression teams and RICs. These could possibly be augmented by later-arriving rescue/USAR companies or truck companies.

Engine companies can also be used for manpower, cutting teams, medical teams, litter teams, and myriad other functions.

USAR/rescue companies. USAR and rescue company apparatus should be placed where the tools they carry can be put to the best use. This may be directly in front of the collapse (out of reach of the collapse zone), near an equipment pool, at the command post, or in a staging area. Reducing the distance to which heavy tools must be carried is important, so a reasonable location needs to be selected (Fig. 2–26).

Fig. 2–26 Leave Room for Heavy Equipment to be Positioned

Paramedic/ALS units. Based on the needs of the incident, ALS and paramedic units may be positioned in a medical group configuration, a mass casualty configuration, or somewhere else away from the collapse zone.

Battalion chiefs. If the battalion chief's vehicle is to be used as the command post, it's desirable to maintain a two-sided view of the incident if possible, while remaining outside the collapse zone.

Helicopters. As a general rule, helicopters should be kept a good distance from (and a significant height above) recently collapsed buildings until they are assessed for stability. This will reduce noise at the scene, eliminate blowing of loose material, and reduce the airborne concrete dust and dirt. Helicopters may be used as aerial vantage points during large-scale collapse disasters, and they may be used to shuttle manpower, equipment, and patients. It's generally recommended to designate helicopter landing zones a safe distance away to avoid the previously mentioned complications.

The placement of lighting units, generator units, air compressor units, commercial vacuum trucks, and other specialized units is incident-specific. The placement of cranes, skip loaders, track hoes, bulldozers, dump trucks, and other heavy equipment is incident-specific (Fig. 2–27).

Fig. 2–27 Placement of Equipment and Teams at the Pentagon Collapse

As shown in Figure 2–27, fire apparatus should initially be deployed in a way that allows cranes and other heavy equipment to be properly positioned without the need to re-deploy hose lines, aerial towers, and other units. This saves time and reduces complications that can lead to preventable danger and delays.

Staging areas. Staging areas should take into consideration such factors as ground vibration, debris-filled streets, the distance that equipment must be carried, etc.

USAR emergency signaling system

The international USAR community has adopted a standardized system of emergency communication at the site of structure collapses and other major rescue operations. While some local systems may differ, there are universally recognized signals that utilize air horns and other audible devices to transmit critical information to many people simultaneously at the site of major rescue sites. These signals include the following:

- Evacuate the area—three short blasts repeated
- Cease operations/all quiet—one long blast (three seconds)
- Resume operations—one long and one short blast

Stage 2: surface SAR

Stage 2 includes searching for victims while working from the exterior of the collapse, using whatever search means are available. All victims trapped at or near the surface beneath structural and non-structural elements (i.e., desks, beds, book cases, furniture, etc.) are rescued. This includes people buried beneath materials readily removable with hand tools, buckets, and other methods that don't require collapse rescue technical expertise and heavy equipment.

Victims found on top of the debris or lightly buried should be removed first. Initial rescue efforts should be directed at removing victims who can be *seen* or *heard*! The next rescue efforts should be directed at reaching victims whose locations are known even if they cannot be seen or heard (Fig. 2–28).

Fig. 2–28 Aerial View of the Pentagon Collapse

It should be mentioned that search-related tasks may be conducted throughout all five stages of the collapse SAR. The gathering of information included in stage 1 operations provides intelligence that will be helpful to the search process. In stage 2, we begin conducting physical searches for hidden victims. During stage 2, there is constant danger of causing secondary collapse or further crushing trapped victims. Therefore, appropriate precautions are necessary to prevent such adverse events.

Scene control is critical here. We should not allow passersby and relatives to scamper over the rubble pile, risking secondary collapse or the crushing of victims beneath the pile. Neither should observers/family members be allowed to pick away indiscriminately at structural members, whose removal may precipitate secondary collapse. They should not be permitted to use torches and other spark-generating tools without appropriate fire protection. One out-of-control fire, especially before the gas and other utilities are controlled, might render victims *unsalvageable* and endanger the lives of firefighters and other rescuers.

We should not allow heavy equipment to operate in any place where surface rescue and void-space search has yet to be conducted. Special caution and discipline are required whenever the use of heavy equipment is contemplated and especially when people may be trapped directly beneath the surface.

Do not hesitate to use the full force of available law enforcement resources to establish and maintain scene control. This will help prevent bystanders from causing secondary collapses, uncontrolled fires, or other events that could make the situation worse. Trapped victims and firefighters/rescuers must eventually crawl into void spaces and other high-risk locations to find and extract victims.

At the scene of any suspected terrorist event, it's doubly important to establish tight security controls around the entire collapse incident. Do not forget the potential for secondary devices to be located around the collapse site, including at or near the surface, where a stage 2 operation may disturb them and cause detonation.

Once the scene has been controlled and secured, caution firefighters/rescuers (including canine search teams) against movements and actions atop the collapse pile that could precipitate a secondary collapse. Every move should be thought-out by each person. One wrong move could have devastating effects on people trapped beneath your feet or adjacent to where you're standing and operating (including your coworkers).

Debris usually must be removed to facilitate the location and rescue of victims during stage 2 operations (Fig. 2–29). We want to move as little debris as possible to get the job done. This may mean working into specific areas where victims have been spotted, or it may mean the beginning of a de-layering operation (transitioning to stages 3 and 4).

Fig. 2–29 Picking through Debris in the Pentagon Collapse

It may be necessary to establish bucket brigades to remove debris from atop the collapse pile. If so, make use of designated open spaces to deposit debris where it will be out of the way of emergency operations. It should be located where dump trucks, front-end loaders, and other equipment can later move it to another site.

This is a function that may be assigned to convergent responders such as police officers, construction workers, and municipal employees. In disaster situations, bucket brigades and other debris-removal tasks may be assigned to passersby and relatives and friends of building occupants. This will help free up rescue-trained personnel to conduct the more technical tasks like cutting, breaching, lifting, and shoring.

As always, maintain a good watch for potential collapse using aerial ladders and other overhead vantage points; watch from the ground level and other appropriate spots. The safety officer should be very involved with evaluating the overall safety of the operation at this point.

As the last victims are removed from atop the collapse and signs of additional victims trapped near the surface diminish, consider transitioning into stage 3 operations or combining stages 2 and 3. The IC should make this determination based on the conditions at hand, after considering reports from those doing the work, and after considering the recommendations of the most experienced rescuers.

Stage 3: void-space search operations

After the victims who are obviously visible, lightly trapped, and readily reachable are rescued (i.e., after stage 2 is completed),[8] all potential survivable void spaces must be searched. This search for live victims should last until it's determined that there is virtually no chance of survival within the collapse zone. This is the beginning of stage 3 operations, which means that firefighters and other rescuers must explore every potential survivable void space that's visible. In many ways, this is the most dangerous part of the collapse rescue operation because rescuers are placed in harm's way, crawling through confined areas that may be unstable in search of victims.

Keep in mind that void spaces in a collapsed structure also qualify as confined spaces, with many of the dangers that are inherent in confined-space rescue emergencies. Consequently, rescuers and supervisors must consider the need to evaluate and address confined-space rescue hazards that may be encountered during rescue operations. Some agencies (including the LACoFD) make it mandatory for members of their regional/state/federal USAR task forces to maintain constant certification as confined-space rescue technicians in addition to other required training.

Exploration of voids and other likely survival places. Search for trapped victims by looking in void spaces and other places that could have afforded a reasonable chance of survival when the collapse occurred (Fig. 2–30). The following places should be considered among the priorities for physical search:

- Between pancaked slabs
- Sheltered parts of structures with high likelihood of surviving the collapse
- Beneath fallen walls
- Basements and other underground locations
- Voids beneath collapsed floors
- Beneath stairs and in stairwells
- Next to chimneys
- Spaces around sturdy furniture, safes, appliances, desks, etc.
- Rooms where access is difficult but rooms have not yet collapsed

Fig. 2–30 Search Cameras

As noted in Figure 2–30, a technical search specialist from FEMA's Florida USAR Task Force I prepares to use a search camera to look beneath layers of collapse debris for victims trapped in void spaces at the collapse of the WTC towers.

It may be necessary to breach walls, floors, roofs, and other structural elements. If so, take care not to further compromise the stability of the structure by cutting through weight-bearing beams and columns. Take care to avoid cutting tensioned cables within concrete slabs.

Breaching and shoring. Breaching and shoring may be required to reach victims. Initially try to avoid breaching walls because it may undermine the structural integrity of the rest of the building. It is generally safer to cut holes in floors and use the vertical-shaft approach than to breach walls. If you must breach a wall or cut a floor, cut a small hole first to ensure that you are not entering a hazardous area.

Shoring may be used to support weakening walls or floors. However, shoring should *not* be used to restore structural elements to their original positions. Rescuers should simply support the structural components to remain exactly as found post-collapse.

Attempts to force beams or walls into place may cause collapse. In the Pentagon operation, USAR task force members pounded wedges between the top of a shoring system and a damaged reinforced concrete beam to snug them up when the beam snapped from upward pressure, sending firefighters/rescuers scrambling. Keep timber shoring as short as possible (Fig. 2–31). The longer the timbers, the less weight they will support and the more unstable the shoring system. As a general rule, the maximum length of a vertical shoring member should be no more than 50 times its width.

Fig. 2–31 Shoring up the Interior of the Pentagon to Facilitate Void Space Search Operations Nearby

The strength of a shore depends on where it is anchored; if it's anchored to the floor, the shore depends on the strength of the floor to hold its load. Shoring should generally be conducted by firefighters assigned to USAR or rescue companies and firefighters on other units trained to perform these tasks. It should also be done under the supervision of a qualified officer.

In some situations, engineered shoring, designed and supervised by structural engineers, including FEMA USAR task force structures specialists, may be required. This was the case in the aftermath of the 1993 WTC bombing, the Oklahoma City bombing, and the 9-11 attacks at the WTC and the Pentagon. In disaster situations, innovative shoring options should be considered. This may include unconventional strategies such as the following:

- Filling air/sea transportation containers with soil and stacking them beneath damaged bridges or against large unstable buildings to prevent secondary collapse.
- Positioning cranes or even aerial ladders against unstable walls to prevent secondary collapse until more permanent shoring can be devised.
- Piling soil or sand against unstable structures to hold them up. In the aftermath of the Kobe earthquake, Japanese public works employees deposited 3-story piles of sand in some streets to protect underground petroleum and natural gas pipelines from the effects of secondary building collapse. Sand was even piled against some heavy concrete structures to prevent overturning during aftershocks.
- Using high-strength cables to stabilize damaged structures. In Kobe, engineers supervised an operation whereby cables were attached to a 10-story apartment house that was tilting and threatening to overturn. The cables were anchored to the foundation of an adjacent apartment building. During SAR operations at the Oklahoma City bombing, engineers supervised cabling operations to secure unstable sections of the Murrah building to the less-damaged front of the structure.

In some disasters there will be a shortage of wood and other materials normally used for shoring operations. This may be especially true in regions of the world where construction-grade wood is a rare commodity and where concrete and steel are the primary components of structures in densely populated cities. Likewise, wood for shoring operations may be severely lacking in the aftermath of a disaster. While it's recognized that wood is among the most effective shoring materials, if none is available, alternatives must be considered. In extreme cases, rescuers have resorted to stacking concrete blocks and other materials to fill voids and shore up unstable collapse zones. These are *not* accepted shoring practices, and they *aren't* taught in recognized USAR training courses. But, pragmatically speaking, rescuers may be forced to consider alternative options to get the job done under extreme conditions when lives hang in the balance and there are no other traditional alternatives.

Generally speaking, shoring should *not* be removed once it is placed in an unstable building (Fig. 2–32).

Fig. 2–32 Note Large Timbers on Left

In Figure 2–32, large timbers were used as shoring to help stabilize damaged areas of the Pentagon after the 9-11 attacks. Now FEMA USAR Task Force members are installing temporary pneumatic shores to support the ceiling.

Currently, the LACoFD USAR task force is exploring a radical new method for stabilizing buildings that uses a composite fabric to *wrap* damaged columns and other structural components. This material was originally designed for use in a revolutionary method of high-rise construction. It was intended to be stronger and more resistant to explosions and earthquakes than steel and concrete. In collaboration with construction, engineering, seismology, and design sciences, the LACoFD USAR task force was attending and teaching at the L.A. Tall Buildings Design Council when they came across the new fabric, almost by accident.

When asked if this new material could help stabilize buildings damaged by earthquakes, explosions, plane crashes, etc., the material's inventors said they hadn't even considered its use in emergency stabilization. They had concentrated only on original construction of high-rise buildings. The LACoFD and the manufacturer soon were embarking on a testing program to determine whether this was a viable alternative to traditional emergency shoring and stabilization methods.

Preliminary results indicate that the composite fabrics could be used to wrap columns and possibly entire buildings. The fabric could then be hardened with polymers that would make the structural components even stronger than their original design specifications. This may represent a new revolution in emergency shoring. This approach may prove especially effective for conducting USAR operations in places where timber for shoring is scarce or nonexistent (especially during disasters). It is yet another example of the benefits derived from expanded collaboration and information-sharing between the fire/rescue services and other applicable sciences and professions. In Figure 2–33, firefighters tunnel through a collapsed building after a train derailed at 90 mph in a residential area. Their objective was to locate and extract any victims who might be trapped alive in a *survivable void space*.

Fig. 2–33 Firefighters Tunneling for Victims in L.A. Train Derailment Collapse Operations

Consider establishing void-space search squads.[9] The void-space search squad is a concept based on decades of collapse rescue experience. The concept is not universally used, but was developed by recognized collapse rescue experts. It holds promise for situations in which it's necessary to probe all accessible void spaces to their ends, looking for signs of trapped victims within a collapse zone. Unlike tunneling or trenching (covered later), void-space probing is more akin to following an earthquake fault or narrow cave through the earth, exploiting weaknesses and openings wherever they may lead. As debris is encountered within the cracks and voids, it is removed to allow forward movement, and shoring is placed wherever necessary to maintain stability for ingress and egress (of rescuers and victims).

A void-space search squad generally consists of six members divided into two three-person teams and identified as a search team and a support team.

In Figure 2–34, advanced or engineered shoring may be required during void space search operations. Here, firefighters and FEMA USAR Task Force members search void spaces with protection from shoring designed by structural engineers attached to the USAR task forces.

Fig. 2–34 Shoring Protecting Firefighters at the Oklahoma City Bombing

As the names imply, the search team performs the initial penetration operation, moving forward along the cracks and voids within the collapse zone, occasionally breaching obstacles. Meanwhile, the support team provides tools, shoring material, expertise, and other assistance as the search team burrows its way toward trapped victims. To make maximum progress, the teams exploit the building's vulnerable (soft) points to move forward, while identifying and taking advantage of protective features like columns and beams to maintain the integrity of the search area as they proceed and (hopefully) remove live victims.

Search team positions and duties. The void-search squad officer supervises the operation and should generally be a USAR-trained company officer. The officer should assign the most experienced members to the search team. The officer triages the building to determine the most promising void spaces where live victims might be found. The most stable void spaces will allow maximum penetration with lower levels of risk to personnel; the officer should be trained and experienced to identify these conditions.

If a victim(s) is located, the officer will be responsible for determining appropriate extraction tactics. It should be noted that the path created by the void search squad might not be the best path of egress for the victim. According to O'Connell and McGroarty, removing larger sections of debris elsewhere may open up better alternate egress routes. This is especially important if the victim must be removed on a litter or backboard.

Void-entry technician is a position given to the first member to enter the collapse void. This position's main duty is to search the voids for the presence of victims and locate other voids within which victims may be trapped. Entering one void may lead to other, larger passageways within the collapsed structure. This is especially true where elevator shafts, hallways, and other access paths may still be partially intact.

As he makes his way into the void, the void-entry technician removes debris and passes it out to waiting members (Fig. 2–35). If possible, all debris should be removed from the void, not just moved to the side. This may require a bucket-brigade operation. Since the debris will likely have to be removed anyway, it makes sense only to handle it once.

Fig. 2–35 Void Space Search Operations Conducted by Tennessee FEMA USAR Task Force at the Pentagon Collapse

In Figure 2–35, canine search teams prepare to search void spaces at the site of a major train derailment in L.A. County.

The void entry technician may find it helpful to use small hand tools to cut rebar and pry debris away. Items such as search cameras may also be needed to look behind debris and into small void spaces for possible victims as the squad burrows into the collapse zone.

If a victim is found, the void entry technician should remove debris from the victim and make physical contact as soon as possible. A physical assessment should be made with emergency treatment as appropriate, including treating for dehydration, shock, crush syndrome, inhalation of concrete dust, etc.

The victim's identity should be obtained and transmitted back to incident command to establish victim-tracking and possibly help determine where additional live victims might be found (using housing or seating diagrams, etc). The following information should also be obtained and transmitted:

- Victim's location at the time of the collapse to help establish victim movement as the structure failed
- Type of victim injuries
- Whether other people were with or around the victim prior to the event
- Whether the victim heard other victims during entrapment and their possible locations
- Possibly, the cause of the collapse (i.e., explosion, etc.).

Shoring technician. This is the second void-space search squad member to enter the structure. The shoring technician is responsible for assisting with debris removal and installing box cribbing or shoring to maintain stability of the collapse void. This position requires advanced skills in determining and then building appropriate shoring as the squad moves further into the void space (Fig. 2–36).

Fig. 2–36 Members of a FEMA USAR Task Force from Tennessee Inspect a Void Space to Locate a Victim at the Pentagon Collapse

Support team positions and duties. The support team consists of three members responsible for supplying physical assistance, tools, shoring materials, patient-packaging equipment, etc. The team is made up of the void-expander technician, the support technician, and the tool and equipment technician. They are also the RIC for the search team.

The void-expander technicians assist the search team with ongoing operations. Their tasks vary according to the needs of the situation. They are primarily responsible for increasing the size of the path created by the search team. This, according to McConnell, enables additional personnel and equipment to enter the void space and provides better egress for the victim after being extracted.

The support technician may also be called a *go-for* or *relay person*. In essence, the support technician is responsible for gathering and moving tools and materials that are needed by the other squad members operating within the void space. He may also assist with expanding the void space, shoring, and other necessary support tasks.

The tool and equipment technician is responsible for establishing an equipment pool and material pool to enable the timely movement of equipment and materials as they are requested by other members of the void-space search squad. The tools must be tested for proper operation before they are passed to members working within the collapse zone. Broken and dirty tools must be rehabbed after use. In some cases, the tool and equipment technician must perform field repair of tools that are badly needed in the collapse. This member must maintain a tool log to ensure accountability of all tools during the operation. This is also a function that can be delegated to the logistics section once it's established at the incident.

Fig. 2–37 Extensive Debris after Oklahoma City Bombing

Stage 4: selected debris removal

After all known survivable void spaces are searched, selective debris removal begins. During this phase, rescuers work with heavy-equipment operators, structural engineers, construction and demolition contractors, and others to de-layer the collapse zone from top to bottom. As upper layers of the buildings are peeled away, additional void-space search operations are generally conducted to check newly accessible parts of the building for potential survivors.

Sometimes the strategy of void-space searches (stage 3) is alternated with selective debris removal (stage 4) until the entire collapse zone is dismantled and all victims are located and extracted. These are high-risk operations because the stability of the building is compromised by the original event and is often made far worse during stage 4.

In Figure 2–37, this view of the Murrah Building shortly after the Oklahoma City bombing demonstrates the immense amount of debris that must be removed in order to locate and rescue any trapped victims.

Prior to this stage, all accessible void spaces have been checked and victims removed. Now it's time to reduce the size of the debris piles, searching for additional victims or accessing victims known to be trapped below.

Selected debris removal must be accomplished based on a *plan*, not haphazardly! Cranes and heavy equipment may be needed under careful direction from USAR-trained heavy-equipment/rigging specialists (HERS) to selectively remove heavy debris. Debris piles are disassembled from the top down, in layers, looking for victims as the operation proceeds. Concentrate primary efforts at removing debris from areas where information suggests victims might be buried.

In Figure 2–38, debris being removed is considered possible evidence. Here Federal agents work in a cooperative effort with firefighters and USAR Task Forces at the Oklahoma City bombing to maintain the chain of custody for debris being removed during search and rescue operations. These factors must be considered by ICs and rescuers operating at suspicious explosions and other potential terrorist attacks.

Fig. 2–38 Maintaining a Chain of Custody

Stage 4 operations may continue for many days and may even exceed two weeks in some large-scale disasters. Even in single-site collapse emergencies, stage 4 may continue around the clock for days. Following the May 2001 collapse of a three-story banquet hall in Jerusalem, Israeli USAR teams conducted stages 3 and 4 for nearly 48 hrs.[10] Work had to be stopped several times to clear widow-makers[11] from the upper reaches of the collapsed building.

Controversy erupted when stage 4 was determined to be complete 42 hrs after the collapse, at a time when the fate of two workers hired for the wedding festivities was uncertain. Some, including relatives of the two employees, demanded that operations proceed in the rescue mode on the chance that the employees might still be alive, or until their bodies could be recovered. But the head of the Israeli rescue operation determined that the risk of continuing in the rescue mode (stages 3 and 4) was too great. This is but one example of the difficult challenges faced by ICs who must make a final determination as to when to proceed to stage 5, which essentially means abandoning hope of making live rescues in collapsed buildings.

In Figure 2–39 structural engineers attached to a FEMA USAR Incident Support Team and a USAR Task Force discuss options for selective debris removal operations at the collapse of the Pentagon. Safely delayering the collapse without causing a secondary collapse that could endanger rescuers working in the building was a complex engineering problem. This is somewhat typical of many large-scale collapse search and rescue operations.

Fig. 2–39 Discussing Delayering Options

Sometimes there is no definitive answer to the question: "When does stage 4 end and stage 5 begin?" In answering this question, the IC must ask himself the following three questions: (1) "Do I have positive confirmation that every possible survivable void space has been inspected for potential victims?" (2) "Have all victims been accounted for?" and (3) "Has the potential window of survivability been exhausted for any victims who might still be trapped within the rubble?"

In Figure 2–40 Stage 4 operations at the WTC included the selective cutting and removal of large steel pieces. Here a pair of workers operates from a personnel basket suspended from a crane while they use torches.

Fig. 2–40 Cutting Pieces into More Manageable Sections

If the answer to all three questions is *yes*, the IC is justified in proceeding to stage 5.

If the answers are *no*, then the wise IC will push the operation in rescue mode as long as it takes for each answer to become a reasonably conclusive yes. In fact, the IC is responsible for doing whatever it takes, within the bounds of reasonable safety and rationality, to continue in modes 3 and 4 until such time as the answers become *yes*. Anything less may be characterized as mere guesswork on the part of the IC, and it's known that guesswork has often doomed viable victims to untimely deaths during technical rescues and disasters.

In Figure 2–41, rescuers and a canine search team search a void space that was uncovered during selective debris removal operations at the Pentagon.

Fig. 2–41 Canine Search Team

Stage 5: general debris removal

Stage 5 generally signals the end of SAR operations and a transition to the recovery phase. These operations should be undertaken only after all other potential life-sustaining voids have been physically or technically explored for signs of victims and after there is reasonable assurance that no survivors remain inside the collapse.

In Figure 2–42, during Stage 5 operations at the WTC collapse, one crane moves into position while other heavy equipment works in the background.

Fig. 2–42 Heavy Equipment at Work

As stage 5 operations proceed, firefighters/rescuers must remain alert to signs of life (or life-sustaining conditions) that require further exploration before the area is demolished and the rubble removed. They must also remain vigilant for human remains and potential evidence of the cause of the collapse. This evidence will help identify and repatriate the deceased with their families and may assist investigators in determining the factors that led to the structural failure.

The aerial view of the WTC collapse site shown in Figure 2–43 shows the immense amount of debris that was ultimately removed during search, rescue, and recovery operations after the 9-11 attacks.

Fig. 2–43 Mountains of Debris

Stage 5 operations may include the use of heavy equipment to bulldoze, demolish, and remove large sections of building, as well as the debris piles left behind. It may involve hundreds of rescuers or laborers participating in the hand-removal of tons of debris not accessible to heavy equipment. A search for deceased victims in the rubble is sometimes continued through stage 5. It may be necessary to sift all the debris for evidence if the event is an act of terrorism, arson, or some other crime.

Applying LCES during Collapse SAR Operations

LCES was developed for wildland firefighting operations and has been successfully used during structure collapse operations since being implemented during USAR operations at the Oklahoma City bombing. LCES was successfully applied during the Pentagon collapse rescue operations. Before committing personnel to the danger zone of a partially collapsed structure, the IC should always ensure that LCES has somehow been addressed and that all members are aware of them. The following is a brief review of LCES as it applies to structure collapse operations.

Lookout. Some member of the team or another reliable responder should be assigned to observe the structure for signs of impending secondary collapse, secondary explosion, fire, or other immediate life hazard. It may be necessary to place the lookout in the basket of an aerial platform, on an aerial ladder, or on an adjacent building to provide a view of the entire collapse zone. It may be necessary to designate multiple lookouts. It may be necessary to use theodolites, plumb bobs, and other tools that can indicate movement of a building toward secondary collapse.

Figure 2–44 shows exactly why the use of LCES is critical during collapse operations and other technical rescue emergencies where personnel are vulnerable to secondary events. In this series of photos taken at the Pentagon collapse operations, two members of the FEMA USAR IST (structures specialist on left, and safety officer on right) are performing lookout duty while FEMA USAR Task Force members are conducting interior shoring and stabilization operations at a critical point in the incident. They noticed a fire erupting in the collapse zone just seconds after a large "bang" was heard and a reinforced concrete beam snapped. Immediately, they made an "emergency traffic" request and ordered all rescuers to conduct an operational retreat from the collapse zone.

Fig. 2–44 FEMA USAR Incident Support Team Safety Officer and Structures Specialist Acting as Lookouts at Pentagon Collapse, Just as a Fire Begins

In Figure 2–45, the fire rapidly grows in intensity, and FEMA USAR Task Force members evacuate the collapse zone while on-scene Arlington County Fire Department (ACFD) units maneuver to attack the fire. Air horns are sounding in the standard patter to reinforce the radio order to conduct an operational retreat.

Fig. 2–45 Operational Retreat and Attack Maneuvers After Lookouts Sound the Alert

In Figure 2–46, all teams conduct a head count and report their personnel accountability report (PAR) to the IC, ACFD units get water on the fire while the IST Safety Officer observes for any additional hazards.

Fig. 2–46 Looking for Additional Hazards as the Fire is Knocked Down

Communication. Each collapse SAR operation should have a communication plan that includes designated radio channels for certain functions and teams. The plan should also include other forms of communication such as voice/hand signals, whistles, and air horns. All personnel operating in and around the collapse zone should be familiar with the communication plan, and each officer should ensure that his charges are using the components of the plan appropriately.

At the command post at a major trail derailment that buried occupied homes (Fig. 2–47), chief officers communicate with division and group supervisors and other units on the scene. Clear communication and a clearly identified command post are important factors during major rescue emergencies.

Fig. 2–47 Effective Command Post Communications are Critical

Clear position designations are also critical to communications (Fig. 2–48). The use of identification vests, marked helmets, armbands, or other identifiers should be mandatory. If none of these are available, use marker pens to hand print designations on shirts, helmets, or even on arms. This is preferable to the chaos that occurs when everyone looks the same and no one can identify who's in charge of what.

For disaster operations, predesignated caches of armbands, helmets, and vests can assist in the process of communication. Communication also includes the use of clear and concise IAPs that coincide with what's actually happening in the collapsed structure.

Fig. 2–48 Face-to-face Communication between Rescue Officers and Incident Commanders at a Major Collapse Operation

Escape route. Every team and team member should have a clear idea of the primary and alternate escape routes, preferably before entering the collapse zone and certainly once they're in it (Fig. 2–49). Each officer should brief his team on escape routes during each entry and as conditions change. Escape routes should be the fastest, safest way out of the collapse zone or to a safe refuge in the event of a secondary collapse, fire, secondary explosion, flood, or other unexpected event. If necessary, the escape route should be identified by fluorescent spray-paint or lumber crayon markings, signs, fireline taping, and/or other method that clearly identifies.

During major collapse rescue operations and other technical rescue emergencies, secondary events should always be of concern to IC, to supervisors, and to all personnel on the scene. Paramount is the establishment of multiple escape routes in the event of an unexpected adverse event that endangers the lives of rescuers.

Fig. 2–49 Using Pre-designated Escape Routes as a Secondary Collapse Starts

In this series of photos shown in Figures 2–49 through 2–52, Stage 4 operations at the Pentagon are interrupted when a mishap occurs. This event resulted in an Operational Retreat, with FEMA USAR Task Force members and local firefighters using pre-designated escape routes to avoid being trapped by the secondary collapse that ensues. Here, an ultra-high-reach multi-processor (known by rescuers at the Pentagon as "T-Rex") takes a "bite" as it selectively cuts and removes debris while attempting to uncover additional void spaces in the layers of collapse. Personnel working in the area have already been alerted and moved to a safe zone, but a number of teams are still conducting interior shoring operations in the perimeter of the main collapse. Lookouts have been placed in multiple strategic locations, communications have been double-checked, and escape routes have been discussed among rescuers. In this photo, we see a FEMA IST structures specialist (on right) watching with concern as the T-Rex bites into a vulnerable beam and slab.

As the T-Rex digs into the debris, a beam snaps and precipitates a secondary collapse (Fig. 2–50). At the moment of this photo, the T-Rex operator is attempting to "hold" the collapse, to slow the progression of the collapse that he can see occurring. The IST Structures Specialist sees that there is trouble. Meanwhile, personnel inside the Pentagon have heard the snap of the beam and are already heading into their designated escape routes.

Fig. 2–50 Equipment Operator Tries to Slow the Collapse

The T-Rex is unable to prevent the collapse, and the slabs progressively fail (Fig. 2–51). This precipitates a sort of avalanche of material that slides into recently shored interior areas, striking sections that rescuers have just evacuated, as the IST structures specialist and another worker scramble to avoid being struck by debris.

Fig. 2–51 Debris Avalanche Occurs as Collapse Continues

After holding back as much of the collapse as possible, the operator places the T-Rex in a resting position and locks it in place as the IST and FEMA USAR Task Forces conduct a head count, transmit their PAR, and begin to assess the results of the secondary collapse and its effect on shoring systems and structural integrity. Some shoring systems were found to be buried in debris, but they basically held and continued supporting the Pentagon (Fig. 2–52).

Fig. 2–52 Some Shores Continue to Hold after the Collapse, Allowing Rescuers to Escape

During SAR operations following an earthquake that shook the Philippines in 1992, USAR specialists from Dade County (FL) and Fairfax County (VA)[12] found that rapid escape through the corridors of an overturned hotel was unfeasible during aftershocks. To expedite egress from the collapsed building, rescuers stacked mattresses outside windows. The agreed upon escape route was through these designated windows. Members would dive out one at a time, each rescuer rolling off the mattresses just in time for the next team member to land safely.

Stacking mattresses might seem comical to some who've never operated inside a collapsed building with recurring aftershocks. However, it was clearly a simple and workable plan that successfully evacuated rescuers from the collapse zone numerous times over a period of several days. When faced with unusual conditions, it's important for team leaders and officers to think outside the box when addressing the safety needs of fellow rescuers.

Safe zone. Team leaders and officers should identify at least one safe zone, or area that is safe from secondary collapse and other hazards. Rescuers can retreat to this zone in the event of an aftershock, explosion, secondary collapse, or other unplanned event. The safe zone may be outside the building and beyond the collapse zone, usually the same distance as the height of the building.

If a secondary collapse occurs, where will rescuers go if they cannot exit the overall collapse zone in a timely manner? These are questions that must be answered by supervisors and IC at major rescue operations (Fig. 2–53).

Fig. 2–53 Where is the Safe Zone?

If escape to the outside will take too much time or is otherwise unfeasible, the safe zone may be within a stairwell or other fortified area within a building. In some cases, a safe zone may be constructed within a damaged building, fortified by shoring or other methods. Everyone entering the collapse zone should be clearly aware of the safe zone(s). In the case of an unplanned event, team leaders or officers should conduct head counts at safe zones to ensure that all rescuers reached safety and to determine whether any are in need of assistance

Spontaneous Collapse

In aging cities across the Untied States and other nations, there is a phenomenon known as spontaneous collapse. Generally speaking, the term refers to a situation where a building has—for no apparent reason—suffered partial or total collapse. Even though there may be no outward signs, such as an earthquake or explosion indicating the cause of the collapse, there is always some reason for a building to fail. It may be a failure of structural members, deterioration of the building materials, inadequate design, or a number of other causes not obvious to the casual observer. The point here is that structures can fail without any warning, even when there is no earthquake, tornado, hurricane, explosion, or other obvious cause.

One example occurred in December 2000 when an aging two-story apartment complex suddenly pancaked down, killing one resident and trapping or injuring 36 people. There was no earthquake, explosion, or other outward cause associated with the collapse. It's possible that the 1994 Northridge earthquake and other tremors that have affected downtown Los Angeles over the past few decades contributed to the collapse.

The Los Angeles Fire Department (LAFD) managed to rescue a number of trapped victims within the first few hours and then spent the next several days shoring up the structure for investigators. The failure, coming six years after the Northridge earthquake, was preceded by typical signs and symptoms of an impending collapse. There were unexplained creaking and snapping noises, leaky pipes, spreading cracks, and doors that suddenly would not close. In fact, one resident was on the telephone attempting to report the problems to the city's building department when the collapse occurred. It soon became apparent that the owner of the building should have been aware that the building was in danger of failing.

In another sign of the times, prosecutors and police in Los Angeles have considered charging those responsible for the building's condition with criminal acts that contributed to its collapse. Building inspectors are trying to determine who actually owns the apartment complex. In a nearly unprecedented move, the Los Angeles Police Department (LAPD) is conducting a murder investigation, attempting to determine whether someone is directly or indirectly responsible for the death of an innocent resident. This type of action hasn't been attempted since the catastrophic collapse (17 fatalities) of the Northridge Meadows Apartments during the 1994 Northridge earthquake.

There's one troubling issue about the current L.A. case; two years prior to the building's failure, a building and safety inspector found a crack in the foundation but deemed that it wasn't an immediate threat to life. After months of wrangling with the owners—who were at that time known—building safety inspectors forced the owner to make repairs, signed off by a city inspector.

Also, at the time of this writing, investigators are seeking a possible link between the cracked foundation, the repairs, and the collapse. There are questions about whether the repairs were adequate and whether or not they should have been approved. As noted by one LAPD investigator, "We're not overlooking anything. If (the investigation) points to some inspector that is negligent, that he knowingly signed off on a building that he knew was a life-threatening hazard that could be a criminal act." The investigator added, "I'm not saying that happened, but a building collapsed and somebody died. And we have to determine if somebody is responsible for that."[13]

In Turkey, recriminations continue as more evidence of shoddy construction methods, failure to follow original design specifications and local regulations, and the use of improper materials continues to surface. Turkish police and prosecutors are considering criminal charges against construction companies and contractors who knowingly violated seismic codes and design specifications to cut construction costs. In some cases, builders have left town to avoid criminal and civil penalties.

Increasing pressure is on Turkish building inspectors to ensure that future construction meets the appropriate standards and to prevent corner-cutting that endangers the inhabitants of office and residential buildings. There's also pressure to upgrade current building standards.

Other Legal Issues

The previously mentioned L.A. case has potential implications for fire/rescue personnel faced with the consequences of buildings damaged by fire, earthquakes, and other causes. Before allowing residents back into damaged structures, it's becoming increasingly important to conduct a valid risk evaluation, preferably after consulting the responsible building and safety authorities. Even allowing residents and business owners into a fire-damaged building to gather records and other goods could have serious consequences if the building's condition hasn't been properly assessed and if adequate shoring and other safety measures are neglected by fire officials. Given the current and future state of the law on such cases, it's entirely plausible to postulate that police and district attorneys will one day investigate firefighters or other rescuers for allowing the reoccupation of damaged buildings that later failed.

Rescuer safety is priority no. 1

The L.A. case has other implications. If the apartment complex failed so catastrophically with no obvious precipitating event, such as an earthquake or explosion, how many more like it are on the verge of failure across Los Angeles and other metropolitan areas?

And how will these already-compromised structures perform during the next major seismic event? For fire/rescue personnel called upon to conduct SAR operations in these structures during a fire or after a major earthquake, it should be evident that the potential for secondary collapse is tremendous. It illustrates the dangers faced by personnel working in these environments and the importance of protecting the lives of rescuers who voluntarily enter them.

To conduct effective emergency operations, the safety of rescuers must remain a prime consideration during the course of structure collapse SAR operations. After the 1985 Mexico City earthquake, it was reported that more than 100 would-be rescuers died in secondary collapses, falls, and even drownings when they were trapped by rising water from broken pipes and ruptured upper-floor water tanks. Would-be rescuers have been killed in other collapse operations, including the Oklahoma City bombing.

The entrapment, injury, or death of rescuers creates effects that extend beyond the personal suffering of the rescuers themselves. It causes other rescuers to stop what they're doing to assist the trapped or injured rescuer, which diverts help from people trapped in the original event. Entrapment or death of rescuers may also cause hesitation on the part of others who may choose not to place themselves at risk of the same fate. Lastly, these mishaps reduce the effectiveness of rescue teams, not just by eliminating the rescuers themselves but also by causing additional mental trauma to their teammates.

In rescue, there are no *acceptable* casualties. If a rescuer is lost, seriously injured, or killed, it usually means something went wrong that might have been preventable. That's not to say there won't be occasional loss of life among fire/rescue personnel. We all know that an aftershock or secondary explosion can bring down a damaged building in toto, killing rescuers and trapped victims alike, regardless of how much shoring was placed or how many other precautions were taken. Even a small and unexpected secondary collapse can kill rescuers within a damaged building. Committing oneself to enter a damaged building to conduct SAR is *always* a calculated risk, but we want to reduce those risks to a reasonable level whenever possible through effective planning, training, equipment procurement, and rational emergency response.

Every rescuer is a safety officer. Whether professional or volunteer firefighters, or USAR team members (or both), our primary duties during collapse SAR are to accomplish the following:

- Respond in a timely manner
- Accurately assess the situation
- Request sufficient resources
- Organize the scene
- Provide the highest level of assistance to the greatest number of victims
- Ensure that all viable victims are located and extracted
- Recover the dead when possible
- Return home safely and in good health

To accomplish this, each member of the team must view himself or herself as a safety officer. Each must be constantly alert to danger, always playing the what-if game, and always prepared to alert other members to take protective action in the event of an unanticipated event. Even when there's a designated safety officer on the scene whose official job is to monitor all operations to ensure that all applicable safety requirements are met, every member is essentially a safety officer in his own right.

Operational Retreat, Head Count, and Rescuing the Rescuers

All rescuers and team members must be familiar with the signal to begin an operational retreat. An operational retreat is the immediate exit of the collapse zone or building when a standard operational retreat signal is given. Operational retreat signals are given when secondary collapses are imminent or when other immediate life hazards have been discovered. An operational retreat system is necessary for the IC to ensure that all personnel are safe and accounted for and to determine whether any rescuers are missing, trapped, or injured.

Supervisors of personnel operating in vulnerable locations, like the USAR firefighters conducting a collapse tunneling operation in Figure 2–54, should be especially cognizant of the need to ensure adequate personnel accountability, and for ensuring members receive word of operational retreat orders if something goes wrong. They should be prepared to conduct a rapid PAR and to initiate a rapid intervention operation if the location of any rescuers cannot be confirmed.

Fig. 2–54 Rapid PARs Can Ensure Safety

With those concepts as a backdrop, imagine now that a major aftershock has occurred or perhaps an explosion caused by a natural gas leak resulting from the aftershock. Imagine that you've suddenly become lost, trapped, or injured by a secondary collapse. Even worse, imagine that the building has also caught fire, threatening to burn you alive while you wait for assistance from your comrades or that broken water pipes are threatening to drown you. You now count yourself among the unfortunate victims trapped within the building. Time is running out; the clock is now ticking.

If as a rescuer, you become lost, trapped, or injured under these conditions, you would (and should) expect that the IC will mount a rapid intervention operation. The operation will attempt to locate and rescue you from harm before normal SAR operations resume or simultaneous to other SAR efforts.

All rescuers and team members should be prepared to conduct rapid intervention operations to rescue other team members who become lost, trapped, or injured during the course of collapse SAR operations. It should be standard protocol for the IC or OIC to designate an RIC and have a rapid intervention plan in place during collapse rescue operations.

The RIC may be another team operating in the same building, a team operating next door, or a team on standby outside the collapse zone. It could even be a team that's rehabbing nearby. Preferably the RIC will be on standby outside the collapse zone, dressed in full battle gear to immediately launch a rapid intervention effort if something goes wrong. In a disaster, the IC or OIC may not have that luxury. All the same, they should have a plan to rescue personnel who become lost, injured, or trapped. The plan should include a radio channel for RIC operations, an RIC team officer, protocol for handing off rescued personnel to the medical group, etc.

There should be a standard radio designation for RIC operations as well as for other notification and communication protocols. When a rescuer is found in need of assistance, a standard radio call such as *rescuer down*, *firefighter down*, or *mayday* (depending on the individual agency's protocols) should be issued. The IC or OIC should call for radio silence to ascertain the downed rescuer's location and predicament, and an RIC operation should be launched immediately, using the designated RIC channel.

Case Study 1: Collapse Rescue Operations at the Pentagon

The 9-11 attacks in New York and at the Pentagon created unprecedented challenges for local, state, and federal fire/rescue agencies charged with finding and rescuing or recovering victims from within the collapse zones. Volumes can be (and will be) written about the WTC attack and subsequent collapse, which killed 343 firefighters and more than 100 officers of the Port Authority and NYPD, the darkest day in the history of the American Fire Service and the nation in general.

Fig. 2–55 The Pentagon in Flames

The USAR operations at the WTC were of such magnitude and scope that we cannot even begin to convey the story within the confines of this book. Suffice to say that the twin towers collapses created every imaginable collapse rescue problem, all in one place and all at the same time. It was the largest and most deadly single structural collapse in history, followed by the largest single SAR operation in history. It took more than seven months to recover all the dead who could be found[14] and to remove more than 1.4 million tons of debris, carried away in more than 98,000 truckloads.[15] The firefighters and rescuers who lived through those dark days on the site of the WTC collapses have stories to tell that we can never convey in this book. To maintain the focus of this chapter on SAR in collapsed buildings, the case study of the USAR operations at the Pentagon collapse is presented. In Figure 2–55, people are trapped on upper floors prior to the collapse as local firefighters attacked the fire, rescued victims, and treated the injured.

Emergency response to the Pentagon. In the aftermath of the plane crash into the Pentagon, the ACFD immediately responded to assist the military crash/fire/rescue unit already on the scene.16 First-arriving firefighters were confronted with an intense, jet-fuel-fed fire engulfing a significant section of the Pentagon between wedges 1 and 2. The ACFD, the jurisdictional fire department for the Pentagon, rapidly requested additional alarms as well as special calls for extra resources. Soon the ACFD would be joined by neighboring fire departments responding to provide mutual aid, including those from Washington, D.C., Alexandria (VA), Fairfax County (VA), and Montgomery County (MD).

Fire attack and rescue operations were complicated by thousands of gallons of jet fuel forcibly injected through all five floors of the Pentagon. This occurred as the 757 aircraft plunged through the first floor of three outer rings (E, D, and C) in a place halfway between lateral landmarks of the building known as wedges 1 and 2. Coincidentally, the plane struck the juncture where an ongoing Pentagon renovation/strengthening project met a section of the building that had yet to undergo structural strengthening.

Since the attack, some engineers have credited the Pentagon renovation project for reducing the level of damage and casualties when Flight 77 struck. According to Doug Sunshine, blast mitigation program manager for the Federal Defense Threat Reduction Agency, the new reinforced walls and roof system, blast-resistant windows, and other improvements probably limited the initial impact damage and reduced the dispersion of fuel and fire.17

Nevertheless, many people were immediately missing and/or trapped by the fire at different levels of the building after impact. The Pentagon, with its thick reinforced concrete floors, heavy concrete-frame and masonry-infill walls, and projectile-repellent windows, contained the heat and smoke much like a massive oven. Standpipes were ruptured, requiring long hose stretches and extensions. Firefighters took a heavy beating during the course of fire suppression and SAR. The structure had not yet collapsed, but the plane had left a large hole in the first and second floors between wedges 1 and 2, with many structural columns blown out or under attack by fire.

To make matters worse, during the initial fire attack, ACFD Assistant Chief James Schwartz (who became the Pentagon IC) received a warning that another airplane was apparently headed for the Capital and possibly the Pentagon. Considering the dual-plane attack just witnessed in New York, Chief Schwartz ordered an operational retreat and a PAR. After notification that the threat had passed, possibly when Flight 93 disappeared from the radar screens over Pennsylvania, the fire attack resumed. Firefighters retraced their lines and found nozzles to begin flowing water onto the fire, continued SAR to locate and extract people still lost or trapped in the Pentagon, and continued the multi-casualty operations in front of the building.

Pentagon collapses during fire-suppression operations. Minutes later, yet another warning of an incoming aircraft was received by Arlington Command. Once again, an operational retreat was ordered. When the all clear was given, the multi-alarm firefighting forces were regrouping and reentering the Pentagon to attack the fire when the heat and flames suddenly intensified. The smoke level dramatically increased, and a large five-story section of the structure collapsed.

Just as in the WTC attacks, catastrophic structural collapse at the Pentagon was not a direct result of the airplane's impact into the building. The collapse was rather a result of structural damage from the impact combined with untenable heat from thousands of gallons of burning jet fuel. This heat ultimately weakened structural members to the point of catastrophic failure (Fig. 2–56).

Fig. 2–56 Firefighters Barely Escape Being Buried by Debris when Burning Sections of the Pentagon Collapse

Just prior to the collapse, live victims had been seen through the windows, attempting to escape smoke and fire in the upper floors.

The confirmed presence of live victims before the collapse made it imperative to knock down the fire and mount an intense USAR operation. Although the fire and collapse encompassed a comparatively small portion of the overall mass and footprint of the Pentagon, it was still a large geographic area. The fire area measured approximately 400 × 300 ft, with all five floors involved. The collapse zone itself was approximately 80 × 240 ft and five stories high.

The collapse, occurring during a major firefighting effort, obviously complicated the situation for the ACFD (Fig. 2–57). Realizing the need for specialized USAR personnel and resources to mount a timely and massive collapse SAR operation, ACFD Fire Chief Edward Plaugher activated his technical rescue team. He requested similar teams from the neighboring Alexandria Fire Department (AFD) and the Military District of Washington (MDW).18 Plaugher also consulted with on-scene members of the Fairfax County Fire and Rescue Department, whose Virginia Task Force 1 (VATF-1) is one of the 28 FEMA national USAR task forces strategically located across the United States.

Fig. 2–57 Reassessing the New Collapse

FEMA USAR task force system activated. FEMA headquarters USAR program staff recommended activation of four USAR task forces for both incident sites within the hour of the attacks. The New York aircraft attacks were seen on television by FEMA senior managers and program staff, and activation authority was authorized verbally at that time. Steve Presgraves immediately activated the emergency support function #9 (ESF-9) cell in the FEMA National Interagency Emergency Operations Center (NIEOC). Four task forces were selected for both New York City and the Pentagon for immediate activation and deployment using non-specific disaster funding until declarations could be authorized. The ESF-9 challenges were to activate task forces and get them on the road and identify IST resources on the east coast to establish an IST presence in two locations, since the commercial air transport system was shut down.

When the IC request for USAR support had been processed through the state to the federal level, the eight task forces had already been given verbal activation orders and were on their way. They began preparing and assembling personnel and equipment. The selection process for the task forces was based on closest-two resources and those that were available through mutual aid. The second set of task forces was selected by FEMA based on the next-closest resource because of ground transportation requirements. Pennsylvania is obviously closer to the Pentagon than NYC. But to send Pennsylvania Task Force 1 (PA-TF1) to the Pentagon would have required task forces from even further away to drive past Washington D.C. to get to NYC as a second wave.

Fairfax County officers assisted ACFD officers with the process of making a formal federal request for USAR resources through the state of Virginia and FEMA (headquartered just across the river in Washington, D.C.). This process quickly opened the door for additional federal resources that extended far beyond the USAR portion of the Federal Response Plan. It included teams to help manage fire, medical/health, and hazmat disasters. Fairfax County USAR task force received its official activation order from FEMA by 1100 hours. The activation order also included Maryland Task Force 1 (MDTF-1), VATF-2, and Tennessee Task Force 1 (TNTF-1).

Not surprisingly, back in Fairfax County, VATF-1 members had already conducted an internal alert and were preparing to respond to the ongoing WTC disaster. Fairfax County firefighters/rescuers not already committed to the Pentagon fire operations prepared to deploy to New York. When the Pentagon collapse occurred, VATF-1 was that much closer to deployment, thereby reducing their response time when they were diverted to the Pentagon. The same thing happened in Virginia Beach, where VATF-2 prepared to deploy to New York and across the river in Montgomery County, where MDTF-1 was in deployment mode for the WTC collapse. In Figure 2–58, as USAR operations are initiated, local firefighters continue to attack the fire and search for victims wherever they can make access.

Fig. 2–58 Attacking the Fire

VATF-1 left for the Pentagon at 1:30 p.m., arriving around 2:00 p.m. MDTF-1 arrived shortly thereafter, and VATF-2 arrived around 10:00 p.m. The Tennessee task force drove all night, arriving in D.C. on the morning of September 12.

Meanwhile, personnel from the FEMA Red IST-Advance (IST-A)[19] began arriving at the Pentagon site around 3:00 p.m. on September 11. Often, IST-A members, who respond from around the nation, report to the disaster site or a designated BoO using civilian airline transportation to get there. On September 11, this was impossible due to the grounding of all commercial flights in the country.

IST-A members converged on both New York City and the Pentagon traveling by any means possible, including fire department vehicles, rented cars, and military aircraft. Some members arrived the following day after driving all night. Others (including this author) arrived via a series of military aircraft cargo planes and Air Force helicopters. Because of the transportation complications, the Red IST-A operated with less than desirable staffing at the Pentagon for the first 24 to 36 hrs after the attack. Because it was designed specifically for disasters (the very nature of which result in unforeseen complications), the FEMA USAR task force system is designed with sufficient flexibility to operate efficiently under less than desirable conditions. In fact, it can be said that the FEMA USAR system is intended to function best under peak duress conditions.

USAR branch established. As part of the upgraded IAP, Chief Schwartz designated a USAR branch to help manage the collapse SAR operations. Key personnel from the local technical rescue teams, VATF-1, and the IST-A were assigned to the USAR branch. The timely division of functional responsibilities proved critical to the effectiveness of round-the-clock collapse SAR operations at the Pentagon over the next 12 days.

Considering the scope of the Pentagon disaster, the integration of FEMA USAR task forces was a smooth process. Several task forces are based in the greater Washington, D.C. area, and Chief Plaugher (a former member of the Fairfax County Fire Department) is familiar with the role, responsibilities, and operation of the task force system. Automatic mutual aid agreements existed for fire and medical response in the D.C. area, and joint training of task force members with both heavy and technical rescue teams provided a nearly seamless integration of the teams into the on-scene operation. In Figure 2–59, USAR task force members and personnel assigned to the FEMA USAR IST evaluate the collapse rescue problem on the evening of September 11, 2001.

Fig. 2–59 Evaluating the Collapse

Fairfax County Battalion Chief Jim Strickland,[20] who became the interim IST leader for the first 30 hrs, noted in his after-action report: "Chief Plaugher had worked with many of the VATF-1 staff as well as some of the FEMA USAR IST members. He immediately assigned the Fairfax USAR task force to the SAR Branch." Plaugher added, "This allowed the USAR teams to begin operations without delay. In other cases, there has proved to be a learning curve by the local jurisdiction about incoming national USAR response system resources that has consumed value time at the beginning of a mission." Plaugher went on to say that, "Fire Chiefs and local jurisdictions could shorten this learning curve and reduce the lag time by locating and interacting with FEMA USAR task forces in their region so that capabilities and requirements are known beforehand."

The main collapse area, in the front of the building was identified as division A, with the standard ICS assignments of divisions B, C, and D made in a clockwise sweep around division A. Battalion Chief Mike Tamillow (IST member and task force leader of VATF-1) and other Fairfax-based members met with ACFD officers at the ACFD command post to plan a long-term strategy to manage the Pentagon collapse operation.[21]

The national scope of the response to the collapse disasters in New York and Washington, D.C. was immediately apparent. Within one hour of the attacks, FEMA headquarters personnel activated 16 USAR task forces (992 rescuers, plus canine search teams) for the New York and Pentagon disasters. FEMA also assigned the White IST-A to coordinate the efforts of the USAR task forces being dispatched to support the city of New York's ongoing SAR operations. The remaining 12 USAR task forces were placed on alert status. It was already the most extensive deployment of federal USAR resources in U.S. history. All but two of the nation's 28 USAR task forces would eventually be deployed to New York or the Pentagon.

Naturally, many other federal resources were deployed to the Pentagon disaster. Among them were three disaster medical assistance teams (DMATs) from Georgia, Maryland, and North Carolina, two veterinary medical assistance teams (VMATs), and disaster mortuary teams (DMORGs).

Unified command. The Red IST-A members were melded into the ACFD's command structure. This is consistent with FEMA guidelines, which emphasize the IST-A's role as a resource to assist local ICs with coordination of federal USAR resources to support local and state agencies.

Unified command was established under the direction of Chief Schwartz. Included during daytime operations were the ACFD (Schwartz), Arlington police (Chief Dan Murray), commanding general MDW (MG James T. Jackson), FEMA USAR (IST-A leader John Huff) and FBI (Special Agent Christopher Combs). This team met every four hours, deciding and coordinating the strategic objectives of the Pentagon incident. During nighttime operations, a mirror of this team continued round-the-clock efforts, with IST-A Deputy Leader Carlos Castillo representing the FEMA USAR system. This unified command structure enabled the Pentagon incident decision-makers to effectively utilize the FEMA USAR task forces to the local jurisdiction's greatest benefit. In Figure 2–60, FEMA USAR IST members evaluate the structural integrity and other factors on the upper floors of the area affected when the airliner crashed through the building.

Fig. 2–60 Inspecting Upper Floors

Early during FEMA USAR operations, Chief Schwartz identified his strategic objective for the task forces. The mission of the USAR system included the following:

- Ensuring reasonably safe collapse SAR operations
- Preventing needless injuries and death among victims and responders
- Locating and rescuing (or recovering) all victims from the Pentagon
- Providing structural stabilization to damaged sections of the Pentagon
- Supporting the recovery of evidence by the FBI, NTSB, ATF, and military by stabilizing the structure
- Helping locate, extricate, and remove bodies and material in a respectful manner

Throughout the incident, the ACFD provided a liaison to the IST leader and deputy leader to allow directed two-way communication.

During unified command meetings, the IST leader provided input from the IST and task force operations leaders that identified operational goals during each 12-hour operational period. Once agreement was achieved by command, the IST conveyed these goals into tactical objectives, which were included in the USAR action plan for each operational period. The IST then distributed the action plan to all USAR resources, conducted briefings with task force leaders, FBI, and other affected parties before each operational period, and provided continuous coordination and supervision of USAR operations and support. In turn, each USAR task force developed and distributed a written IAP specific to that task force's operations during every operational period. This is a standard FEMA USAR system approach to ensure that the task forces are operating in a manner that's consistent with the needs of the local IC, and to ensure that all resources at the disaster are on the same page. This is particularly important at large-scale and/or complex disasters where different resources can potentially find themselves employing opposing tactics because the action plan isn't transparent, or effectively communicated.

To assure that these strategic objectives were completely understood, IST-A leaders followed FEMA protocol by drafting a memorandum of agreement for USAR operations with the ACFD. The agreement specified the purpose of the USAR task forces, established performance benchmarks, and spelled out demobilization protocols. This is an example of the extent to which the FEMA USAR system is designed to support the needs of local entities (Fig. 2–61).

Fig. 2–61 Rescuers and Investigators at the Area where the Airliner Punched through the Inner Wall of the Outer Ring of the Pentagon

USAR Command and Control. Based on collective experiences from the United States, the Philippines, Turkey, Taiwan, Greece, Armenia, Latin America, etc., VATF-1 and the IST members were able to determine the scope of the Pentagon operation. It was clear to them that locating and extracting all the victims from the Pentagon collapse would require approximately two weeks of nonstop, round-the-clock operations. The personnel required would include at least four 62-person USAR task forces: two assigned to daytime operations and two assigned to nighttime operations, in 12-hr rotations.

They recognized the effects of the intense fire caused by jet fuel and the loss of structural stability when the aircraft plowed through the fifth (outer) ring of the building. The impact had obliterated columns and other support structures on the first and second floors. The IST determined that the main objective was to locate and extract an unknown number of victims from within the large combination lean-to/pancake collapse in division A without causing additional casualties. The collapse involved five heavy reinforced concrete floors topped by a 2-ft thick concrete roof.

The IST also understood that the combined effects of the plane crash, fire, and collapse essentially required the task forces (assisted by technical rescue teams) to re-engineer the first and second floors in divisions B, C, and D. This would be accomplished through a massive and innovative shoring operation coordinated by IST-A. The IST-A structures specialists were highly experienced in advising USAR task forces during emergency shoring operations at disasters.

The IST-A members recognized the need to reduce the potential for secondary collapse and overhead hazards before committing personnel into the main collapse area. They realized that it might take several days to shore and re-engineer divisions B, C, and D before rescuers could start the job of de-layering the collapse area in earnest. And they understood that it might take up to two weeks to locate and remove all victims, even with the commitment of four FEMA USAR task forces, three local technical rescue teams, and hundreds of local firefighters, military personnel, and heavy equipment resources.

With those factors well established and understood, the IST-A and the USAR task forces (in tandem with technical rescue teams from the ACFD, AFD, and MDW) dug in and aggressively continued round-the-clock operations. The USAR action plan was implemented with the unwritten understanding that additional terrorist attacks, a USAR disaster elsewhere, or even a secondary disaster at the WTC might necessitate alternate strategies and redeployment of USAR task forces.

The daily routine included 0700- and 1900-hr shift changes. Two USAR task forces and one-half the roster of each local technical rescue team were staffed throughout each 12-hr operational period. Approximately 180 USAR personnel were working at all times.

For members of the IST-A day shift, work began at 0500 with a shift change briefing from the night shift and a tour of the collapse area. The status of strategic and tactical objectives was reviewed and observed, complications noted, and action plan changes discussed. This was followed by a formal IST planning meeting at 0700 hrs, attended by all IST-A members, USAR task force leaders, the incident management team (IMT), representatives of the ACFD, AFD, MDW, FBI, ESF-9, and others with operational responsibilities.

Prior to each night shift, the same briefing process and planning meeting occurred from 1700 to 2000 hrs. Consequently, each 12-hr operational period actually consisted of a 16-hr shift for IST-A members and others with operational command/control/coordination responsibilities. This is an example of the standard daily routine during long-term USAR operations when FEMA task forces are deployed (Fig. 2–62).

Fig. 2–62 USAR Task Force Members Begin Shoring Operations inside the Pentagon

Collapse SAR operations. Viewed from a distance or from television news cameras, the 80-ft wide gash and the fire/smoke damage to the west face of the Pentagon (the only obvious external signs of damage) were somewhat misleading. In truth, damage to the structure and contents—and of course the carnage—were far worse and more extensive than could be seen from outside the Pentagon. Hidden within the building was a 240-ft deep, five-story collapse zone. The interior of the building affected by the attack stretched more than 400 ft, with collateral damage all around. In terms of size and complexity, some FEMA USAR task force members at the Pentagon noticed similarities between this collapse problem and those they encountered at the Oklahoma City bombing.

Pentagon IST-A was charged with coordinating the overall USAR operation to ensure that the many stakeholders could fulfill their purpose without additional loss of life. Victims of the airliner crash and subsequent structural collapse were discovered throughout the 12-day USAR operation.

The USAR operations were predicated on the five standard phases of structure collapse SAR: (1) size up and recon (2) surface rescue that includes the primary search (3) void-space search that includes the secondary search (4) selected debris removal that includes carefully dissecting the building with heavy equipment and rescue personnel and looking for live victims and the deceased, and (5) general debris removal that includes the clearance of all material using heavy equipment and other methods.

Each USAR task force consists of 62 to 70 persons who make up four 6-person rescue squads. Task forces also include two 2-person technical search squads; four canine search teams; two medical squads; and specialists in heavy equipment, logistics, communications, engineering, hazmat, and other skills.

Naturally, USAR operations were preceded by standard fireground primary/secondary searches. Searches covered all floors of the affected area of the building accessible during fire attack operations. These searches were conducted diligently by firefighters from the local agencies during the intense firefighting operation that took place during the first 12 to 18 hrs of the incident. Yet, the magnitude of the disaster left thousands of square feet on all five floors chest-high in incinerated debris. This debris required a third round of thorough searching by hand to ensure that no bodies (or body parts) were left behind or ended up in dump trucks during stage 5.

Re-engineering parts of the Pentagon. From day one, it was clear to engineers (structures specialists) attached to the IST-A and USAR task forces that a major stabilization operation was required. Stabilization was needed to prevent division B and parts of divisions C and D from collapsing and burying rescuers in division A, where the combination lean-to/pancake collapse had occurred.

Another complication was the 2-ft thick concrete roof structure topping division A, which was designed to prevent bombs from penetrating into the structure. Because of the angle of the lean-to collapse, the concrete roof appeared poised to start a massive slide. With no easy access to conduct inspections below the roof, engineers were unable to determine exactly what was holding it in place. Some of the structures specialists suspected that a small number of rebar strands might still be connected, while others postulated that it was a combination of inertia and internal connection between the collapsed/uncollapsed sections of the building.

The engineers were concerned that the entire upper deck might avalanche into division B, much like a layered cake slides in layers if tilted sufficiently. Such a cascade of hundreds of tons of material striking division B might in turn cause that part of the structure to fall back onto division A. Because of the intense fire that raged for half a day on September 11, the survivability profile for victims trapped in the division A collapse was considered relatively low. With these factors in mind, IST-A agreed it would be less than prudent to commit personnel to penetrate and de-layer division A until the most vulnerable portions of divisions B, C, and D were properly supported. In addition, they needed to figure out a solution for the roof problem.

So the first USAR action plan specified that initial efforts would include a massive shoring operation on floors 1 and 2 of division B. At the same time, other USAR task force rescue squads and search teams would conduct a thorough and methodical secondary search of the upper floors, looking for victims who might have been missed during firefighting operations. At this point, the number of missing people taken from the Flight 77 passenger/crew rosters and Pentagon employee lists hovered around 200. Although dozens of bodies had been located and marked for later removal in various parts of divisions B, C, and D, approximately 150 more people were unaccounted.

Shoring operations. In USAR, a generally accepted strategy for shoring large unstable buildings is to work from the unaffected areas of the building toward the affected area. This allows rescuers to stand in relatively safe areas and methodically work their way into non-shored areas that would otherwise be unsafe to enter. This was the approach adopted at the Pentagon. The USAR task force rescue squads, with expert advice from the structures specialists, began at the front of division B and constructed a variety of shores to replace missing and damaged structural columns. As they worked their way into the building, they built a kind of forest consisting of immense shoring combinations to support damaged and broken beams (Fig. 2–63).

Fig. 2–63 Interior Shoring Operations

To reach spots where shoring was to be placed, they methodically worked their way from shored areas to non-shored areas, placing temporary pneumatic pipe shores and other short-term measures. Working by hand and with hand tools, rescue squads removed tons of debris, locating bodies and body parts later documented by the FBI and removed by military teams. When a section of slab of sufficient size was cleared down to *clean* concrete, a base was built and the shoring set up. It was painstaking work, requiring rescuers to cut and remove tons of debris and pass it out bucket-brigade style.

Tons of lumber cut to specified dimensions by cutting teams outside the building were passed into the building in the same manner. The rescue squads called out the required dimensions, which were relayed to the cutting teams who quickly sized the timbers and cut them to the specified dimensions. Within one day, a virtual shoring factory had been established. It was complete with its own lumberyard, constantly restocked by local lumber trucks, and various cutting stations built on-site to expedite the sizing of shoring lumber.

Essentially, the affected parts of the Pentagon were re-engineered by firefighters, USAR task forces, and various teams to affect the removal of all victims and provide a relatively safe crime scene investigation (Fig. 2–64).

Fig. 2–64 An Example of the Shoring Systems Placed inside the Pentagon

Differences and similarities. Although the Pentagon collapse required a significant response of USAR resources, it paled in comparison with the WTC disaster that occurred at the same time. Still, the Pentagon collapse was a large one, and it created significant challenges for rescuers.

The five-story, reinforced concrete frame/masonry infill building with Kevlar-armored walls had been struck broadside by a fully fueled 767 airliner. Traveling more than 400 mph, the subsequent impact caused a tremendous fire that burned through the night of September 11. Many victims were still missing; finding them would require the commitment of a significant number of rescuers in a high-risk structure collapse SAR operation (Fig. 2–65).

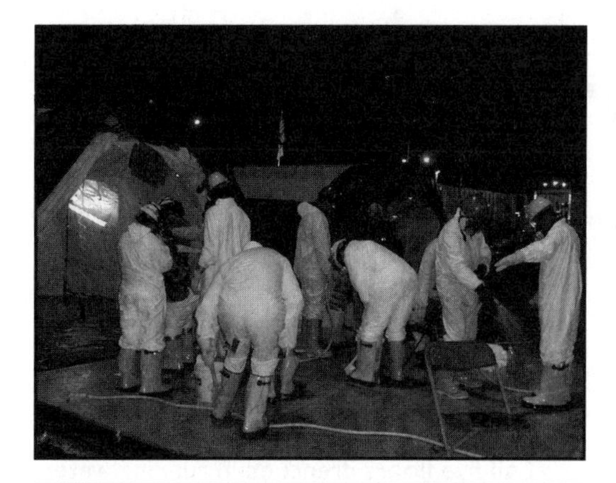

Fig. 2–65 Decon Operations at the Pentagon

Some task force and IST members (including this author) who had been deployed to the Oklahoma City bombing likened the Pentagon collapse in some ways to the damage encountered at the Murrah building. In terms of technical difficulty, carnage, overall size, and personnel hazard, there were similarities, including the following:

- Multiple secondary bomb scares that prompted the operational retreat of personnel
- Heavy, reinforced concrete buildings with constant potential for secondary collapse
- Potentially lethal overhead hazards, including sections of concrete slab overhanging the rescue site and bowling-ball-sized concrete chunks (widow-makers) that threatened to fall on rescuers
- Office furniture and other heavy items perched on the edges of slabs with the potential to become lethal projectiles
- Extreme carnage and other disturbing sensory elements
- Typical, difficult operating conditions, including confined spaces, darkness, heavy equipment operating in close

proximity, trip and fall hazards, entrapment hazards, and concrete dust possibly laced with harmful substances like asbestos

- Sheer physical labor of moving and operating heavy tools and materials into the collapse zone and removing hundreds of tons of debris by hand and with buckets and shovels
- Flooded conditions. In fact, at the Pentagon, burst water mains/piping kept all five floors drenched in running water for several days
- Routine difficulties (i.e., communications, regional variation in use of the ICS system, etc.) expected when hundreds of people from dozens of agencies from across the country converge on a single incident site for two weeks
- Constant safety hazards due to multiple and simultaneous SAR operations in different parts of the building involving cutting, breaching, lifting, and removing of major structural elements

USAR action plan. The USAR action plan, developed during the 9-11 attacks, was revised twice daily by the FEMA USAR IST. The plan provided a blueprint for the operation of all USAR resources for each 12-hr operational period and was synched with Arlington's IAP throughout the incident. Liaisons were established between the IST, the ACFD, and other agencies with roles in the collapse SAR operations. Throughout the USAR operation, IST leaders and operations chiefs consulted with Arlington command about all major USAR issues. This was to ensure that ACFD chiefs would not be surprised and so ACFD personnel could keep the appropriate decision-makers (as well as the news media) accurately informed about the progress of the operation.

In an after-action report on the Pentagon incident, IST leader John Huff explained how this worked:

> *Early during the FEMA USAR operations, Assistant Chief Schwartz (ACFD) identified his strategic expectations of the USAR teams. These expectations were: to provide search for victims in the affected areas of the building, provide structural stabilization, as needed in areas of the building that sustained damage (specifically structural collapse areas) and to support the victim recovery efforts as victims were located.*
>
> *During the unified command meetings, the USAR IST leaders (Huff and "Night" shift IST leader Carlos Castillo) provided input from the IST and task force operations leaders that identified operational goals during each 12-hr operational period. Once agreement was achieved by command, the IST conveyed these goals into tactical objectives, which were included in the Action Plan for each operational period. In order to assure these USAR strategic objectives were understood, the IST drafted a written Memorandum of Agreement to the Arlington commander. This written agreement specified the purpose of the presence of the USAR task forces, and established criteria to be utilized to identify when the task forces had successfully completed their mission, and could be disengaged, or recommitted to other needs.*

The IST ensured that the Arlington County, Alexandria, and MDW technical rescue teams were fully integrated into the USAR operations throughout the incident. This was an important factor in maintaining local presence in the USAR operation, a key element of FEMA USAR operations.

Long before the Pentagon attack, ACFD commanders had been familiarized with the FEMA USAR system through training and preplans; they were therefore familiar with the role of the IST in such collapse-related disasters. This provided a good comfort level and allowed them to place the appropriate levels of responsibility on the IST to manage USAR operations under the umbrella of the Pentagon incident command structure. This structure included the ACFD and a host of local, state, and federal agencies all working closely in the unified command mode. Special USAR-related problems were addressed rapidly and jointly by the IST and Arlington command, and solutions were generally arrived at by consensus. As the incident progressed, some observers commented that the Pentagon incident was an example of how a major terrorism event that results in a major firefighting/hazmat/mass casualty/WMD/USAR/investigatory response, can be effectively managed.

"There were many different stakeholders in this operation," said Huff. "This was not only a major disaster with early reports of nearly 800 people missing, but it was also a crime scene on federal property, in the nation's military command center." "Therefore," continued Huff, "stakeholders included the local fire departments, the military, local law enforcement, the FBI, the FEMA teams, and many volunteer agencies. Each had a legitimate purpose for being there and had needs and wants to be met." The real tragedy is that this model of efficiency and inter-agency cooperation evolved in the shadow of one of the worst terrorist attacks on American soil.

USAR operations continue. The USAR task forces used practically every tool in their 50,000-lb equipment caches. The wide variety of work assignments required to locate and extract victims of the Pentagon disaster proved to be a substantial test of the equipment and logistical resupply system developed for the national USAR response system.

Shoring. Under direction of the IST, USAR resources were engaged in major round-the-clock shoring operations. Shoring was needed to prevent the division B side of the building and parts of divisions C and D from collapsing and burying rescuers searching void spaces and de-layering division A. Division A was a combination lean-to/pancake collapse. It took several days to re-engineer these areas of the Pentagon to enable rescuers to proceed with de-layering and searching of void spaces. These searches included canine and technical search teams in the main collapse area. Shoring operations continued in various parts of the Pentagon until September 21.

USAR task force rescue squads assigned to shore and stabilize divisions B, C, and D worked from the unaffected areas of the building toward the affected areas. This is standard protocol for penetrating into a heavily damaged and highly unstable building. This process enabled USAR rescue specialists (task force rescue squads) to work in relatively safe areas while they methodically proceeded into non-shored areas, otherwise unsafe to enter. The group received expert advice from IST structures specialist Dennis Clark and VATF-1 structures specialist Kent Watts. Structural engineers concentrated most of the shoring in division B, levels 1 and 2, to replace missing and damaged structural columns. Division B took on the appearance of a harshly lit forest consisting of immense shoring combinations that supported damaged columns and broken beams.

A solution for the monolithic slab. A huge section of reinforced concrete slab hung precariously above the rescuers for days. It was suspended by just a few strands of rebar. This 2-ft thick reinforced slab roof topping division A of the Pentagon collapse presented a significant danger to rescuers. A similar slab had existed in the Oklahoma City incident and became known as the *mother slab* after Ray Downey of the FDNY coined the term.

On the evening of September 12, the IST held a meeting to discuss solutions for the problems presented by the concrete roof structure. Included were IST structures specialists, USAR task force leaders, task force structures specialists, HERS, and others. It was agreed that the concrete deck could be broken up into manageable sections that could be lifted off with cranes in a controlled manner to prevent the whole thing from avalanching.

Then a member of the IST recalled a machine known as a Caterpillar 345 ultra-high reach multiprocessor with shear. This machine has a multi-purpose claw called a multi-processor that swivels 360° mounted on the end of an articulating boom. The multi-processor can pulverize concrete slabs, pinch large pieces of slab between a sort of metal thumb and forefinger, and even grasp steel I-beams to sweep debris from a rubble pile. The machine's imposing appearance eventually earned it the nickname T-Rex, which became the moniker used by rescuers through most of the rest of the operation. The T-Rex was thought to be exactly the answer to the problem of how to section the monolithic slab into manageable pieces and get it out of the way so rescuers could proceed with the search in division A.

IST logistics members were assigned to track down the closest Cat 345 multiprocessor, order it, and get it on the road to the Pentagon with police escorts to expedite its response. The closest machine was found in Baltimore, operated by a heavy-equipment firm called Potts and Callahan. Plans were made to get it to Washington, D.C. with a code 3 police/state trooper escort. Meanwhile USAR task forces continued shoring around the perimeter of division A. By the afternoon of September 13, the T-Rex was sitting in front of the Pentagon, ready to work.

Heavy-equipment operations. The HERS attached to the FEMA USAR task forces were responsible for interfacing with local heavy-equipment and crane operators. Task force specialists educated local operators about their role in the SAR process at the site. Operators were informed about their place in the ICS and their interaction with the USAR task forces. They were informed about communications issues and the tactics/strategy being used to locate and clear victims from the Pentagon. Task force specialists also performed the following functions:

- Coordinated emergency activities of heavy equipment at the rescue site
- Made recommendations about heavy equipment issues to the local IC, the IST, task force leaders, and various task force components as necessary
- Assisted in the planning process
- Determined tactics and strategy for heavy equipment use
- Dealt with the myriad heavy-equipment issues and problems that can arise during SAR operations

The Pentagon SAR operation was heavily dependent on the capabilities of local crane operators and other heavy-equipment resources. Throughout the incident, HERS from the FEMA USAR task forces played a vital role in maintaining safe and effective operations.

As the T-Rex went to work on the evening of September 13, the USAR task forces were pulled back to a safe distance from areas designated by engineers as subject to secondary collapse from the multiprocessor. The task forces continued their massive shoring operation on floors 1 and 2 of division B and elsewhere. They conducted a methodical search of upper floors, looking for victims who might have been missed in earlier sweeps. But they were kept out of the areas thought to be most vulnerable to secondary collapse if the division A roof were to avalanche into division B as a result of the multiprocessor's work.

Just as promised, the T-Rex was capable of dissecting a 2-ft thick reinforced concrete slab. It carefully picked up each piece off the top of the collapse zone and deposited it on the ground or in a dump truck. Coordinating with IST structures specialists and HERS from the USAR task forces, the operator was able to pulverize the slab at will, then grab large pieces and place them wherever he chose (within reach of the articulating boom). He was able to grab large items like safes and computers from upper-floor debris and deposit them gently on the ground where they were taken into custody by FBI agents strategically positioned around the building.

The T-Rex operators demonstrated the practice of grabbing steel I-beams and using them to sweep debris from atop the collapse pile as they took turns de-layering the building 24 hrs a day. They could cut rebar from which large chunks of concrete were suspended, thereby eliminating widow-makers from a safe distance.

With the T-Rex in operation, it quickly became apparent that the collapse could be delaminated and dissected far more quickly that way than by cutting slabs by hand and rigging and slinging them for lift by cranes. The latter is a standard skill taught to FEMA USAR task force rescue specialists and coordinated by the HERS who are part of each task force. It was also much safer; the T-Rex operator could essentially perform the same tasks from the safety of the cockpit.

In short, the T-Rex proved safer and faster at dissecting the collapse of division A during stage 4 operations than standard crane rigging and lifting operations. Under these high-risk circumstances, it proved itself by helping to keep rescuers out of the direct collapse zone during critical moments. Many observers agreed that the T-Rex cut several days off the time expected for the division A de-layering operation.

One drawback to the T-Rex was the limitation imposed by the operator's line-of-sight vision, complicated by debris, fatigue, and darkness. These factors inhibit a human's depth perception. As the operation proceeded around the clock, it was increasingly difficult for the operators to discriminate debris from bodies trapped within it, in part because everything began to take on the color of concrete dust. It was also necessary to provide instant guidance regarding the selection and order of pieces that could be cut and removed safely to prevent secondary collapse and other safety hazards.

Recognizing this, the IST established a 24-hr lookout for victims, evidence, and signs of secondary collapse. Each lookout crew consisted of one USAR task force search team member and one structures specialist stationed in a personnel lift basket suspended high above the Pentagon. They were suspended beneath a 200-ft crane and rotated periodically to ensure that they remained sharp and alert. It was their job to constantly look for victims in the rubble, spot potential safety problems, and communicate problems to the heavy-equipment operators. This avoided situations where victims might be mistaken for debris and loaded into dump trucks and where secondary collapses caused by the machines might endanger lives.

The only exception was when a secondary collapse occurred on the evening of September 15 while the T-Rex was working (see Figs. 2–52 through 2–55). Given the potential for secondary collapses throughout this incident, it's a tribute to the skill of the heavy-equipment operators and effectiveness of the shoring operations that only one significant secondary collapse occurred during the entire 12 days.

Combination stage 3/stage 4 operations proceed in division A. The de-layering of division A required a mixed stage 3 (void-space search) and stage 4 (selected debris removal) strategy. As the building was delicately dissected by the T-Rex and the bucket crane, there was a constant watch for victims, their remains, and void spaces that might contain additional victims (potentially live ones). The searchers' observers were watching from the ground as well as from lifts suspended above the Pentagon. Members of the USAR task force search teams were stationed in strategic locations where they could watch the building and each load of debris being deposited on the ground in front of the Pentagon. They were on the lookout for any sign of victims unburied in the collapse; in which case, all operations there would cease and personnel would identify the object/remains and determine whether further extrication work was needed. They were also watching for signs of remains being deposited with debris on the ground.

Periodically new void spaces would be opened up during the de-layering process. For every new void space, USAR task force search team members physically searched every accessible void area for victims, crawling into them when possible. They used snake-eye and search-camera/remote-search devices, trapped-person locators, search dogs, and direct visual/voice contact to search for victims in void spaces. Some of the deceased victims were mostly intact in void spaces that might have offered a chance of survival had it not been for the intense fire that raged on September 11 and 12.

When victims were discovered in void spaces, search teams would pause and call for the FBI to photograph the scene for evidence. Then a task force rescue squad would come in to shore the area, breach floors or walls, cut rebar, and generally make the victim ready for extraction. The FBI agents would take additional photographs before the victim was removed, and the military morgue detail would make the actual extraction by hand.

After a victim was removed, the T-Rex would move back into place and dissect the building to the limits of its reach, or until a floor was removed. The pulverizer would pull back to allow a clamshell crane to move in and remove debris and then the pulverizer would eat its way further into division A. The clamshell would deposit materials on the ground in front of the Pentagon for further searches, then move out to enable the T-Rex to come back in. These rotations sometimes took hours; it was a meticulous process that continued around the clock for nearly a week and a half.

Debris was spread out by heavy equipment and on the signal of a IST USAR specialist, the equipment would stop and canine search teams from the USAR task forces would deploy across the material in search of human scent, indicating remains. Then, USAR search team members conducted a physical search for remains, crawling and walking over all the debris. Finally, after being searched three or more times, the debris was loaded into trucks with skip loaders and taken to a Pentagon parking lot to be further combed for human remains and evidence by the FBI, ATF, and military units.

The strategy of alternating void space searches with selective debris removal continued until the entire collapse zone had been dismantled and all victims were located and extracted. Heavy equipment was used to expedite the removal of hundreds of tons of rubble in a safe, effective manner. Structural specialists from the IST and USAR task forces made critical decisions about removal, breaching, shoring and cutting of large debris throughout the operation.

FEMA USAR task force search team managers effectively coordinated all canine search operations. They established a canine search plan for operational periods, evaluated requests to search, kept the dogs operating at optimum, and generally looked after their welfare. One of the IST priorities was to assure safety for the canine specialists and dogs that often found themselves working in some of the most precarious positions.

The attention paid by IST USAR specialists and safety officers to safety aspects of the canine search function enabled dogs and handlers to concentrate more fully on the actual work. Their knowledge of structural issues, search methodology, and other aspects of the operation was an important factor.

One USAR search team member stated, "It was evident (after several days) that the probability of finding live victims was very low. This did not deter the canine search mission because we felt we could locate deceased victims and help the families with closure." He went on to say, "The dogs were able to direct rescue squads to almost the exact point where victims were located. This allowed the squads to focus their efforts and helped maximize the available manpower."

As the days went by, this process was streamlined to find bodies in the most timely manner, prevent deterioration and loss, allow better identification, and reduce the time families had to wait for word on the ultimate fate of loved ones. Operation of the heavy equipment and personnel were choreographed assembly-line style by the IST USAR specialist, the HERS, and the IST structures specialists. This coordination shortened the time of discovery of victims and the removal of debris to find more victims.

Canine search teams were used extensively throughout the operation and proved extremely accurate in pinpointing the location of victims buried in the rubble. This helped the rescuers focus their efforts on critical areas for rescue/recovery and reduced the danger to rescuers by limiting their exposure to victimless areas.

Victim recovery. A time-tested approach was applied to the recovery of Pentagon incident fatalities. When a victim was located, work in the area was halted to protect the body, personal belongings, and evidence. An FBI evidence team (one of several on constant standby in front of the collapse) was called in to photograph and gather victim-related evidence. If physical extrication was required, a rescue squad from the assigned USAR task force was given this task. Next, a military mortuary team collected and removed each victim from the building, handling them with military dignity and respect. Some of the victims were clearly military personnel because of their attire. Others were in civilian attire, and still others could not be identified as either. This was a military facility, full of both civilians and uniformed people doing military work.

Victim recovery ensured that all victims encountered in the Pentagon collapse were treated with military respect befitting their activities at the time of the terrorist attack. It helped ensure that all available evidence was being located and collected, including top-secret documents, computers, safes, etc. It provided documentation that might answer later questions about the mode of death, their location within the collapse, and other factors.

In at least one case, a victim could not be freed without severely endangering the lives of rescuers due to the potential for secondary collapse that threatened to bury them. After deliberating, IST and USAR task force members decided to perform a surgical amputation of the deceased victim's shoulder, which was pinned beneath dozens (perhaps hundreds) of tons of reinforced concrete slabs and material. A medical team manager (emergency room physician) from one of the USAR task forces was selected to perform a surgical amputation, which was completed without delay. The trapped arm and shoulder were covered and marked with fluorescent orange spray paint and flagging tape to ensure rapid location after the debris was de-layered. Later, the remains were reunited in the same manner previously described and sent along for autopsy and identification.

This is just one example of the unusual and often unpleasant tasks associated with disasters of this nature, something that must be considered by the IC and other decision-makers. One can extrapolate the potential for victims to be trapped alive under similar conditions after an earthquake or terrorist attack that causes structures to collapse. It's happened before and it will happen again. With good planning and command, even unpleasant tasks like field amputation can be accomplished in the most humane way possible.

IST operations. IST Operations Chiefs Buddy Martinette (daytime ops) and Ruben Almaguer (nighttime ops) were assigned to help develop SAR strategy with the IST leaders. They were also required to coordinate the tasks assigned to FEMA USAR task forces, USAR teams from the Arlington and Alexandria fire departments, a military rescue team, and civilian heavy-equipment contractors. They were further tasked to accomplish USAR objectives established by the ACFD command and IST leaders in divisions A to D (floors 1 to 5). To accomplish these tasks, they assigned the IST USAR specialists to provide direct supervision and coordination.

In strict ICS terms, the role of USAR specialists at the Pentagon incident approximated that of USAR group leader, one of the potential tasks listed for FEMA IST USAR specialists. In this case, the job of the USAR specialists included coordination with elements of the ACFD, the FBI, ATF, and the NTSB. Because of the risk of secondary collapse and other life hazards, it was critical for them to work closely with USAR structures specialists, USAR safety officers, and task force HERS. They were to keep the IST leaders and operations chief informed of the current situation, progress, and needs.

Although there is no HERS assigned to the IST, the HERS assigned to the FEMA USAR task forces were a critical part of the Pentagon incident. During stage 4 collapse search operations, at least one HERS was assigned to work directly with the heavy-equipment operators. As the USAR operations progressed to stage 4, the HERS were pressed into service to assist the IST USAR specialists by performing the following tasks:

1. Determine and communicate clear direction to civilian heavy-equipment operators—in understandable terms—regarding the USAR action plan and individual objectives assigned to the pulverizer, bucket cranes, skip loaders, dump trucks, and the other heavy equipment employed at the Pentagon collapse.
2. Communicate to rescue personnel the capabilities and limitations of various types of heavy equipment.
3. As necessary, communicate to rescue personnel the procedures, protocols, and techniques (in understandable terms) related to heavy-lifting applications and other heavy-equipment operations.
4. Develop and maintain a comfortable level of assurance for firefighters and other rescuers regarding the competency and abilities of local heavy-equipment operators to perform specific tasks throughout the incident (e.g., "Yes, that crane is capable of lifting that piece of debris.").

Operations near completion. The operational area was painstakingly reduced through combination stage 3/stage 4 operations in division A and ongoing stage 5 operations in divisions B, C, and D. Shoring operations also continued as the building was de-layered. By September 18, the need for four full FEMA teams no longer existed. Recognizing this, the IST developed a plan to relieve the first two FEMA USAR task forces to arrive (VATF-1 and VATF-2) and to work one task force on both the day and night shift.

A fifth USAR task force from New Mexico (NMTF-1) was activated to provide new personnel for the balance of the Pentagon incident. By September 20, it was decided to demobilize MDTF-1 and TNTF-1, as the physical workload was reduced to a level manageable by a single USAR task force and local technical rescue teams. Therefore, NMTF-1, the WMD, and local teams were split into day and night operations.

On the evening of September 21, the IST advised Arlington command that all areas of the building had been searched several times, all shoring operations were complete, and USAR goals had been met. It was agreed that at 0800 hrs on Saturday, September 22, all SAR efforts would cease, and the FBI would assume responsibility for on-scene command and control to complete their crime scene efforts.

This marked the end of the fourth largest deployment of FEMA USAR task forces in the history of the program. The results of this mission were bittersweet. Deceased victims of the second-worst terrorist attack on continental U.S. soil were recovered. However, four hours away, thousands of rescuers continued to toil around the clock in a desperate search for possible survivors at the smoldering WTC collapse.

Lessons learned

In a disaster of the magnitude and duration of the Pentagon incident, there naturally are too many lessons to list in one place. What follows is an attempt to capture some of the most pressing lessons that might be applied at future collapse disaster operations, including those resulting from terrorist attacks.

Local fire/rescue agency chiefs and all officers will be at a distinct advantage in future disasters *if* they are familiar with the Federal Response Plan (including ESF-4, ESF-9, etc.). They should also understand the FEMA USAR task force system and the FEMA IST concept. They will have an advantage if they have developed local plans to integrate these vast and highly experienced federal resources seamlessly in the event of a large or complex disaster. Even company officers, who may be in charge of certain scenes during the course of large-scale disasters, should be familiar with these plans and resources. Information about the Federal Response Plan and the FEMA national USAR response system may be obtained from a number of sources, including the official FEMA Web site (www.FEMA.gov/usr), regional FEMA offices, and state OESs.

Collapse SAR operations frequently last days or weeks. They require round-the-clock operations to locate and extract all victims within the window of survivability and to recover the dead in a timely manner. Emergency officials should plan to pull live victims from the rubble up to three weeks after catastrophic earthquakes, and they should be prepared to sustain nonstop SAR operations until all hope of locating viable victims has passed. The public has the right to expect this level of response to devastating events in the United States.

One advantage of FEMA USAR task forces is that they are designed to conduct nonstop operations without a great deal of local support. Each task force is prepared with sufficient food, water, and supplies to operate for 72 hrs without resupply. One of the initial functions of the IST is to resupply the USAR task forces to continue a mission indefinitely, without requiring anything of the local jurisdiction. This level of self-sufficiency is critical in large disasters because it reduces the impact on local agencies that have their hands full dealing with the disaster itself.

Another advantage is the tremendous amount of collapse SAR (and overall disaster rescue) experience that FEMA USAR task force members bring to an incident. Still another key advantage is the USAR task force mission of supporting the local IC's incident objectives. This keeps the local IC in charge, yet provides him with an advanced federal USAR capability that may become part of his command and operational structure, depending on local needs.

Generally, all five stages of structure collapse rescue should be conducted, in order. For example, skipping stage 3 (void-space search) and going directly to stage 5 (general debris removal) may doom live victims trapped but undetectable from the surface who might otherwise be saved alive. With few exceptions, all five stages of collapse rescue should be accomplished.

In large-scale or complex collapses, it may be necessary to simultaneously conduct various stages of collapse rescue. In the Pentagon, void-space search (stage 3) was happening in one section of the building while selected debris removal (stage 4) was occurring in another. One advantage of the FEMA USAR IST is that the IC can assign coordination of large-scale collapse rescue operations to it (e.g., create a USAR branch with the IST leader as branch director, or assign the IST to supervise a rescue or USAR group). The IC can do so with confidence that the IST can dictate and manage multi-stage collapse operations.

USAR ISTs have demonstrated that they are in their element when assigned to manage multiple, simultaneous collapse SAR operations at major disasters involving more than one FEMA USAR task force and local teams. It is a positive sign that these teams, consisting of selected members of USAR task forces from around the nation, can be dispatched within minutes of a disaster and arrive from every corner of the United States. They can then function as an effective management team to help coordinate and maximize the efforts of the USAR task forces to help rescue trapped people who might not otherwise survive a disaster.

The Federal Response Plan works. This was evident at the Pentagon disaster, where numerous federal resources supported the needs of local and state agencies in charge of handling the consequences of the attack. It was evident at the Oklahoma City bombing, at the Northridge earthquake, and a number of other disasters around the United States in recent years. While there is always room for improvement and while efforts are underway at this time to further streamline the system, the Federal Response Plan is one of the true success stories of the federal government.

The FEMA national USAR response system has been fulfilling the mission that was given it in 1989. The FEMA USAR system has proven itself repeatedly at disasters ranging from hurricanes to earthquakes and floods to terrorist attacks. The system works as it was designed, fulfilling the federal government's promise to provide timely and expert response to collapse-related disasters and other disasters that leave people trapped anywhere in the United States or its territories.

The FEMA national USAR response system is *expandable*. The mission of this system had already been expanded to include WMD response before the 9-11 attacks. This is one reason why FEMA USAR task forces were strategically deployed at the 1996 Atlanta games and the 2002 Winter Olympics in Utah, as well as other events of national and international importance. Currently there is discussion about expanding the role of the national USAR task forces to cover flood rescue operations and other disasters involving trapped people. In California for example, all eight of the federal USAR task forces are already equipped and trained for flood rescue operations (part of an initiative by the state OES). Task forces have already been successfully deployed to rescue people during flood disasters in the Sacramento area, Yosemite National Park, and elsewhere.

Fire, heat, and smoke from the jet fuel-stoked fire killed many occupants of the Pentagon who might otherwise have survived the initial crash of the 757 into their building. If not for the fire, it's quite possible that live victims might have been located and rescued from the collapse of division A in the days following 9-11.

As a matter of principle, rescuers should assume—until proven otherwise—that there may be live victims in a collapse event of the size and nature of the Pentagon incident. Victims should not be written off until their survival is demonstrated to be impossible because of conditions such as intense fire, lethal radiation, or non-survivable submersion, etc.

FEMA's decision to immediately mobilize ISTs and USAR task forces to the Pentagon and WTC disasters was prudent and provided an added layer of protection to trapped victims and rescuers alike.

One of the critical needs in a collapse disaster is to prevent secondary collapses that kill or trap rescuers. The response of FEMA USAR resources helped ensure that this need was met. At Oklahoma City, the WTC, and the Pentagon, not a single rescuer was killed while conducting SAR operations in the aftermath of these events. It becomes clear that the FEMA USAR system is an effective means of ensuring rescuer safety and effective USAR operations at disasters of this nature.

Based on the resource needs at the simultaneous disasters in New York and Washington on 9-11, it's clear that fire/rescue officials need to consider the potential for even larger (or multiple) disasters. Bigger disasters could require the commitment of all available federalized USAR resources and possibly USAR teams from other nations. This would be an unprecedented event in the annals of American history. To those who study these matters, it's quite clear that a catastrophic earthquake in the New Madrid fault zone, in San Francisco, Los Angeles, Seattle, or New York, could cause damage and fatalities that exceed the 9-11 attacks. And the American fire/rescue services must keep in mind the potential for terrorist attacks that exceed the level of damage experienced on September 11.

A variety of non-traditional rescue resources proved effective at the Pentagon. The use of tools like the T-Rex, lift truck, Bobcat, and man lift reduced the time and manpower required to perform many critical tasks and improved the level of safety for personnel operating in high-risk areas.

At major collapse incidents, ICs are well advised to consider requesting structural engineers to advise them on the safe mitigation of complex collapse problems. Each USAR task force and each USAR IST includes two structures specialists, civil engineers specially trained to help rescuers conduct high-risk USAR operations. This is a huge advantage when developing tactics and strategy, evaluating the effectiveness of stabilization operations, and determining the advisability of high-risk tactics. The value of the IST and USAR task force structures specialists at the Pentagon cannot be overstated. The following lessons were learned.

Experience counts. The efficiency with which members of the IST and various USAR task forces managed the range of challenges presented by this disaster was indicative of their training and experience. The depth of experience was an undeniable advantage at the Pentagon incident. This lesson emphasizes the responsibility of participating agencies to ensure—as much as possible—that all USAR task force members are trained and prepared to perform all necessary tasks related to their positions. Clearly, experience is a key component of readiness for unusual missions like the Pentagon and WTC incidents. With that in mind, the single-most important thing for some agencies may be raising the threshold of real-life USAR/collapse rescue experience among members assigned to these special units.

Potential for dispatching all layers. There may be a need in the future to deploy full USAR task force rosters, which are three deep in each position on every task force. Since the 1994 Northridge earthquake, there has been consideration of the need to dispatch all three *layers* of USAR task force rosters to triple the capabilities of the system. In disasters of a scale necessitating the mobilization of all task forces, this would allow FEMA to transport additional layers of rostered personnel, carrying only their PPE and other personal gear, to bolster on-scene task forces. Instead of rotating different USAR task forces in and out of the scene each operational period, it would be possible simply to rotate personnel from other platoons of the same task force. Rotating different USAR task forces necessitates the time-consuming process of exchanging task force equipment at each shift change. It would also allow equipment from each USAR task force cache to be used in wider operations by infusing additional personnel from the same USAR task forces.

Such a protocol may be used for earthquakes that devastate large urban areas in such a way as to require the response of all FEMA USAR task forces. Few of us suspected the possibility of a terrorist attack so large as to necessitate the response of 20 USAR task forces. But after the events of 9-11, we must be cognizant of the potential for one or more simultaneous attacks or a terrorist attack at the same time as another unrelated disaster to cause draw down of the nation's FEMA USAR task forces.

Prepare for unexpected disasters. There is a need to consider using FEMA USAR task forces for disasters that may not have been expected at the program's inception in 1989. Terrorism resulting in a collapse disaster on U.S. soil is the first example of events that were not on the radar screens of those charged with developing the FEMA national USAR response system in the late 1980s. Today, it's important for fire/rescue officials to consider the potential for events that would have been deemed implausible just a decade ago. The destruction of a dam through a terrorist attack, for example, might necessitate the mobilization of FEMA USAR task forces to help conduct SAR in areas decimated by sudden flooding/inundation in urban, suburban, or rural areas.

In the case of dam failure from any cause, people caught near the outer perimeters of the flood may be trapped alive in trees, on homes, on bridges, and within collapsed structures. Likewise, avalanches, landslides, mud/debris flows, and other events may create conditions requiring the mobilization of FEMA USAR task forces, or components thereof (e.g., search and recon teams, search teams, canine teams, etc.).

Familiarity with the system. One of the keys to the relative timeliness and operational safety for rescuers and investigators of the Pentagon incident was the professionalism and preparedness of the ACFD. The ACFD personnel were familiar with the FEMA USAR task force system. This was due, in part, to the close proximity of VATF-1, based in the neighboring Fairfax County Fire and Rescue Department. Not only was the Arlington command staff familiar with the USAR system, they were also on a first-name basis with members of the FEMA USAR IST and some members of VATF-1, VATF-2, and MDTF-1. This allowed the ACFD to manage a relatively smooth USAR/investigative operation that involved local, federal, and military resources operating with the highest level of cooperation in a very hazardous environment, with few injuries and no fatalities to responders.

Soon after the five-story collapse occurred during fireground operations on 9-11, the ACFD commanders essentially told the Fairfax USAR staff: "You handle USAR operations, using local, state, and federal USAR resources, and we'll handle the fire, EMS, and hazmat operations." This early division of labor and responsibility led to a smooth transition to a large USAR operation that proceeded around the clock for two weeks.

In this sense, the Pentagon incident carries a lesson for all fire/rescue agencies. It pays to ensure that all chief officers and company officers are sufficiently familiar with the local, state, and federal USAR systems and that they are prepared to effectively utilize these resources whenever the need arises. Regardless of what type of disaster occurs that traps people beneath collapsed structures, it is counter-productive for highly trained FEMA USAR task forces and their specialized SAR capabilities to sit on the sidelines. They could and should be used to expedite the discovery and extraction of live victims. They should be used to make the scene safer for first responders through the implementation of time-tested strategies and tactics. They should be there to help facilitate the needs of the FBI, ATF, NTSB, and military. They should also help ensure non-stop, round-the-clock USAR operations until the last victim is recovered, or until local agencies are sufficiently recovered to assume responsibility for remaining USAR operations.

Combine collapse/NBC responses. Modern fire/rescue agencies should give serious consideration for the potential for combined collapse and NBC terrorist attacks. For years, some terrorism experts and fire/rescue authorities have warned about the potential for a terrorist attack that would trap people in collapsed structures *and* involve NBC agents. This worst case scenario could place firefighters in the unenviable position of being confronted with live victims trapped in various places within a collapsed structure *and* in proximity to strategically placed NBC agents dispersed by explosion or collapse.

Some people have dismissed the suggestion of dirty bombs and other worst-case scenarios as preposterous. Others state flatly that we would simply have to write off anyone trapped within the perimeters of such an event. This author strongly disagrees. Response would depend on the extent and type of contamination and whether trapped people are actually exposed to toxic levels. It would also depend on whether personnel could safely enter uncontaminated sections of collapse zones. No one would automatically be written off without confirmation of their actual condition.

The criticality of experience and training in collapse disasters cannot be overstated. After the Murrah building was bombed, quake-seasoned USAR task forces were dispatched to assist the OCFD. The experience of USAR task force members told them that in the three-story mound of debris surrounding the nine-floor, there was a chance that people might be alive in survivable void spaces in sections shielded from the blast wave. As it happened, no survivors were found after the first 24 hrs, although SAR continued nonstop until almost the entire collapse area was cleared of debris. This was one of the most complicated and dangerous rescue operations in U.S. history to that point.

In retrospect, the absence of survivors after day one of the Oklahoma City bombing was mainly a function of the damaging physiological effects of the blast wave. In addition, the particular way that the building collapsed in pancake fashion after its support systems were blown away also contributed. The low survivability of trapped victims should lead no one to conclude that SAR operations are futile after the first day in future bombing- or earthquake-induced structural collapses. To the contrary, experienced rescuers understand that long-term survival of trapped victims should be expected until determined otherwise by physical or visual examination of potential survivable void spaces.

These principles were applied at the Pentagon collapse, leading to two weeks of round-the-clock USAR operations to give the best chance of survival to any victims trapped alive in the rubble. Although there were no survivors found after day one, it remains true that anyone trapped alive there would have received the best possible change of rescue, given the current level of experience and technology in the United States today. The same approach (with improvements resulting from lessons learned in the 9-11 attacks) will be directed to victims of the next national collapse disaster, regardless of its cause.

Realistic assessments. Effective disaster planning is based upon a *realistic* assessment of the types of disasters that are possible in a particular locale. It used to be that extensive disaster planning was conducted only in areas prone to earthquakes, hurricanes, floods, tornadoes, and other natural disasters. Clearly, every community needs to reevaluate its exposure to potential disasters, whether natural or manmade. Based on the events of 9-11, the time has come to begin plans that will ensure a timely and effective response to terrorist attacks. This planning should include preparations for the effective management of worst-case scenarios, not just the ones that we know can be managed with existing resources and experience.

The 2002 national intelligence estimate, an annual assessment of U.S. intelligence agencies, states that "ships, trucks, airplanes, and other means" are likely modes for delivery of terrorist WMD in coming years. According to the report, these unusual delivery methods using common objects are "less expensive, more reliable and accurate, and more effective" for producing explosions or disseminating biological/chemical agents. It's clear that fire/rescue agencies should plan for worst-case scenarios. Disaster planning should include worst case scenarios, including such things as high-impact times of the day, multiple simultaneous events, adverse weather, etc. to avoid being caught flat-footed. Disaster plans should be intended as *living* documents, subject to periodic updates that reflect the latest intelligence.

Local fire agencies affected by disasters should update their pre-attack plans for firefighting and rescue operations in damaged buildings. Some buildings (like the Pentagon) may need major shoring operations to make them safe for entry, SAR, and firefighting that may be needed *after* the disaster. Fire alarms and extinguishing systems may be inoperable in buildings damaged by explosions and other destructive events. Structures with damaged fire systems may require 24-hr fire watch or evacuation, and some commercial processes and occupancies may need to be closed down until repairs can be made.

Tools and equipment. The Pentagon SAR operations placed high demands on tools. It is necessary to consider additional equipment and resources to cut, lift, and break reinforced concrete in future collapse operations at concrete buildings. This is especially true with regard to tunneling through reinforced concrete. The need for rescuers to crawl into small spaces to break and breach their way through layers of reinforced concrete makes for an interesting challenge in terms of tools. One device that has proven helpful in these situations is called a rivet buster. It resembles a small jack hammer, but it has such a powerful punch that it has proven effective for firefighters and USAR task force members to use while lying at odd angles in confined areas.

In this sense, we should also consider the lessons learned during SAR operations at the collapse of the WTC towers, where the destruction was so complete that there was very little reinforced concrete left to cut. At the WTC collapse–which consisted largely of mountains of steel beams and metal structural components mixed with other debris—the overwhelming need was for metal-burning and cutting tools and rescuers trained in their efficient use. The

Puget Sound-based FEMA USAR task force (Washington Task Force-1) brought with them a tool that proved useful for metal burning and cutting at the WTC operations. Known as the *Petro-Gen*, this device has a cutting tip that mixes compressed oxygen with gasoline from a pressurized can to create a torch that readily cuts very thick steel and metal at a rapid rate. The Petro-Gen proved so effective during the WTC operations that dozens were ordered to be used in the collapse.

Because leaded or unleaded gasoline and compressed oxygen are so readily available at, or transportable to, the scene of most disasters, the Petro-Gen has proven very effective for collapse SAR operations. As of this writing, other USAR task forces are procuring Petro-Gens, and a number of fire departments (including LACoFD) are buying them for their rescue companies and USAR units. This is simply one example of the evolution of rescue tools in real time, based in part on the fire/rescue service's ability to quickly adapt to ever-changing conditions and new challenges.

Security. FEMA USAR BoO were secured by law enforcement officials and the military, which was on high alert throughout the Pentagon incident. It's a reminder of the importance of proper security for rescue resources to prevent theft, sabotage, or secondary attacks.

Self-sufficiency. The feeding and other support provided by the Red Cross and other agencies (including many volunteers) was tremendous. This level of support will not always be available in the immediate aftermath of long-term disasters. For this reason, local fire/rescue agencies should be prepared to be self-sufficient, including food and water, for at least 72 hrs.

Signaling devices. It's advisable for every firefighter and rescuer working in a large collapse site to carry a personal signaling/alarm device (i.e., whistles, boat air horns, PALS, etc.) at all times. This will allow them to warn of impending adverse events such as collapses, and to call for help if they become trapped.

Filters. Filtration respiratory filters were in high demand throughout the Pentagon incident. The filters required changing daily, and the masks themselves required decon every time rescuers left the red zone.

Decon. All personnel were required to be deconned every time they exited the red zone. Consider the implications of these requirements on future disasters, where hundreds (or perhaps thousands) of firefighters, rescuers, and investigators may have to be deconned.

Conclusion

The timely recovery of victims without serious injuries to rescuers can be attributed to a number of factors, including: (1) the ACFD's rapid and effective initial response (2) the familiarity of local fire commanders with the FEMA USAR task force system (3) the ACFD's readiness to utilize FEMA USAR resources effectively, including delegating responsibility for USAR operations to the FEMA USAR IST early in the incident (4) extremely close coordination between local fire commanders and the FEMA USAR IST (5) excellent cooperation between the ACFD, military, FBI, NTSB, ATF, local and federal police, and private heavy-equipment contractors (6) the IST's ability to develop and implement an IAP that supported the needs of the ACFD and all local and federal agencies on the scene, and (7) the overall readiness and professionalism of the FEMA USAR task forces.

Given the incomprehensible number of innocent people who lost their lives, it can be said that USAR operations at the Pentagon were conducted effectively and with as much success as could be hoped for. Despite the complications caused by the shutdown of all commercial air transportation for the first week, the FEMA national USAR response system worked as designed.

The Pentagon USAR operations were practically textbook in many respects. This was due, in part, to the close proximity of VATF-1. It was also a result of the rapid on-scene arrival of FEMA USAR IST members who helped pave the way for the arrival and timely engagement of all FEMA USAR task forces assigned to the Pentagon. The FEMA USAR task forces moved directly into position, deployed their equipment and personnel, established BoO, and began working without delay. As a result, the Arlington County fire command was able to concentrate on firefighting, EMS, and hazmat issues, with confidence that the FEMA USAR system was managing in a manner that met the objective set by the ACFD's Pentagon IAP.

Endnotes

1. Written by an unnamed 7th century bard upon observing the ruined stonework of the Roman town of Sulis (modern-day Bath, England).

2. If all the daily collapse incidents in the typical year in New York City were to occur on the same day from a single event, it could legitimately be called a disaster. It's no wonder that FDNY firefighters, particularly those assigned to the rescue companies, are among the nation's most experienced experts in structure collapse rescue operations. And now New York has the unfortunate distinction of being first-due on history's worst single collapse disaster, the most deadly single terrorist act, the most devastating failure of any high-rise building (twice in the course of one hour), and the worst loss of firefighters and other emergency responders, all in a single event.

3. Within days of the collapse, Israeli police arrested the four owners of the building, the contractor who built the structure, the owner of a renovation company that did work in the structure, and a structural engineer who worked on the project. A judge ordered them to stand trial for various charges related to the cause of the collapse.

4. Jim Day, structures specialist (structural engineer), LACoFD USAR task force (California USAR Task Force-2).

5. Proceedings from the Management of Complex Civil Emergencies and Terrorism-Resistant Civil Engineering Design (New York: City University of New York, June 24-25, 2002).

6. These precautions should be taken whenever personnel are engaged in collapse SAR operations and other high-risk tasks.

7. Until the first four Stages have been completed (leaving stage 5 [general debris removal] as the last option), officials should generally refrain from declaring that the recovery phase has begun.

8. As previously mentioned, stage 3 operations are sometimes initiated even as stage 2 functions continue, depending on conditions at the collapse site.

9. Void-space search squads are an alternative tactic taught by John O'Connell (FDNY Rescue Company 3) and Mike McGroarty (California OES), leading instructors on shoring and collapse rescue operations who have conducted SAR operations at disasters ranging from the Oklahoma City bombing to the WTC collapses. Both have conducted extensive work to define and teach "what it takes to manage collapse rescue emergencies." Although not officially taught as a universal approach for all collapses, void space search squads remain an option for agencies confronted with circumstances that make this approach advantageous. Once again, ask the magic questions: Is it safe? and Will it work?

10. Several Orthodox Jewish members of SAR teams were given permission from rabbis to continue their efforts through the Sabbath, the traditional day of rest as prescribed by the Bible. It's part of Jewish tradition to override religious mandates when lives are at stake, such as during times of disaster and war. For international USAR teams (including those deployed by the USAID/OFDA), this is an example of the kind of cultural, religious, and traditional issues that must be considered at the scene of a major disaster.

11. Widow-makers is a term given to large chunks of concrete suspended by rebar or other means, or other debris that remains perched over the top of rescuers during the course of SAR or firefighting operations. It is often necessary to take extra time to remove these immediate hazards before other work is completed to enable rescuers to conduct their operations with a reasonable degree of safety.

12. The Dade County and Fairfax County rescuers are members of two U.S. USAR task forces dispatched under the auspices of the USAID/OFDA, which coordinates the operations of American USAR assets during disasters on foreign soil. The third international task force used by OFDA is from the LACoFD.

13. "A Different Kind of Police Investigation," *Los Angeles Times* (December 17, 2000).

14. As of this writing, only about 800 bodies of the approximately 2,830 victims of the WTC attack had been recovered.

15. "In the Last Piles of Rubble, Fresh Pangs of Loss for Ground Zero Workers," *New York Times* (March 17, 2002).

16. Later, during mop-up, firefighters came across numerous half-emptied fire extinguishers scattered across floors on all levels of the building surrounding the fire. This indicated that many Pentagon employees attempted to fight the jet fuel-fed fire with fire extinguishers, but were forced to abandon their fire-control efforts because of intense fire and/or smoke.

17. Robert Patrick, "Shared Data on Bomb-Proofing Buildings Proposed," *Los Angeles Times* (November 22, 2001).

[18] The MDW's technical rescue team consists of engineer/soldiers.

[19] FEMA fields three national USAR IST-As, identified as the Red, White, and Blue IST-As. These specially trained teams, each comprised of 20 highly experienced members selected from among the nation's 28 USAR task forces, are dispatched whenever task forces are deployed, to assist the local IC with command and control/coordination of USAR task forces assigned to their incidents. They normally arrive ahead of the task forces, establish liaison with the local IC, determine the mission and scope of the USAR operation, develop a written memorandum of understanding with the local commander, and perform a wide variety of supervision, coordination, operational, and logistical functions to ensure effective, round-the-clock USAR operations that support the local responders and meet the overall incident objectives.

[20] Strickland, normally the leader of the FEMA USAR Blue IST, functioned as the USAR White IST leader for the first 36 hrs until the arrival of Lincoln (NE) Deputy Fire Chief John Huff, who drove directly from Nebraska, and Miami-Dade Assistant Chief Carlos Castillo, who responded in a fire department vehicle from Florida with IST Deputy Operations Chief Ruben Almaguer, also from Miami-Dade County.

[21] ESF-9 (USAR) is one of 12 emergency support functions listed in the Federal Response Plan. Under ESF-9, FEMA is the lead federal agency for coordination of USAR operations, and other federal agencies are listed as assisting agencies. When federally declared disasters strike, an ESF-9 emergency support team (EST) is dispatched to assess the need for federal assets to assist local and state authorities. The EST reports to the local IC and provides an immediate source of expertise and counsel for the IC, and is prepared to expedite the process of supporting the local and state agencies through the deployment of federal resources.

3

Water Rescue

Introduction

Every year, firefighters and other would-be rescuers are killed attempting to pull victims from rivers, streams, lakes, ponds, flood-control channels, flooded rock quarries, flood tunnels, the ocean, and other bodies of water. The vast majority of water rescue operations are successful, but there are significant numbers of close calls. The inherent dangers associated with water rescue demand attention to strategy, tactics, training, SOGs, and other factors related to operational safety. Rescuer safety must be of paramount concern in all cases because injured, lost, or dead rescuers can help no one.

The history of the fire service is replete with stories of ill-prepared, improperly equipped would-be rescuers who were killed trying to help coworkers. In many of these cases, firefighters and other first responders entered fast-moving water or large ocean waves without the appropriate equipment. To make matters worse, they may not have been trained for the conditions they faced, or they were simply overwhelmed by the force of moving water. Regardless of the reason, these rescuers ended up in the same predicament as the original victim(s).

We've all heard that the force of moving water is deceiving and that it can overwhelm even the strongest swimmers and rescuers under the right (or wrong) conditions. This author, a lifelong swimmer assigned to an engine company, once found himself fighting for his own life after jumping into the raging sea during an on-duty rescue. During a 1982 El Niño storm, a fisherman was washed off a rocky reef at the base of a 250-ft cliff in 12- to 15-ft waves. There was nothing but rock and reef along the coast.

The author, who like most firefighters at the time, had no formal training in surf and ocean rescue, was able to make his way to the victim and began swimming the man back into the surf line. Quickly, however, the situation turned into a life-and-death struggle to keep the victim and rescuer afloat. Not only were both dashed against the rocks by big waves, but they were also pulled beneath the surface in the venture created by a canyon of submerged rocks. They were washed back and forth 25 yards at a time by the storm swells.

After several minutes of struggle, the rescue was ultimately successful. However, more than 20 years later, this author is still left with strong impressions from the experience. One understands the true worth of *air* when repeatedly forced beneath the surface to the point where just one more breath of air becomes the most important thing in the world. It's at this point when one must decide whether to hold on or release the victim to get a breath of air before blacking out. One also understands the true value of a rapid intervention system (RIS) to provide emergency assistance when things start to go wrong and the rescuer gets into trouble. Fortunately, through youthful determination and luck, this author was able to hang onto the man and everyone escaped alive, albeit with various levels of hypothermia, cuts and scrapes, and a renewed respect for the power of water.

For people who put on a badge, there is an implicit understanding that each of us is willing to take these kinds of calculated risks to save the life of another; it is, in some ways, the essence of our job. We can say that such sacrifice is honorable, but we can also say that the need for self-sacrifice can often be prevented through appropriate planning, training, equipment procurement, and exercises. These activities can be learned from the study of many firefighting and rescue operations and especially those that become problematic or where unexpected hazards are encountered.

In localities where water rescue is either a constant or seasonal hazard, operational effectiveness and personnel safety can be improved if both factors are considered integral to the response system. An IC who has the advantage of access to well-trained, properly equipped firefighters and rescue teams can manage these situations aggressively and safely to ensure the best chance of rescuing victims. To achieve this goal, fire/rescue departments must undertake solid preparation *before* their personnel are confronted with potentially lethal water rescues.

A New Emphasis on Water Rescue in the Fire Service

For many years in some regions of the United States, water rescue was not considered a fire department role. Fire departments would respond to water rescue situations when requested, but there was little or nothing in the way of water rescue equipment, training, planning, and preparation for firefighters to conduct effective water rescue operations. Whereas many eastern fire departments traditionally performed water rescue operations, some parts of the West and the Southwest were slow to recognize the necessity for fire department-based water rescue systems.

Until the mid-1980s, there had been little emphasis on training firefighters to manage the complexities of swift-water rescue operations. The situation was generally the same across most of the nation, with the exception of progressive swift-water rescue training programs in places like Ohio, Colorado, Texas, and New Mexico.

There were no generally recognized national standards for swift-water rescue training, and there was no mandate from government to develop one. The situation was doubly complicated for firefighters and other rescue professionals in southern California. Flood and swift-water rescue conditions are among the most diverse and challenging to be found anywhere. In addition, the corresponding training and equipment requirements were daunting for funding-strapped fire departments struggling with the results of Proposition 13. This tax-relief law placed a heavy financial burden on many public safety agencies that had relied on property taxes for basic funding.

As a result, local firefighters frequently found themselves fighting for their own lives while trying to save victims from fast-moving water. Their ability to assist victims—and even to save themselves—was severely hampered by their structural firefighting gear. Although it successfully warded off the cold and rain during traffic collisions, flooding, and other storm-related incidents, heavy clothing and rubber boots filled with water became anchors during swift-water rescues.

Some firefighters who decided to enter the water to attempt rescues quickly found themselves dragged down and nearly drowned by their heavy clothing in the churning water. In other cases, they attempted maneuvers that exceeded the boundaries of their swift-water rescue training and equipment. Fortunately, there has been a sort of revolution in the development of water rescue capabilities among many fire departments that previously did not boast an emphasis on water rescue operations.

As the swift-water rescue revolution continues, placing further emphasis on water rescue operations, the public has come to expect the fire department to perform a multitude of special skills such as swift-water rescue, dive rescue, marine rescue, and other water rescue disciplines. It's a sign of the times that many fire departments have absorbed municipal lifeguard services, creating lifeguard divisions, marine divisions, and water rescue teams that include professional lifeguards, sometimes mixed with firefighter-staffed units.

It is a distinct advantage that water rescue training and preparedness is challenging and interesting; it makes participants think on their feet while performing arduous tasks like swimming, climbing, and boat handling. Water rescue training, planning, and exercises force firefighters and other rescuers to think on their feet, often under rapidly changing conditions—factors that lend themselves to more effective fireground operations.

Fire/rescue agencies that have developed formal water rescue programs will find that their firefighting forces tend to be well-rounded, physically fit, and enthusiastic about their jobs. They are ultimately better able to deal with the rigors and challenges of structure firefighting, wildland firefighting, USAR operations, and other emergencies. Fire chiefs will find that many operations run more smoothly, more efficiently, and above all, more safely. The public will recognize that tax monies are being spent to save lives and that the fire department can be relied upon even when unusual rescues and disasters strike.

Swift-water Rescue Operations

Assessing swift-water rescue hazards

The process of water rescue hazard evaluation starts with knowing what to look for in terms of potential rescue problems. This skill is closely related to the experience and training of firefighters and other rescuers. There are often good indications of potential swift-water problems if one knows where to look. It's often helpful to review past incidents and to discuss trends and patterns with dispatch personnel and others who have a broad sense of your department's jurisdiction. The simplest way to identify potential problems is for fire/rescue personnel to survey the district with their own eyes. If there are natural rivers and streams, flood-control channels, aqueducts, or areas subject to flash floods, your agency has potential for swift-water rescue emergencies.

Sometimes, swift-water rescue hazards may not be obvious, especially in regions that are normally arid. An oft-heard comment from officers in these areas may be, "We've never had a swift-water incident. Why are we going to all this effort?" The fact is, most firefighters will spend an entire career without actually rescuing a victim from moving water. But if there are identifiable swift-water hazards, the question of past incidents is moot. If a hazard exists, there is potential for incidents. Firefighters need to address that potential through planning, training, procuring the right equipment, and establishing a multi-tiered response system.

Dynamics of moving water

As firefighters, we devote a great deal of time to learning about fire behavior. We want to know what conditions cause fire to develop and how it reacts in different situations. It is this knowledge that allows the IC to properly size up a fire, a critically important task. A proper size-up sets the stage for all the operations that follow. To perform a proper size-up, firefighters must have an understanding of what they are seeing. The ability to look at a given situation and understand what is happening is usually related to training and experience.

Similar comparisons may be drawn with respect to water rescue. Decisions made by the IC and others at the scene should be based on an accurate understanding of what is happening in the water and what the water is likely to do. The initial size-up and all subsequent decisions will affect the safety and effectiveness of the entire operation. Fortunately, water (like fire) follows specific laws of nature 24 hours a day, 365 days a year. With training and practice, rescuers can even observe the features of a dry river bottom and have a pretty good idea what will happen when it fills with water during a storm.

A slow-moving river may give the appearance of continuity from shore to shore. This is deceiving. The speed of the water is really not constant across the profile of a river. Water along the shore is slowed by friction with the sides, and water in the center moves faster. This is an example of *laminar flow* (Fig. 3–1).

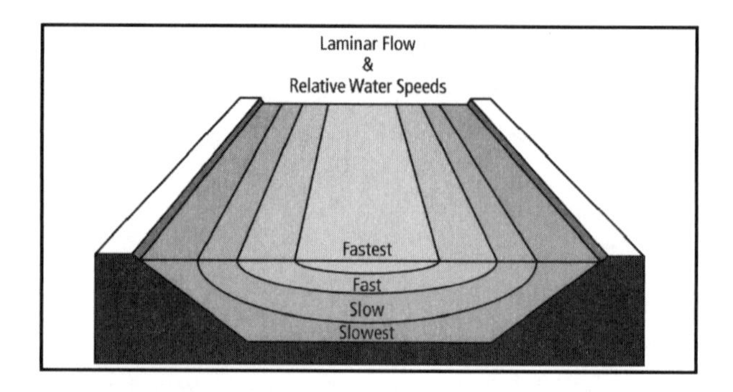

Fig. 3–1 Laminar Flow

A similar thing happens along the bottom. The deepest water is slowed by friction with the bottom, while the fastest water is found near the surface, riding over the top of the slower current below. This is also an example of laminar flow.

The slower-moving water along the shore tends to circulate with water in the middle in a spiral motion, which we call a *helical flow*. The water along the shore is pulled to the middle and dives down, returning to shore along the bottom and then moving to the surface in a corkscrew motion. This explains why people and other floating objects are often pulled to the middle by the current and may have difficulty getting back to shore. Flood-control channels and aqueducts are frequently built to increase the effect of helical flow (Fig. 3–2).

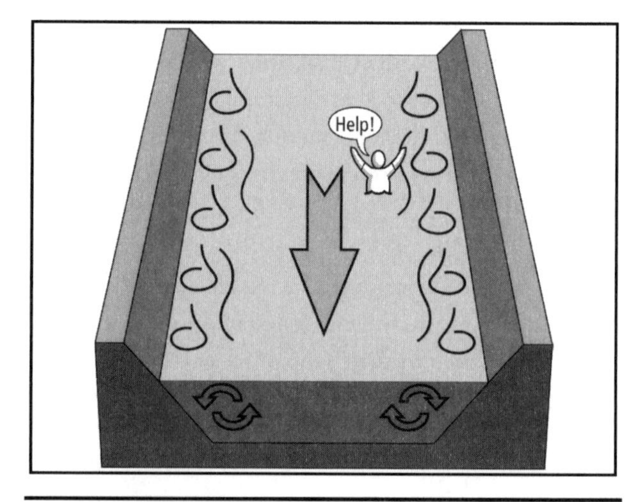

Fig. 3–2 Helical Flow

The purpose is to create a self-cleaning channel; debris is not allowed to pile up in the channel because the helical flow continually sweeps the bottom and sides. This creates an extremely dangerous situation for victims and rescuers in the water. The strong helical flow pulls them away from the shore and to the middle, where the water moves fastest. The combination of swift-moving water, strong helical flow, and slick, moss-covered cement shores makes it difficult (if not impossible) to escape without assistance (Figs. 3–3 and 3–4).

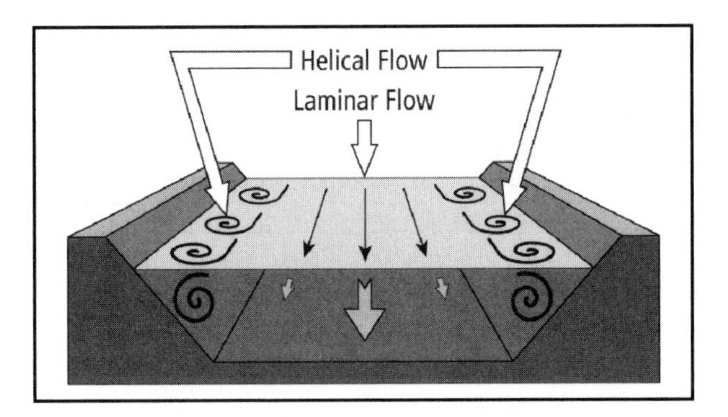

Fig. 3–3 The Combined Effect of Laminar and Helical Flows

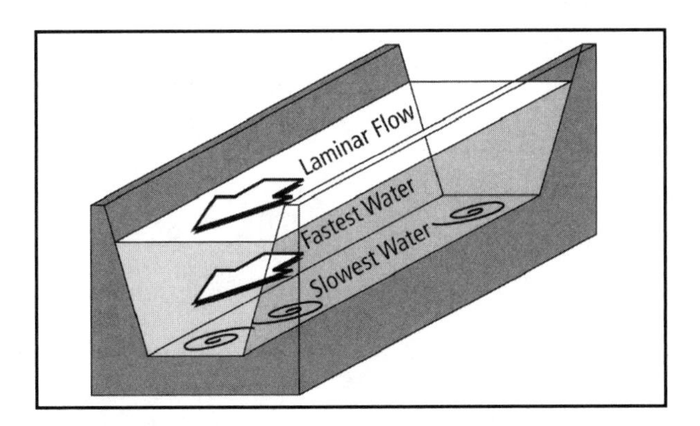

Fig. 3–4 The Fastest Water is Usually Found Midstream Just below the Surface

The result of all this is that the fastest water is usually found midstream and just below the surface. Water in a smooth channel may move deceptively fast. This is especially true in aqueducts, which are built to move large volumes of water with the least turbulence. Typically, the surface may be smooth and give the appearance of slow-moving water. However, a swift and dangerous current is often found just below the surface. This combination results in tragedy when rescuers enter the water expecting only a slow current. Frequently, they are dragged into the center of the channel by the helical flow and find themselves moving quickly downstream.

When a river encounters a curve, a standard set of changes can usually be observed. The water moves straight ahead until it strikes the curve. As it strikes the shore at the outside of the curve, the water tends to carry away material and undermine banks that are not lined with cement or other material. The water deflects off the outside bank and moves downhill, once again in a straight line.

If the curve is S-shaped, the current again deflects off the outside shoreline and heads downhill in a straight line. Even though the current appears to make a graceful arcing change of direction, it is actually trying to move in a straight line until it is deflected by the curving shoreline. The fastest-moving water is found moving from the middle of the stream to the outside of the curve, then back to the middle (if the river straightens out). If there are a series of turns, the fastest water is found moving from the outside bank of the first turn to the outside bank of each subsequent turn.

The slowest average speed is found at the inside of the turn, just downstream of the bend. In fact, the difference between the fast and slow currents often creates an *eddy* effect in this area. These shore eddies can be used as escape routes.

The power of moving water

Water moving downhill is forceful and relentless. Table 3–1 illustrates the relationship of force and speed in a river. As indicated in Table 3–1, rescuers or victims who become pinned against a rock in water moving 12 mph must be able to bench press the equivalent of 540 lbs to free themselves.

Table 3–1 Force of Moving Water

Current Velocity (mph)	On Legs (lb)	On Body (lb)	On Swamped Boat (lb)
3	16.8	33.6	168
6	67.2	134	672
12	269	538	2688

* Courtesy Ohio State Division of Natural Resources

The importance of water depth

The water level generally has the greatest influence in the look of a river. It is important for firefighters to go out and look at waterways in their districts both wet and dry, keeping in mind that dangers obvious in the dry season may be hidden by water at high levels. Many small riffles indicate shallow water. This is particularly true where the current passes over gravel bars. As the water gets deeper, the surface tends to look smoother. Submerged obstructions, such as rocks, cause a characteristic swelling on the surface just downstream of the obstacle. These bulges are an indication of objects below the surface (Fig. 3–5).

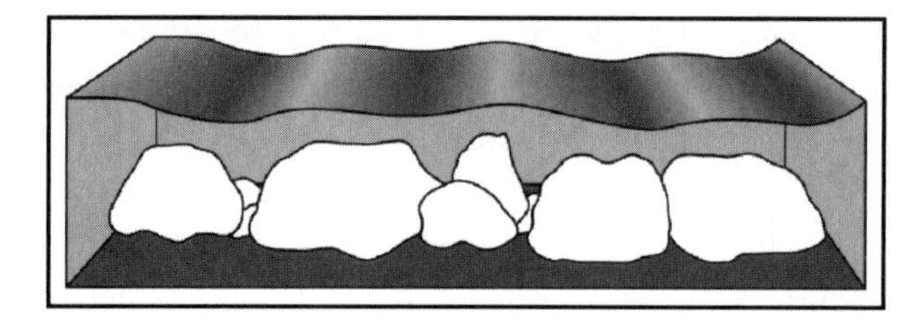

Fig. 3–5 Surface Riffles Indicate Shallow Water and/or Hidden Obstacles

Obstacles

Where the current is parted by an obstruction breaking the surface, a *cushion* or *pillow* appears on the upstream side, and an eddy is formed on the downstream side (Fig. 3–6).

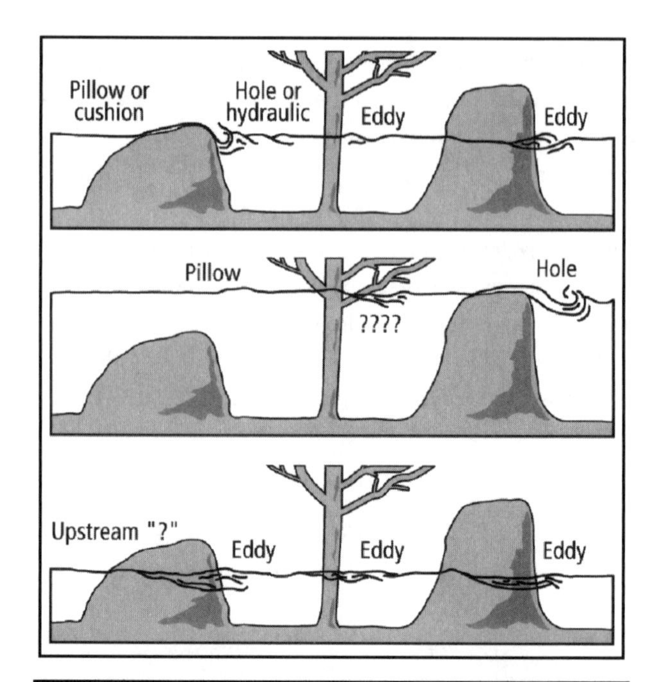

Fig. 3–6 Different Water Levels Equal Different Swift-water Hazards

The cushion or pillow is the result of water piling up on the upstream side of the object. In swift water, the cushion may pin a person or even a boat to the upstream side of the object (Fig. 3–7).

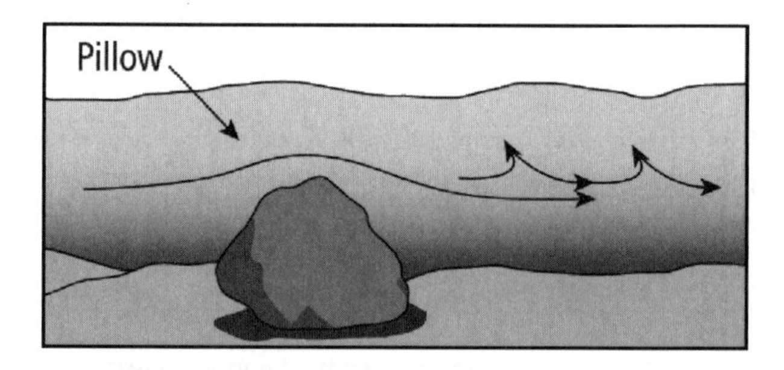

Fig. 3–7 A Cushion is Water Piling Up

Eddies are calm areas separated from the main current flowing downstream. As the obstacle upstream of the eddy breaks the current, a gentle reversing current is often the result. This reversing current is an area where large amounts of floating debris (and victims) may be found. Tired rescuers can also use an eddy as a resting area.

When the current is broken by a protruding object, the water parts into a V that points upstream. This is called an *upstream V*. The line of white water separating the main stream from the eddy flow is called the *eddy line*. Swimmers approaching from upstream will see the eddy line in the shape of a V pointing at them. This indicates an obstruction to avoid. The downstream point of the eddy lines, where the main current resumes, is called the *eddy tail*.

Where two eddy tails meet, or where the channel narrows and pushes the current together from both sides, a V pointing downstream forms. *Downstream Vs* are good indicators of deeper, faster water with few obstacles (Fig. 3–8).

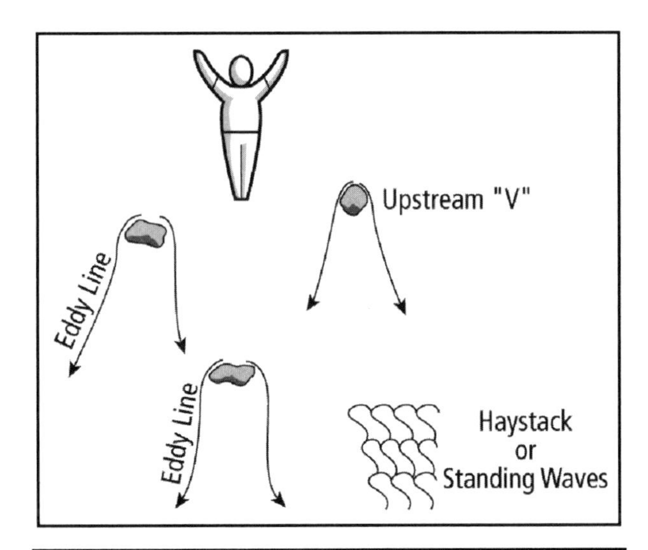

Fig. 3–8 Upstream *V*s and Downstream *V*s.

Another important feature of a river is the *haystack wave* or *standing wave*. These waves are often found at the downstream end of downstream Vs. Water accelerates as it is forced through the downstream V (much like wind through a narrow canyon). At a certain point, the acceleration of the water reaches terminal velocity, and the water has nowhere to go but straight up into the air. Gravity pulls the water down, causing a wave that breaks back on itself again and again. As long as the water velocity and depth remain constant, the wave will remain (hence the name standing wave or haystack wave).

Another type of eddy is formed along the shore, either by obstructions or current differentials found just downstream of curves on the inside bank. These are called *shore eddies*. Some shore eddies are noted for dramatic current reversals, with water flowing back upstream along the bank at a rapid speed. Other shore eddies are calmer, with the water creating a gentle reversal along the shore. A tired swimmer or rescuer should try to use shore eddies to exit swift-moving water (Fig. 3–9).

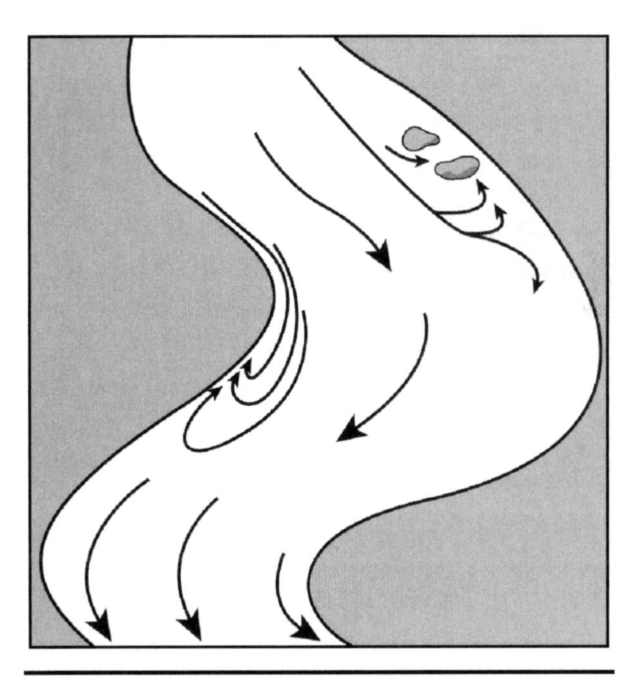

Fig. 3–9 Current Flow and *Shore Eddies* on a Curving Stretch of Waterway

Another common place to find eddies is where two channels meet. Normally, the eddy will be found just downstream of the convergence of the currents, toward the side of the weaker current. All eddies should be considered potential rescue points because they are often where the slowest water is found. Eddies may provide the only escape from fast-moving channels. Where possible, preplans should include rescue points where eddies are known to exist.

Hydraulics and holes

Whenever water flows over a steep vertical drop, the force of the water creates a depression just downstream of the drop. The further the drop and the greater the volume of water, the larger the hole. At a certain point, the depression is so deep that water flowing downstream is forced to move back upstream to fill the hole. Gravity has overcome the force of the current. This sets up a continuous recycling of the water that we call a *hydraulic* or *hole*.

The line where the recycling water is separated from the main downstream current is called the *boil line*. This line is easily recognized for its constant boiling white turbulent water, which wells up from below and moves back upstream toward the drop. The turbulent white water found in a hydraulic is about 60% air. Even buoyant objects may not float in white water.

In a hydraulic, the water is constantly recycling between the drop and the boil line. An object caught in this action must somehow end up on the downstream side of the boil line to escape into the main current. Hydraulics can be powerful enough to hold people and even boats for seconds, minutes, or hours.

The configuration or shape of the drop has an effect on the holding power of a hole (Figs. 3–10 and 3–11).

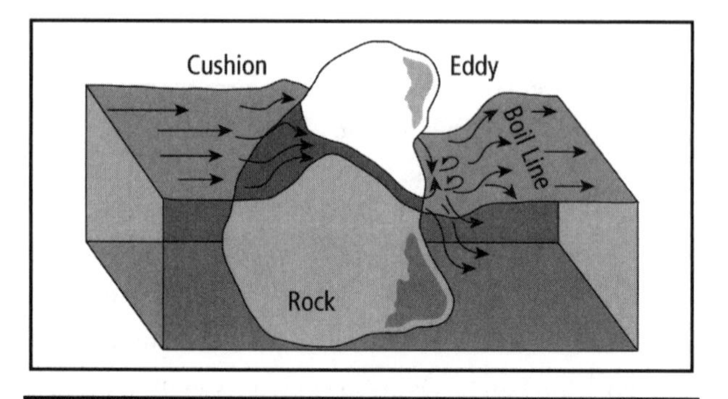

Fig. 3–10 Possible Hazards of the Rock Are Affected by the Water Level

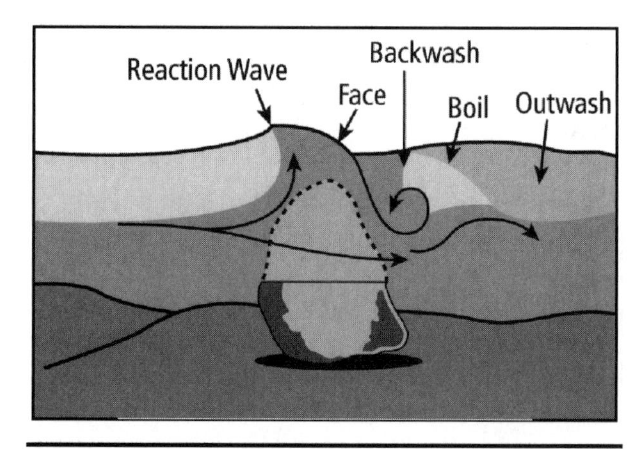

Fig. 3–11 How Water Level and Its Relationship to Submerged Objects Affects the Appearance of the Surface of Moving Water

In Figure 3–12, a *smiling hole* is flushing floating objects out to the side and into the main current. Notice that a swimmer approaching from upstream will notice a boil line forming the shape of a smile.

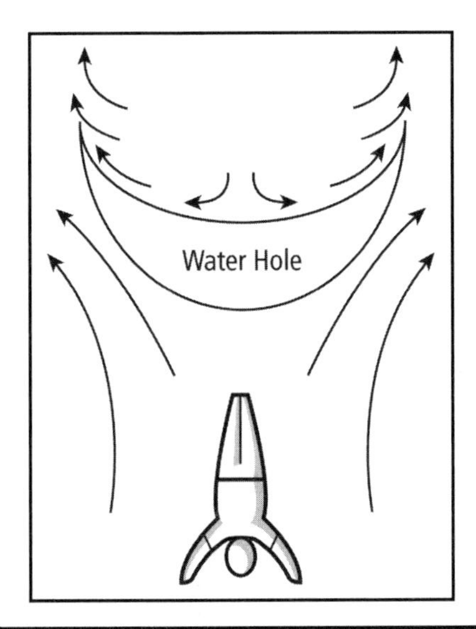

Fig. 3–12 Smiling Hole

In smiling holes, a floating object (including humans) may recirculate several times but will usually be flushed out to the side. Therefore, smiling holes are usually safer than other types of hydraulics (Figs. 3–12 and 3–13).

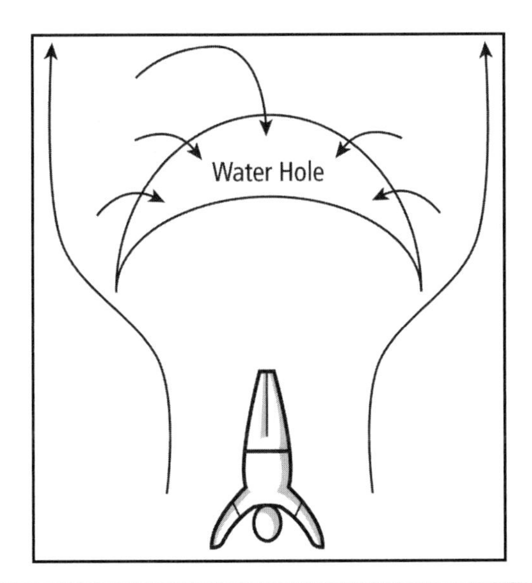

Fig. 3–13 Recirculating in Smiling Holes

Frowning holes, on the other hand, tend to flush objects back into the middle of the hole over and over again. Avoid frowning holes at all costs.

Lowhead Dams

Let's suppose someone decided to create a large vertical or near-vertical drop that stretched all the way across the river. The result would be a perfect hydraulic from shore to shore with no downstream Vs, eddies, or other river features normally used to escape from fast-moving water. This type of structure is found on waterways all across the United States. They are commonly known as *lowhead dams* or *drop structures*.

A typical lowhead dam is built from shore to shore with a drop of between 2 and 12 ft. Some lowhead dams are built at angles ranging from around 45 to 70°, but most are vertical drops. Lowhead dams are called *drowning machines* for good reason. These structures slow the current while providing elevation changes and backup water for conservation projects. They often result in vertical drops approaching 40 ft (Fig. 3–14).

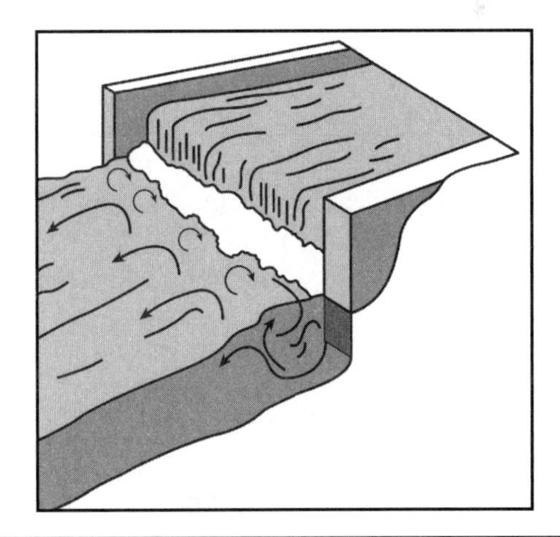

Fig. 3–14 Lowhead Dam

General characteristics of lowhead dams

Lowhead dams have the following general characteristics.

- *Construction.* Usually poured cement or stones laid in cement. Some are vertical drops and some are built at a high angle.
- *Width.* Almost without exception, they span the entire width of the channel.
- *Sidewalls.* Almost without exception, vertical walls are constructed along the riverbanks to prevent the tremendous hydraulic action from undercutting the banks. These vertical sidewalls can doom victims and complicate rescue efforts (Fig. 3–15).
- *Pans.* Cement pans are normally poured on the riverbed on the downstream side from shore to shore to prevent the hydraulic action from undercutting the dam.
- *Locations.* Lowhead dams are found all across the United States in areas ranging from desolate canyons to rivers and flood-control channels running through densely populated areas.
- *Deaths/Injuries.* Numerous firefighter/rescuer deaths and close calls have occurred in lowhead dams throughout the United States (Fig. 3–16).

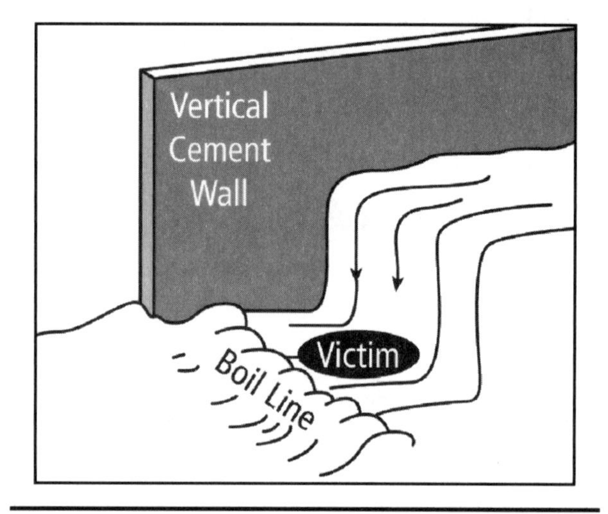

Fig. 3–15 The Boil Line in a Lowhead Dam Enclosed by a Vertical Concrete Wall

Fig. 3–16 Do Not Enter the Water Upstream of a Lowhead Dam

Personnel should not be placed in the water upstream of lowhead dams unless they are fully trained swiftwater rescue team members using precisely controlled means that will prevent them from being swept over the falls.

Since these dams stretch from side to side, so do the boil lines. It is impossible to escape by moving laterally, because there is no main downstream-moving current to get into. In this area, the only main current is the hydraulic. Victims caught in the hydraulic of a lowhead dam usually are helpless and disoriented. After being forced down to the bottom at the base of the dam, they are torn through debris along the riverbed until they reach the boil line. Pushed to the surface momentarily, victims are pulled upstream with the backwash. Many victims never have a chance to catch a breath of air before they reach the face of the dam and are dragged under again. The boil line at a vertical drop may be so close to the dam that the victim never surfaces and instead is left spinning in circles under the water.

Victims lucky enough to get to the side are confronted with a high vertical sidewall. There is usually no escape unless rescue is available from shore.

Escape from lowhead dams

Some victims have survived after a hydraulic has spit them out on the downstream side of the boil line. It has been said that a person able to maintain orientation might be able to purposely swim through the boil line. However, there aren't many volunteers willing to test this theory. *No one should enter the water to attempt a contact rescue from a lowhead dam.* If the hydraulic is powerful enough to trap the victim, it's likely to leave rescuers in the same predicament. Without shore-based help, the chances of survival in a powerful hydraulic are often zero.

Working near lowhead dams

Extreme caution should be used when working anywhere near a lowhead dam. Entering the water anywhere near the boil line is asking for trouble. Often the boil is not well defined, and a powerful hydraulic can even pull a powerboat upstream into the hole. Personnel should not be placed in the water upstream of lowhead dams unless they are fully trained swift-water rescue team members using precisely controlled means that will prevent them from being swept over the falls. These means of rescue include helicopters, inflatable rescue boats controlled by high-line systems, live bait rescues, or other positively controlled systems. To prevent personnel from being committed into the water upstream of lowhead dams, it's necessary to know the characteristics of the waterways in your jurisdiction. The only way of knowing about downstream hazards is through preplanning or by scouting the channel during the incident.

Strainers

The term *strainer* refers to bridge abutments, trees, fencing, and other conditions that strain out floating objects (including people) from water moving past them (Fig. 3–17).

Fig. 3–17 Typical Strainers

In the dry months, brush and trees grow along streambeds and river bottoms. Barbed-wire fences, debris piles, bridge supports, and downed trees are also found in channels and drainage. When the water level rises, these objects tend to strain out floating objects. Victims pinned against strainers by the force of the water may drown before being discovered. Strainers are commonplace at the outside banks of turns in a channel. In these areas, the bank may be undermined, exposing roots of trees and brush. Debris piles acting as strainers are common at turns in the river.

Rescue is especially dangerous when victims are trapped in rivers and channels that are full of strainers. Water rescue experts consider strainers among the most lethal of hazards to be found in any swift-water rescue situation.

The lethal power of strainers was demonstrated on November 26, 1997 during a fatal swift-water rescue incident in a flood-control channel in the San Gabriel Valley. Firefighters at LACoFD Fire Station 4 were alerted to a flash flood that was ripping past in the flood channel directly behind the fire station. To their surprise, they noticed five youths being swept downstream, right past the backyard of the fire station, which was fenced off from the channel. There was no time to mount any type of rescue attempt from the back of the fire station, so the firefighters alerted their captain, and they responded to a point at which they thought they could intercept the victims. The captain radioed a still alarm and requested a full swift-water rescue response, consisting of five additional engines, one truck company, one paramedic squad, two fire/rescue helicopters, one USAR company, and a battalion chief.

Two miles downstream of the point at which the children first became trapped, the flood-control channel drained into the 200-ft wide Rio Hondo River. At this confluence of two powerful rivers, the Rio Hondo channel is choked with 60-ft high palms and other trees, thickets of brush, and large piles of debris deposited during previous storms. This section of river is a virtual obstacle course of strainers that quickly trapped all five children when they swept into the Rio Hondo from the Alhambra Wash. There, they were washed through a nightmare thicket of trees and brush.

Two of the children managed to climb out of the water, where they were stranded for nearly an hour. Tragically, their three friends, two of whom were apparently trying to keep their non-swimming friend afloat, became trapped against the upstream side of the obstacles, below the water level, where they died and were later found by firefighters.

Several LACoFD swift-water rescue teams were dispatched to assist first responder units in searching for the five children from the first moments of the accident. It was clearly evident that all five children were probably trapped within the thick bramble of strainers that characterize the confluence of the Rio Hondo River at the Alhambra Wash. The helicopter-based rescuers were equipped and ready immediately to be suspended below the helicopters to snatch any victims from the flood once they were found. The worst problem facing them was the difficulty of spotting the trapped children against the backdrop of the roiling brown waters, whose surface was largely hidden from view by the thick canopy of trees growing in the riverbed.

Later, one of the LACoFD pilots, who flew MedEvac missions during the Vietnam War, likened the Rio Hondo rescue problem to searching for downed airmen in the thick jungles of Vietnam. It came as no surprise that all five children were found within 900 ft of the Rio Hondo/Alhambra Wash confluence. Tragically, under the dangerous conditions of the day, it was possible to rescue only two of them alive.

Rocky Shallows

Shallow, rocky areas are dangerous when water is moving swiftly. A swimmer trying to stand may find an ankle wedged between submerged rocks. If the water is powerful enough, the victim may be forced under the water (Fig. 3–18).

Fig. 3–18 Rocky Shallows Can Trap Feet and Ankles

The strength of the current and lack of adequate footing may prevent rescuers from reaching the victim in time. In these areas, it is best to crawl toward shore until the water is no longer deep enough to cause a problem, or until an eddy is reached. This is known as the *safe eddy rule*.

Flash Floods

Mountainous areas, drainages, desert areas affected by intense monsoon storms, and regions affected by hurricanes are especially prone to *flash flooding*. The conditions leading to flash flooding can sometimes be deceiving. A heavy downpour at higher elevations can produce flash-flood waves that wash downstream for miles along natural drainages, even when areas downstream are dry and sunny. Sometimes, the only warning for people downstream is the roar of the approaching flood wave as it carries boulders, trees, and other debris with it. Flash floods can develop so rapidly that rescue operations may not be able to keep up with the rising water and the volume of trapped people.

Fast-rise flooding

Fast-rise flooding is related to a number of factors. It could be a storm so intense that there is simply too much water for the waterway to carry, so the river spills over its banks. Another cause of fast-rise flooding may be related to the development of alluvial plains. River courses change drastically with the passage of time. In mountainous areas, these changes can happen in a distinctive pattern when the mountains are slowly worn down by rainfall that deposits sedimentary material along riverbeds and canyon drainage. This activity is especially evident at the base of mountains, where the majority of sedimentary or alluvial material is left along the shores of the drainage. Over the course of many years, the drainages may move back and forth across canyon bottoms and basins in a fanning motion that creates an alluvial fan.

As the drainage in the alluvial fan fills up with sediment, additional storms may force the water over its banks. The floodwaters cut new paths in their journey to the sea or to the middle of the desert, where they pool and eventually soak into the sand or create lakes. When this happens, the result can be fast-rise flooding (Fig. 3–19).

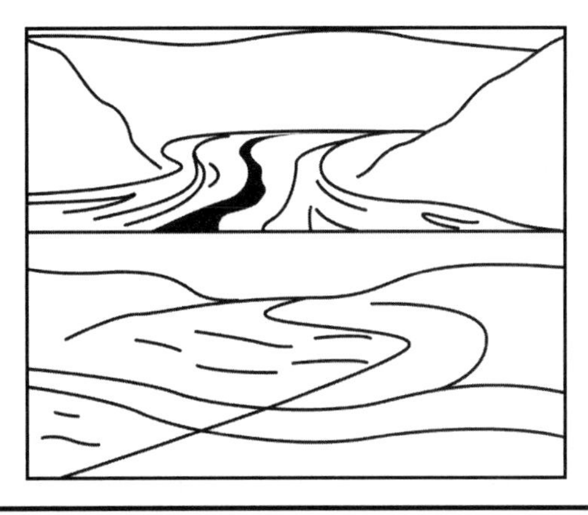

Fig. 3–19 Conditions Ripe for Flash Flooding and Fast-rise Flooding

Fast-rise flooding can be related to development of land on either side of a river, which tends to increase runoff and may squeeze the banks of the waterway between city streets. This reduces the ability of the river to stay within its banks during and after heavy storm systems.

One example is found in the Los Angeles River. For years, there has been concern that this river could overtop its banks and cause massive, violent flooding in an area populated by hundreds of thousands of people. Within 20 miles of the Pacific Ocean, the river is hemmed in by levees. These levees were created because of the large volume of water the river carries (during flood stage, it exceeds the volume of the Mississippi River) and the river's course through the flat plain of the Los Angeles Basin. These levees create a river that is mostly carried above grade.

The collapse of a levee in this densely populated area would be similar to the collapse of a dam; a wave of water would sweep through more than a dozen cities, finding the path of least resistance to the ocean. In these rapidly developing, high-intensity flood episodes, firefighters and other rescuers are faced with a number of problems. The most imminent problem is the sheer force and speed of the water, which may catch the populace by surprise and result in death and widespread destruction before any rescue efforts can be mounted. This would complicate the process of gaining access to victims. Under these conditions, rescue would require the use of boats, helicopters, hovercraft, and personal watercraft (PWC).

Another problem with fast-rise flooding is the tremendous amount of debris found moving with the water. Floating debris can include trees, automobiles, refrigerators, and other material that accumulates in rivers and canyons during the dry months. This material can be pulled under the surface and carried long distances under water.

Implementing a River and Flood Rescue Program

After gaining an understanding of the dynamics of moving water and after assessing the local swift-water rescue hazards, it's time to develop a response system that will allow rescuers to overcome any problems posed by moving water. Again, there is no single best swift-water rescue system, and each program must be tailored to the hazards of the local area. But there are a number of models (like that described in case study 2) that can be used as examples.

River and flood rescue, like many technical rescue disciplines, is undergoing a tremendous upswing in the national and international learning curve. New techniques and equipment are being advanced at a record pace for swift-water rescue. As this process mushrooms, the variety of new tools and methods will continue to expand.

There are relatively few regulations governing swift-water rescue. NFPA 1670 includes a water rescue track that finally sets one national standard for water rescue readiness, and organizations like NASAR have training and information related to this topic.

To take advantage of this information and training, some fire/rescue departments assign a core group of personnel to conduct research and development and provide recommendations for establishing a swift-water rescue system. In some major departments, existing technical rescue units are assigned the duty to develop swift-water rescue training, SOGs, and formal response plans.

Training sources include some private companies that specialize in technical rescue or swift-water rescue. Some of these companies began by training natural river guides and then branched into the rescue services. Therefore, some existing courses teach solutions that are mainly intended for wild river scenarios. As awareness of problems faced by municipal fire departments expands, training is being developed for rescues in flood-control channels (which are frequently more dangerous than natural rivers) and flash-flood zones. In California, Ohio, and other states, state fire marshals are increasingly involved in swift-water training.

The extent to which training should be conducted depends on the local hazards and the local response system envisioned. In most areas with hazards, basic training should be given to all personnel because it is difficult to predict just who will be faced with a swift-water situation. Advanced rescue capability may be established by assigning that responsibility to particular units, such as truck companies, technical rescue units, and other units located in high-hazard areas. Personnel with advanced skill levels may be used to provide basic training for the rest of the department.

It's important to provide strategic and tactical training for all officers who may function as IC, division/group supervisor, or other critical position during a swift-water rescue emergency. All officers should have a working understanding of the hazards and dynamics of moving water, what equipment and tactics may be most effective under given conditions, and how to use a waterway rescue preplan (WRP).

In some places, swift-water rescue training sites may be in short supply, especially during dry months. However, it's not always necessary to get in the water to practice rescue techniques. For first responders, dry land drills have proven successful in teaching basic skills like tossing throwbags or setting up tensioned rope systems. Safe, clean, moving water is required for in-the-water training. Sometimes, innovative ideas must be used. For example, the LACoFD uses a man-made river at a local amusement park and even fast-moving segments of the California aqueduct system as swift-water rescue training sites. Before this type of training is conducted, make sure instructors are sufficiently trained and that the training plan ensures safety. Personal safety equipment is a must for all participants training near moving water.

Recommended equipment

Personnel operating in close proximity to moving water should be provided a minimum level of protection. The equipment should be suited to the level of exposure to danger. A wide variety of rescue equipment, ranging from line-throwing devices to inflatable rescue boats, is available in today's market. Fortunately, new equipment for swift-water rescue is being developed all the time. Finding room on apparatus for this equipment is often difficult. Therefore, specialized equipment must be as compact, lightweight, and durable as possible while still meeting the requirements of the job.

Whenever possible, many fire/rescue agencies find it advantageous to use tools already carried by most engines, trucks, rescue companies, and USAR task forces. There are, however, some items exclusive to swift-water rescue (Fig. 3–20).

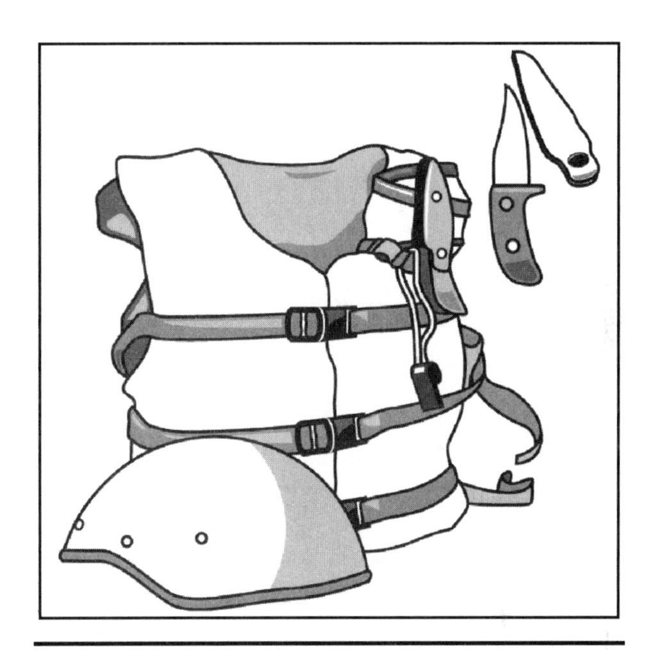

Fig. 3–20 Standard PPE for Swift-water Rescue Includes Rescue Helmet, Personal Flotation Device (PFD), Rescue Knife, and Whistle

PFDs. PFDs used by the department are Type III/V U.S. Coast Guard-approved for use in rough water conditions. They can also be used to provide flotation in calm water. These vests will keep an unconscious person upright and tilted slightly back in the water to maintain an open airway. A foam collar keeps the wearer's head out of the water. PFDs also provide limited torso protection.

Rescue helmets. Rescue helmets provide full head protection and do not impair the wearer's head movement and vision. Rescue helmets differ from structure firefighting helmets in that they are lighter and do not have a rear bill, which is a danger in moving water because of drag. This excess drag may rip a structure firefighting helmet off the wearer's head, tangling the chinstrap around the wearer's neck (Fig. 3–21).

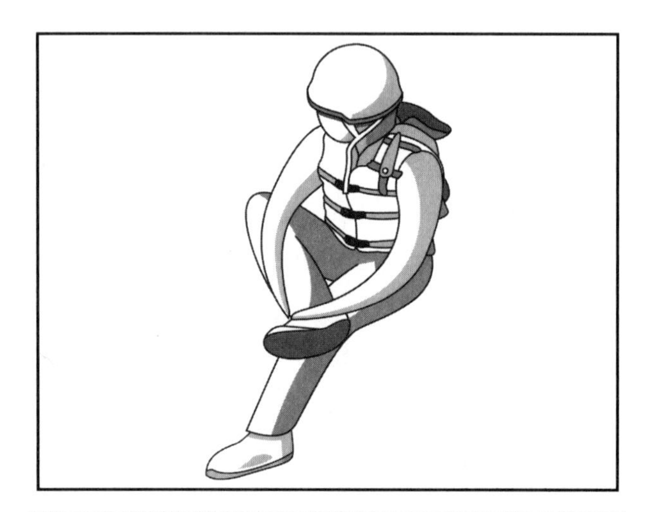

Fig. 3–21 Firefighter in Work Uniform with Standard PPE for Conducting Shore-based Swift-water Rescue

PFD accessories. Whistles, rescue knives, and strobe lights are often attached to PFDs. Whistles serve as communication/warning devices in the water. Rescue knives have a special blade for cutting rope safely and quickly to free victims from ropes or debris. Strobe lights are high-intensity lights used to monitor the location of personnel in the water.

Wetsuits. Wetsuits are provided to selected units with rescue potential in the ocean and in lakes at higher elevations. These types of rescues frequently require contact rescue in the water because victims may be far from shore with no other way to extricate them from the water. Wetsuits provide protection from hypothermia and also protect the body from rocks and abrasive surfaces. They may be worn in swift-moving water for both purposes.

Drysuits. Fire/rescue personnel increasingly use drysuits for swift-water rescue operations. They are especially important to provide a barrier between the skin of rescuers and contaminants carried in floodwater and even in some natural waterways. New models of drysuits include materials and construction suited for the rigorous conditions of river and flood rescue operations. Obviously, drysuits that become ripped or otherwise damaged can be a liability to rescuers. However, new materials and construction help safeguard against this, and modern drysuits can truly be said to provide a level of warmth and protection that was not available just a few years ago (Fig. 3–22).

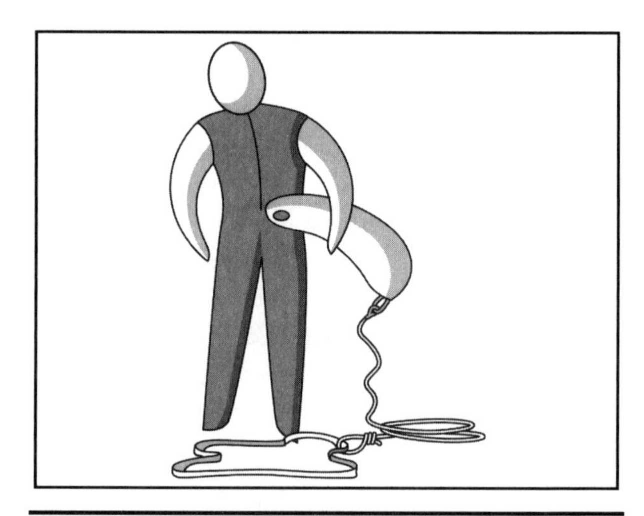

Fig. 3–22 Wetsuits and Drysuits are Common for Swift-water Operations

Footwear. Most fire department activities require safety shoes with steel toes. Water rescue incidents clearly have different footwear requirements. The use of heavy boots can be a hazard because they provide little traction on wet riverbanks. A firefighter who falls into the water wearing boots is at a disadvantage from the start.

Most firefighters and rescue personnel have a pair of running shoes at their worksite for participation in physical fitness programs. Running shoes are much more effective than work boots or turnout boots when working near moving water. They are the recommended footwear for all water rescue incidents. There are a wide variety of specialized rescue and outdoor recreation-related footwear items being developed and marketed all the time, and they are too numerous to list here.

Whenever water is flowing in sufficient quantity to create water rescue conditions, it is recommended that firefighters and rescuers carry their swift-water footwear on the units to which they are assigned. This may sound too obvious to mention, but it's especially important to mention to firefighters, who may not be inclined to carry their running shoes on the fire truck, ladder truck, or other units. Obviously, some regions have water rescue hazards year-round, and public safety personnel should carry appropriate gear at all times (Fig. 3–23).

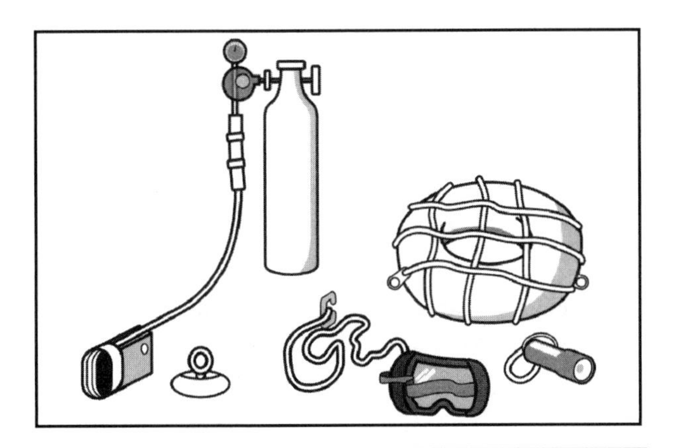

Fig. 3–23 Gear for Inflating Fire Hose with Compressed Air from SCBA bottles, Flotation for Victims, and Other Common Gear

Life floats. *Life float* is a very general term that we will use here to describe any floating object used for rescue. Fire/rescue agencies use a variety of devices as life floats. Some are commercially made, while others are as simple as inner tubes wrapped with bungee cord netting (Fig. 3–24).

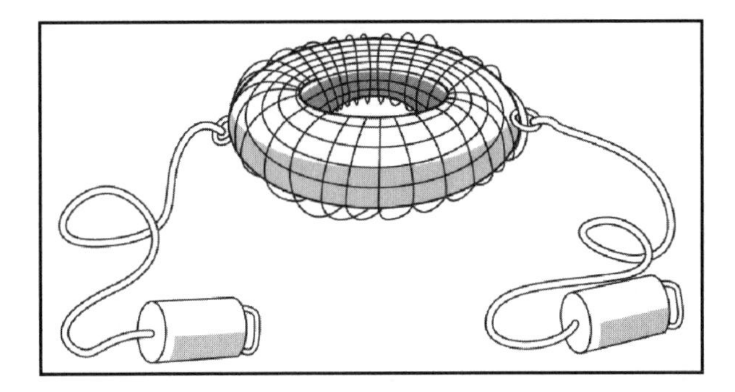

Fig. 3–24 Improvised Flotation Device with Two Throwbags for Control

Throwbags. *Throwbags* are the most basic water rescue tool. The rope of the throwbag is nylon mixed with polypropylene and sometimes other materials that allow it to float. The bag itself is usually fitted with a foam ring for flotation. Throwbags are tossed to victims who are then pulled to safety. The floating rope is not rated for vertical life safety. It should not be used for any purpose other than water rescue.

Hose inflator rescue kits. These kits allow engine companies to inflate 2½-in. fire hose with air from SCBA bottles. The inflated hose is used to effect rescue from bridges, lowhead dams, and other water rescue situations (Fig. 3–25).

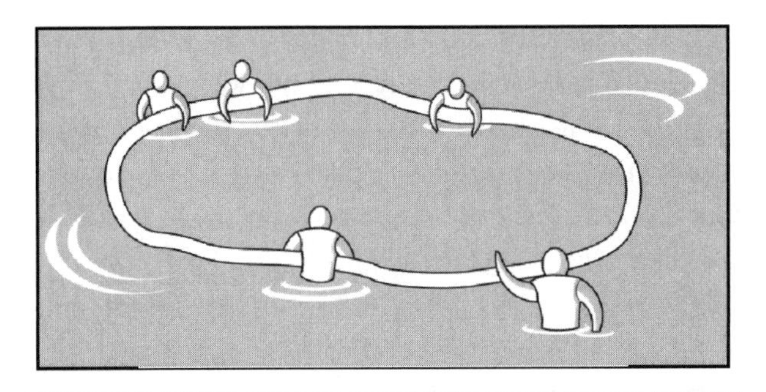

Fig. 3–25 How a Fire Hose Inflated with Air can be used in Unusual Configurations for Flotation

A different version of this device will allow personnel to build floating 2½-in. hose rings for multiple-victim situations, such as capsized boats or plane crashes.

Swim fins and equipment bags. Swim fins are used to provide extra propulsion through the water. Equipment bags come in many shapes and sizes. They help segregate and protect water rescue equipment. Backpack-type shoulder straps are helpful for cases where equipment must be hauled long distances to the rescue site. These bags may be strapped onto the hose beds or equipment bays of engines, trucks, rescue companies, and other units (Fig. 3–26).

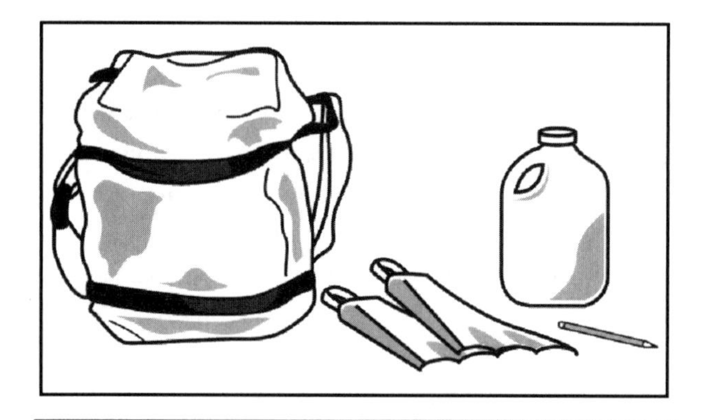

Fig. 3–26 Swift-water Gear Bag, Swift-water Fins, and Marker Float

Equipment and methods to avoid in swift-water conditions

Some commonly used equipment and methods have been found hazardous to rescuers and victims alike. They have been widely used and accepted by the fire service in past years because they appear to be logical on first appraisal. The hazards associated with these practices are of sufficient severity to justify this instruction to discourage their use. The following equipment and techniques have been deemed dangerous or potentially dangerous for use in swift-water rescue situations.

Structural firefighting helmets. Structural firefighting helmets are generally not designed for use in water rescue situations. While they provide excellent head protection from falling objects, they can cause more problems than they prevent when worn in moving water. The bill on the rear of structural firefighting helmets tends to catch water and create sufficient drag to rip the helmet off and tangle the chinstrap around the wearer's neck (Fig. 3–27).

Fig. 3–27 Firefighting PPE is not Designed to be Worn in Swift-water Rescue

Structural firefighting helmets are heavy and awkward in the water, and they can obscure the vision of the wearer. In many cases, the head protection provided is outweighed by the dangers they can create. Although it's acceptable to wear structural helmets in support positions along the shore, the decision to wear a structural firefighting helmet for entry into the water should be left to the discretion of the rescuers at the scene.

Turnout clothing. Unless some other hazard exists (e.g., structure fires near the water), there is no reason for firefighters to wear turnouts near moving water. Turnouts are absolutely dangerous near moving water. In recent years, firefighters have been taught to safely float in calm water with full turnouts. This technique works only in calm water where firefighters may remain motionless in the water, trapping air in the turnouts. However, in swift-moving water, it is impossible to remain motionless. Once the firefighter begins to swim, the turnouts fill with water, preventing effective swimming.

Turnouts and bunker gear are often worn against cold, wet weather. This is acceptable when swift-water hazards are not involved, but turnouts should be discarded whenever personnel are working next to moving water. Turnout boots and work boots can also be dangerous on slippery banks and in the water.

Ropes around the body. There is a great temptation for rescuers to tie themselves into rope systems when working in moving water. This may be inviting tragedy. If the line becomes tangled in debris or if the rescuer is unable to get free of a static line, he can be forced to the bottom and drown. The effect is similar to a water skier who falls off the skis and attempts to hold onto the towrope. Although ropes may be used as *travel limiters* to prevent rescuers from contacting the water, they should not be attached to rescuers in the water. The only exception to this rule is if the rescuer is wearing a PFD that allows immediate release if the line becomes snagged. Even then, rescuers should attach rope only if fully trained in its use (Fig. 3–28).

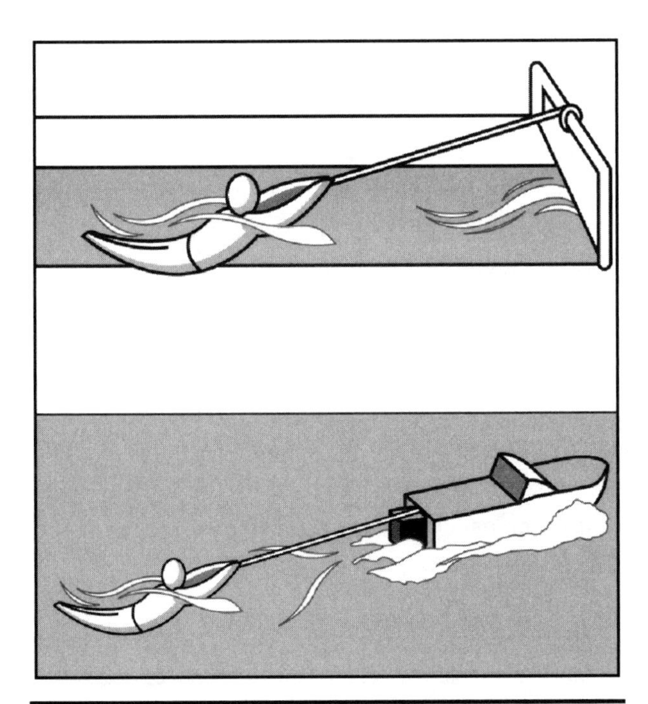

Fig. 3–28 Tangled or Static Lines Can Drown Much Like a Water Skier Trying to Hold the Tow Rope after Losing the Skis

If ropes are to be used for support in the water, a hand loop can be made for the user to grab. This is acceptable because the rescuer can drop the line if it becomes tangled. Hand loops should not be placed around the wrist; if the line quickly becomes taut, the loop can tighten around the wrist, trapping the user (Fig. 3–29). Appropriate use of ropes includes methods like throwbag tossing.

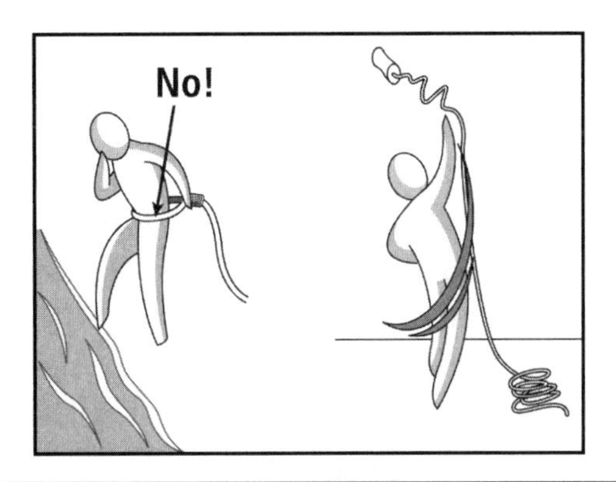

Fig. 3–29 No Ropes around Your Body in or near Moving Water!

Dangling firefighters from bridges and vertical-sided channels. For many years in some regions, it was an accepted practice for firefighters to tie rescue knots around themselves and for other crew members to lower them off a bridge above (or into) the water to snag victims from moving current. Unless a rescuer is specially trained and wearing the proper gear, this can be both ineffective and deadly (see case study 3). There is little chance of placing the firefighter in exactly the right spot to reach the victim and less chance that the rescuer will be able to pull the victim up and out of a fast-moving current. The rescuer may be forced to the bottom by the current, which will form a cushion on the upstream side of the rescuer's body, keeping it down. In a swift current, personnel on the bridge may find it impossible to pull the rescuer to the surface. Consequently, the rescuer may drown, submerged and helpless while trapped at the end of a taut rope.

When moving water is involved, there is generally only one situation appropriate for a vertical rescuer lower. This is when a victim is stranded on a stationary object or island below a bridge or other overhead access. Even in this case, the rescuer should not be attached to the line unless there is a positively controlled lowering and raising system that will keep the rescuer out of the current altogether.

If there is any possibility that the rescuer could enter or fall into water, do not tie personnel to ropes near moving water! An alternative method is the use of an *etrier*, or foot loops and hand loops tied together to form a rope ladder. A properly equipped firefighter can either stand in the loops while being lowered, or climb down using the loops as a ladder (Fig. 3–30). In this way, if something goes wrong and the rescuer ends up in the water, rope escape can be achieved.

Fig. 3–30 Dangers of Rope System Misuse in Moving Water

Self-rescue

As a general rule, firefighters and other rescuers should strive to complete swift-water rescues without actually entering the water *if* there are effective shore-, boat-, and helicopter-based options. As always, there are exceptions based on emergency conditions, the victim's predicament, the level of experience among the rescuers, the availability of equipment, the local SOGs, and other factors.

Although each of us understands the concept of risking a life to save a life, which can be applied to any emergency situation, we should also recognize the concept of risk vs. gain. Fast-moving water is a foreign environment to a significant (but fortunately, ever-smaller) percentage of fire/rescue professionals and volunteers. It does little good to sacrifice well-meaning but ill-equipped and inadequately trained rescuers to contact rescue attempts if we can conduct equally effective rescues working from shore, from boats, or from helicopters. The first impulse of many would-be rescuers is simply to jump in and grab the victim, even when the victim is being ripped along at 25 mph under swift-water conditions likely to overwhelm both victim and rescuer. So we have to be smart about who we put in the water, and when.

We can say that only rescuers trained in self-rescue and contact rescue—and properly equipped with PPE—should enter fast-moving water. We can say that those who are untrained and lack the minimum PPE should refrain from attempting contact rescue in fast-moving water. Firefighters and other public safety officials should find themselves in moving water for only three reasons: (1) to conduct a contact rescue in the water as a last resort to save a life, (2) to perform flood-control work in or near moving water (i.e., sandbagging operations, etc.), and (3) because they have accidentally fallen or slipped into the water. We also must acknowledge that unusual and dire circumstances compel firefighters and other rescuers to take extraordinary risks.

The main things then are to: instill a healthy respect for the power of moving water, be able to read the waterway to reduce exposure to the most dangerous conditions, and be able to use the water to advantage.

We must also teach firefighters and other rescuers to protect themselves if they choose to enter the water.

It should be noted that any manipulative swift-water rescue training, including self-rescue, should be conducted under the direction of qualified instructors. The following information is provided to help demonstrate the wide variety of river and flood rescue methods and equipment available to the modern fire/rescue practitioner. This information is not intended to replace manipulative instruction from a qualified instructor.

Four rules of self-rescue. When performing a self-rescue, keep the following four rules in mind:

1. *Remain calm.* It's second nature for experienced rescuers to remain calm and level-headed under stressful conditions. However, this is not easy when you are in true danger of death or serious injury. It helps to be properly trained and equipped. Calm, deliberate (but rapidly employed) actions are the key to survival in extremely dynamic situations like swift-water rescue.
2. *Use energy efficiently.* Hypothermia and exertion take a toll on energy in cold, rough water. Don't fight an unbeatable force; learn how to use the water's power to your advantage.
3. *Minimize exposure to the water.* The longer you are in the water, the greater the chance of hypothermia and injury. Don't linger in the water any longer than necessary.
4. *Wear personal safety equipment.* Wear your PPE! Need we say more about that?

The self-rescue position. In most situations, the proper position in moving water is supine, with the feet pointing downstream. This is sometimes called the *self-rescue position.* This position allows you to kick away from rocks and other obstacles. Another advantage is that this position allows you to lift your head above the water and look downstream (Fig. 3–31). Keep your feet at the surface to avoid submerged objects and foot entrapment. Flatten out and slip over submerged objects.

Fig. 3–31 The *Self-rescue* Position

Power swimming. The crawl stroke is effective when making an effort to get into a particular spot on the river. It allows the swimmer to move quickly through the water while looking straight ahead at the intended destination.

Strainer approach. Under some conditions, swimmers in swift-moving water may not be able to avoid hitting strainers. The current may be so powerful that the swimmer is pulled into trees or debris piles. This is one situation where the swimmer does not want to be in the normal swimming position (Fig. 3–32).

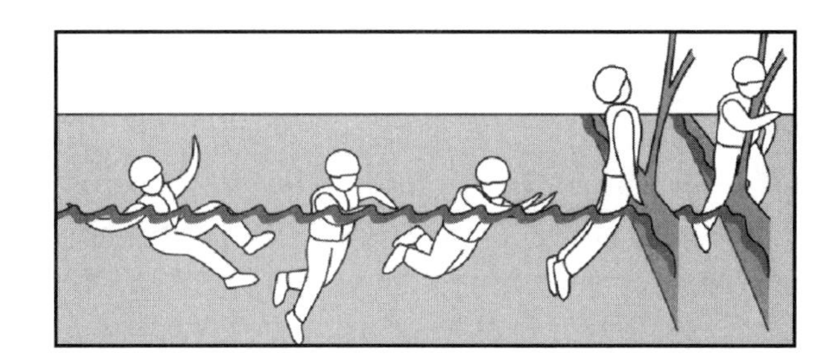

Fig. 3–32 The Proper Approach to Unavoidable Strainers

If you hit the strainer feet-first, you may find your legs around a branch or wedged in debris. A powerful current can then force your torso under water and drown you. This is a common cause of death in rivers and streams. When strainers are absolutely unavoidable, the best position is prone with your head facing the strainer. Just as you are about to hit the object, *explode* out of the water, climbing the strainer and getting your torso out of the current. The remainder of your body may be pinned, but at least you may be able to breathe until help arrives.

Ferry angle. The *ferry angle* got its name from the practice of stringing a cable across a river at a 45° angle to the current. A barge is then attached to the cable via a pulley. The action of the current against the upstream-side of the barge propels it to the opposite shore.

The same principle can be used to swim laterally in the current. Put your body at a 45° angle to shore, with your head pointing at the desired shore. The force of the water against your side will push you laterally toward shore. To increase this effect, backstroke against the current. The crawl stroke can be combined with the ferry angle for rapid lateral movement in the current.

Use moving water to your advantage. In swift-moving water, it is not only important to know *how* to swim, it is equally important to know *where* to swim to avoid hazards and make a safe exit. Correctly reading the water can save your life if you have the ability to use that information to your advantage.

Upstream Vs. Upstream Vs are caused by obstructions to the current. The characteristic V or white water pointing upstream indicates the obstruction. Avoid the point of the V to prevent being pinned on the upstream side of the obstruction (See Fig. 3–8).

Downstream Vs. Downstream Vs are normally the safest path of travel in a rough stretch of river. They usually mark the deepest water with the least obstacles. If you are faced with the choice of going into a downstream V, an upstream V, a strainer, or any other hazard, the obvious choice is the downstream V. If you cannot make an immediate exit from the water, look for downstream Vs as an alternative route.

Haystack/standing waves. Haystack waves are often found immediately downstream of downstream Vs (or other areas where the water is accelerating). They are recognized by churning white water in a continuous, stationary wave in the current. Haystack/standing waves are typically found in series, with the largest ones upstream. They get smaller as you move downstream and the water slows or the channel widens. While they may look dangerous, haystack or standing waves are usually safe, and a swimmer or boat can normally plow right through them and continue downstream.

Midstream eddies. Midstream eddies are found immediately downstream of obstructions to the water's flow (these obstructions normally create upstream Vs). They often make good resting places. The technique used to get into an eddy is simple. While in the self-rescue position, allow yourself to float past the upstream V (but not so close as to risk pinning on the upstream side of it). Just as you pass the obstacle, turn over and crawl stroke hard into the eddy. You may have to stroke upstream to get into the eddy because the main current will continue to affect you until you are in the eddy. Now you can wait for help or make your way laterally to other eddies en route to shore. This is called *eddy hopping*.

Shore eddies. Shore Eddies may be excellent exit points in swift-moving rivers. Shore eddies are usually found just downstream of turns on the inside bank, or downstream of rocks or other obstacles along the shore (See Fig. 3–9).

Because of the gentle reversing current that hugs the shore, shore eddies may be the only way out. The technique for exiting via a shore eddy is called *eddying out*. To eddy out, use the self-rescue position or the crawl stroke to get close to shore at the upstream part of the shore eddy. Approach the reversing shore eddy current and swim right into it. Just getting your torso into the eddy is usually enough to get away from the main downstream current.

Dealing with hazards encountered in moving water. In addition to using moving water to your advantage, you must also contend with various water hazards.

Rocky Shallows. Rocky shallows are typically recognized by the surface riffles that accompany them. The rocky bottom may be visible through the water, but the rocks may be a foot entrapment hazard (See Fig. 3–5).

If the current is swift, get on your belly and crawl to shore. Do not try to stand up until you are certain that the water is shallow enough that you can free yourself if foot entrapment occurs (See Fig. 3–7).

If you get a foot caught on the downstream side of a crevice, it may be easy to free it. But if your foot enters a jamb from the upstream side, the force of the current may prevent you from pulling it out. Try to kick back against the current to free your foot. If unable to free your foot, maintain your composure and keep your head out of the water. Call, whistle, or hand signal for help.

Pillows. Pillows themselves are not dangerous; they are merely the result of water being directed to the surface, creating a bulge in the surface of the water. The danger is the submerged object that is redirecting the water to the surface. In most cases, the hidden object is just upstream of the visible bulge (Fig. 3–33).

Fig. 3–33 Horizons Mean Extreme Danger: Quickly Exit and Scout the Shoreline to Avoid Being Pulled over a Lowhead Dam or Waterfall

Avoid traveling directly into a pillow to prevent being pinned on the upstream side of the submerged object. Move to the side if possible. If you are unable to move laterally in time to avoid the bulge, it is important to flatten out on the surface, keeping both feet up to avoid being pinned on the upstream side of the submerged object.

Horizons. Seeing a horizon appear ahead of you on a river should elicit only one response: Don't enter the water. And if you're in the water, get out. Horizons are evidence of an elevation change, such as steep rapids or a lowhead dam.

Holes. Holes should generally be avoided, whether they are of the frowning or smiling variety. Frowning holes are dangerous because they tend to flush objects back into the middle of the hydraulic over and over (Fig. 3–13). Smiling holes are usually much safer, but it is often hard to distinguish between the two. Avoid holes altogether if possible (See Fig. 3–12).

Strainers. Many victims die when, caught in a swift current, they try to climb onto trees, partially submerged branches, and other objects midstream (Fig. 3–34). In a swift current, it is easy to become pinned on a strainer or tangled in associated debris. Victims may find themselves pinned with the water rising. Others are struck by floating debris that slams into the strainer. It is almost always safer to make an exit at the shoreline than to get into a strainer situation. If it's necessary to use a midstream obstacle to get out of the water, it is usually best to float past the object and drop into the eddy behind it. This avoids potential entrapment or *pinning* situations. Now you can float safely in the eddy or climb onto the object to rest or await help.

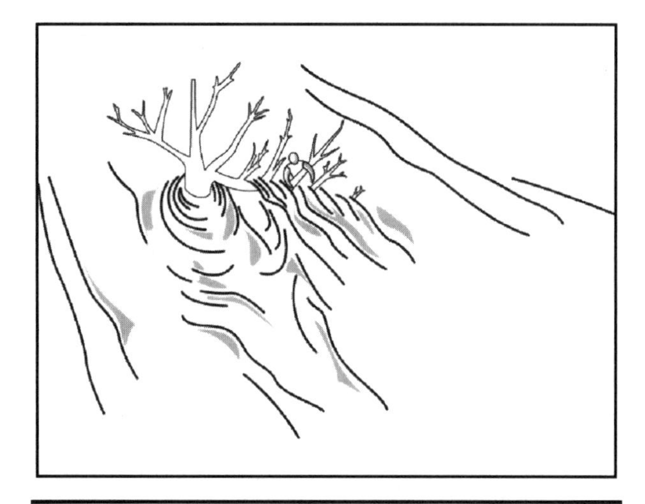

Fig. 3–34 Avoiding Strainers

Lowhead dams. Stay clear of the danger zone created by lowhead dams. This danger zone includes the following areas:

- Area upstream of the dam (where you could be swept over the face of the dam into the hydraulic)
- Area between the face of the dam and the boil line (this is the actual hydraulic at the base of the dam)
- Area immediately downstream of the boil line (where you could be pulled upstream across the boil line and into the hydraulic) (Fig. 3–35)

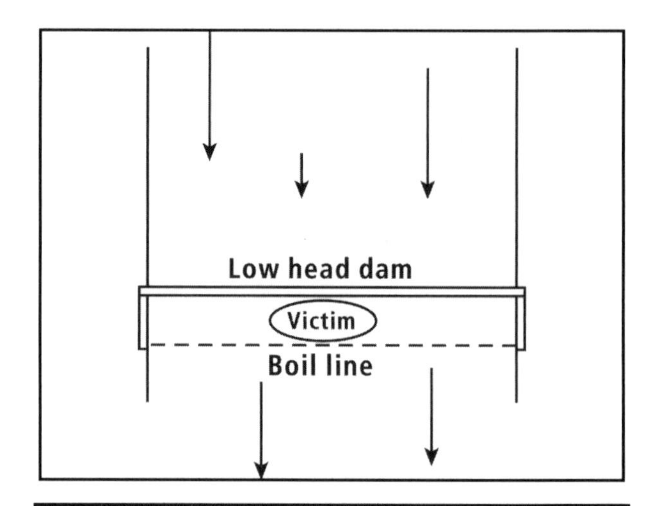

Fig. 3–35 Lowhead Dam Entrapment

Escaping from lowhead dams may not be possible in all cases. However, you may increase the chance of survival by doing the following. First, try to move laterally to the shoreline. If there is no vertical wall, escape may be possible by grabbing objects on shore and pulling yourself out. If there *is* a vertical wall, stay close to the wall and await help.

If you have insufficient flotation, you are likely to be pulled under at the face of the dam and (hopefully) resurface at the boil line. If you end up on the downstream side of the boil line, you will be released into the downstream current. If you end up on the upstream side of the boil line, you will be pulled back to the face of the dam and under the water again. At this point, try to swim downstream with the hydraulic and plow through to the downstream side of the boil line.

Arizona crossings. The term *Arizona crossing* is used to describe roads or driveways that cross streambeds and river bottoms. Arizona crossings are usually cement pads. The downstream side is commonly a vertical drop into the streambed. Arizona crossings appear to be harmless. However, water pouring over the downstream side of the pad can create a hydraulic identical to the type found at a lowhead dam (See Fig. 3–19).

SOGs

As in other rescue operations, SOGs are designed to support the operations that training and continuing education have made possible. SOGs formalize the use of particular tactics and techniques for given generic situations. They are guidelines that should be flexible to allow informed decision-making at the scene. SOGs are very important in setting minimum standards. Swift-water rescue SOGs should include the following topics as a minimum:

- The use of personal safety equipment should be mandatory for all firefighters, officers, and other rescuers operating in positions/locations vulnerable to moving water. No equipment/apparel should be used that is deemed a danger to rescuers in the water. This would include turnout gear, structure fire helmets, and ropes tied around the body of rescuers intending to conduct operations that require them to enter the water.
- Upstream safety spotters should be used to warn of debris approaching the rescue scene.
- Downstream safety should be ensured by assigning personnel to stage downstream of a rescue site in case a rescuer or victim is swept away. Downstream safety personnel should be properly trained and equipped for immediate rescue attempts.
- Low-risk methods should always be employed first. Consider increasingly high-risk options only when necessary. Be prepared to back off if conditions are too dangerous to conduct rescue operations with reasonable safety.
- Only highly trained and properly equipped personnel should operate around severe hazards such as lowhead dams or concrete flood channels with vertical walls or channels that go underground.
- Standard communications, techniques, and tactics should be used. Mutual aid frequencies should be considered when multiple agencies are assigned to a swift-water incident.
- WRPs should be used as the initial IAP whenever possible. This provides a basis for the initial actions of all responders.
- The safety of each individual rescuer, the rescue team, bystanders, and the victim should be considered in that order.

WRPs

WRPs are a concept advanced by the Ohio State Department of Natural Resources and the Ohio State Fire Marshal in its widely renowned river rescue training program. The Albuquerque, New Mexico Fire Department added a new dimension to the concept with its *code RAFT* preplanned response system for rescues in cement-lined flood channels.

In L.A. County and other regions with similar hazards, this concept has been widely adopted by fire/rescue agencies and refined to cover flood channels, natural rivers and streams, and flash-flood zones. The LACoFD worked closely with Albuquerque through the 1980s to integrate WRPs with computer-assisted dispatch (CAD) response matrixes. This allows the dispatch of numerous units downstream along a linear pattern following the channel in which a rescue occurs.

This is an important consideration because water in flood channels can exceed 35 mph. If units are delayed, or if they are not dispatched far enough downstream, the victim(s) will pass rescue points and may never be seen again. This was a fairly common occurrence before these systems were implemented.

These new approaches have repeatedly proved successful under dire circumstances. In Albuquerque, more than a dozen firefighters and police officers have been swept into flood channels attempting rescue during the past decade. Each one of these rescuers was retrieved at downstream rescue points specified in response matrixes and WRPs. Many children and adults have also been saved by these systems. Without WRPs and their corresponding networks of predesignated rescue points, the results might have been tragic.

In some departments, it is common to see the simultaneous dispatch of up to 20 units from different local fire departments that participate in automatic aid agreements and in the WRP program. Within a few minutes of dispatch to a swift-water rescue, engines, trucks, paramedic squads, USAR units, battalion chiefs, swift-water rescue teams, lifeguards, and helicopters can be set up for rescue at strategic points or for conducting searches.

While it may seem like overkill to the casual observer, to those who understand the problems associated with flood-control channels and other high-speed waterways, it's a sign that the victims are being given the best opportunity to be rescued. Every year, victims are rescued alive from high-speed flood-control channels under conditions where their deaths would have been likely if not for these very elaborate swift-water rescue systems. The WRP program, begun in Ohio more than two decades ago, is now being used in many parts of the United States with great success. Combined with proper training, equipment, and inter-agency cooperation, this type of integrated response program can be adapted for use almost anywhere there is a swift-water rescue problem.

Incident Command for River and Flood Rescues

Whether you are commanding an incident in a 25-ft deep flood channel moving 35 mph, a natural river, or a flash flood in the middle of the desert, there are basic parameters you should maintain as the IC, division supervisor, group leader, or other command assignment. If the preparations covered in the preceding pages have been made, many of these parameters are already established and everyone on the response team should know them.

However, if none of these preparations have been made, the IC will already be behind the eight ball because it's an open question about whether anyone responding to the incident knows even the most basic safety precautions. There may be no controls in place to prevent firefighters from jumping into fast-moving water wearing full turnouts or something equally ill-fated.

Without these preparations, there may be no SOGs requiring personnel to wear safety gear; there may be *no safety gear to wear*. With no formal plan, everything will be on your shoulders. In addition, there's a good possibility that if one of your personnel suffers an unexpected mishap (like falling into a raging flood-control channel and drowning), no one may be prepared to help. Worse still, everyone might start jumping into the water without the proper equipment and training to assist the would-be rescuer, thereby creating a secondary rescue problem that could result in disaster.

Sound far-fetched? Think again. In a single swift-water rescue incident in Albuquerque, nearly a dozen firefighters and police officers ended up in water at the same time while attempting to rescue two teenagers. In addition to the original victims, each rescuer was being swept away by water moving more than 30 mph. Some firefighters purposely entered the water to attempt contact rescue. Others accidentally slipped on the wet, sloped concrete and found themselves in the water. Still other rescuers were pulled into the water while using rescue ropes near the channel.

The IC was faced with a situation where nearly a dozen firefighters and police officers were being swept along in different segments of a 20-mile long flood-control channel. As the situation grew exponentially worse, with more firefighters and cops being reported in the water, the IC had no way of knowing how many were actually in the water, who they were, where they were, and their exact status. During the height of the incident, there was no way to get an accurate head count.

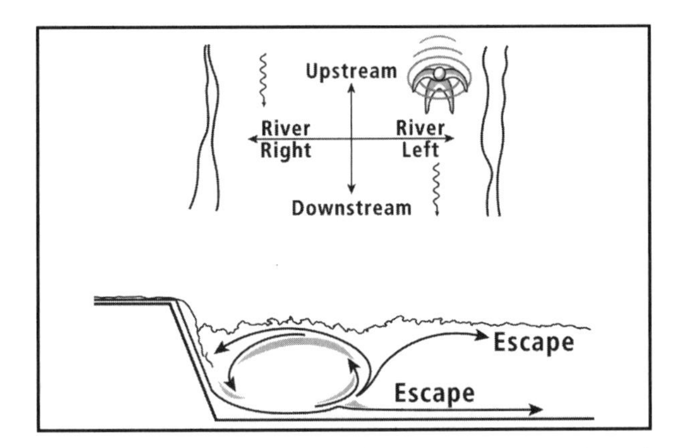

Fig. 3–36 Escape from Lowhead Dams

Fortunately, the firefighters were wearing PFDs, which allowed them to stay on the surface (and probably saved lives). None of the police officers had PFDs, and several received serious injuries. Because the Albuquerque Fire Department had taken the time and effort to equip and prepare its personnel, it guaranteed that rescue points were set up downstream to rescue their own people. The situation could have been far worse if the same thing had happened to a fire department that had not been as progressive as the Albuquerque FD.

River orientation

When discussing directions with regard to rivers and other waterways, four basic terms are used:

- Upstream
- Downstream
- River-right (the right side of the river as one is looking downstream)
- River-left (the left side of the river as one is looking downstream)

Waterways often twist and change direction as they flow. Thus the terms north, south, east, and west may be confusing and should not be used. Always use the terms listed instead (Fig. 3–37). We use the terms like "River Right," "River Left," "Upstream," and "Downstream" to describe directions in swiftwater rescue operations. Terms like "North Shore," "South Shore," "East Shore," and "West Shore" can be unnecessarily confusing on a twisting waterway.

Fig. 3–37 Minimize Confusion by Using Better Terms

Communications

It's often necessary to work long distances apart or from opposite sides of wide rivers or floods. It's not uncommon for rescuers to be out of the line of sight, especially when waterways bend or when foliage or terrain is in the way. These problems threaten to complicate an already difficult situation and may compromise the safety of rescuers operating in vulnerable positions. Good planning, equipment, and communication protocols can help overcome these problems.

Handi-talkies. Consider purchasing commercially available waterproof handi-talkie pouches or other protection against water damage for handi-talkies. While this may not protect them from total immersion, it may be possible to limit the exposure of radios to rain, mud, and moisture.

Whistle signals. SAR teams have long relied on whistles for communication in remote areas. When ordinary voice contact is impossible due to background noise or distance, whistles can be heard clearly for long distances. This is particularly important when visual contact is impaired. Whistles should be attached to PFDs with a cord or lanyard in a way to allow easy use.

There are four standard whistle signals:

- *One long blast:* stop all action; pay attention to me
- *Two short blasts:* start action; move upstream; pay attention to upstream action
- *Three short blasts:* move downstream; pay attention to downstream action
- *Three long blasts:* help; rescue in progress

Important—Three long blasts, repeated over and over, are used when a life-threatening situation is in progress, or whenever help is needed *immediately.* When this signal is heard, halt all activities and assist the firefighter blowing the whistle.

Arm signals. When visual contact is possible, arm signals can be used to communicate basic messages. They can also be used in combination with whistles when necessary. For example, one blast on the whistle will get everyone's attention; now the rescuer can use arm signals to point out a danger or to direct movement. The following are some standard arm signals:

- Hands pointed in a particular direction—go that direction; look over there
- Both arms crossed in front of body forming an *O*—OK; I don't need help. This signal can also be used by personnel on shore to *ask* if the rescuer needs help. The rescuer in the water should return the OK signal if no assistance is required.
- One arm in the air—Help! One arm in the air is the easiest motion for someone who is injured or struggling in the water. Any time a rescuer in the water gives the *help* signal, there should be an immediate rescue effort.

Bullhorns, public address systems, and other forms of communication. Bullhorns and PA equipment, if available, will be helpful to facilitate communication with victims and other personnel. Some situations, however, may require the use of imagination to get the message across. Examples include the use of handmade signs to give directions to victims when voice contact is impossible.

Case Study 1: Rescue at Triunfo Creek

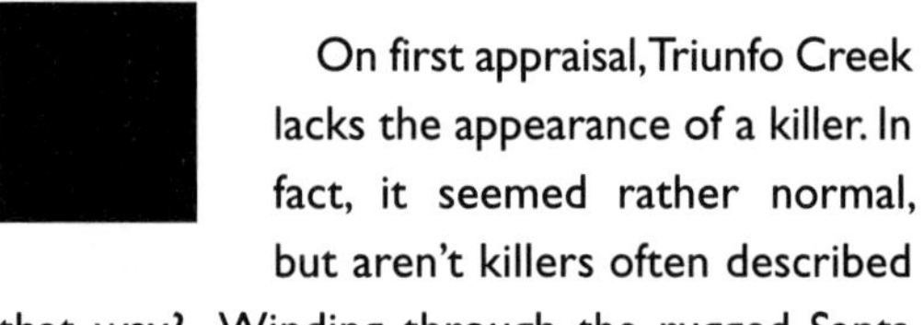

On first appraisal, Triunfo Creek lacks the appearance of a killer. In fact, it seemed rather normal, but aren't killers often described that way? Winding through the rugged Santa Monica Mountains on its way through Malibu to the Pacific Ocean, Triunfo Creek is home to multi-million dollar estates, horse ranches, and private vineyards. Across the creek from Triunfo Canyon Road near a famous riding stable is a small preschool. The only access to the school is a concrete drive that crosses the creek, a situation known as an Arizona crossing, a reference to the many dry rivers crossed by roads in the neighboring state.[1]

It was the winter of 1982, and in those days the LACoFD had recently begun its formal swift-water rescue program, one of the first in any urban area of California. Rudimentary training courses in swift-water hazard awareness and shore-based techniques were being conducted, and PPE for swift-water rescue was being purchased incrementally and distributed to units in the highest-risk areas. This was part of a revolution in swift-water rescue and other technical rescue capabilities beginning to sweep California's fire/rescue services.

Back at Triunfo Creek, automobiles lined up on a gravel road crossing the normally placid stream,[2] their windshield wipers swiping rainwater from the glass. The drivers patiently waited their turn to cross the creek one at a time. The water was flowing deeper and faster than normal, as a serious storm passed across southern California. Headlights pierced the darkness and lit the falling rain.

In a Jeep Wrangler, the wife of a Los Angeles city firefighter picked up her four-year-old son, waited in line, and started back across the creek toward the main road. Visibility was poor, with a low-hanging mist pierced by rain. The woman never saw the wall of water roaring down the canyon toward her. Halfway across the creek, she caught a glimpse of something roiling and black coming from the side, but by then it was too late. Most of the witnesses waiting in line behind her didn't even see the wave of water rush out of the darkness, pushing trees and debris in its path.

The water struck the Jeep and swept it downstream. It spun several times as it floated, and finally lodged itself on some rocks in the middle of the creek. Flood waves surged over the hood and the water rose quickly as the driver climbed out of the window, pulling her child with her. She had nowhere to go. As water flooded the vehicle, she wrapped one arm around the roll bar and the other around her son. Bystanders on shore lost sight of the Jeep. They could only see her bobbing in the water with her son in one arm, screaming for help. A man jumped into the water to assist but was immediately swept past the vehicle into the dark canyon below. He was lost from sight.

Personnel at the LACoFD dispatch center were already scrambling to cover a deluge of emergency calls for rescues, vehicle accidents, and floods related to the intense storm. Fire department units were hopscotching from one emergency to the next in the most heavily impacted areas. Moments after receiving a frantic 911 call from the preschool, the LACoFD dispatched one engine, a paramedic squad, a brush patrol unit, and an ambulance to the rescue on Triunfo Creek. There were no USAR units or swift-water rescue teams in those days, so first responders were expected to manage incidents like this with the resources on hand.

Engine 65 arrived on the scene seven minutes later, and Captain Jerry Dwyer knew this would be a troublesome rescue. Citizens frantically waved their arms and pointed at the creek, now flowing white water over 60 ft across. The water was fast and deep; the firefighters had trouble making out the form of the Jeep's roll cage, to which the female driver was clinging with one arm; she still held a small boy in the other. She was screaming she could not hang on much longer. It was cold and raining, and the current was buffeting the two victims. Refrigerators and other large items were observed floating past in the darkness.

Dwyer knew he had only seconds before these two victims would be washed away. As he stepped out of the cab of the apparatus, he was told that a male victim had already been swept downstream and was nowhere to be seen. Dwyer radioed for additional units to begin searching downstream, but he knew he and his crew would be on their own for an extended period of time due to the rural nature of their district, the heavy rain, the accompanying traffic jams, and the high volume of incidents being reported.

Engine 65 had responded from a two-engine station near Malibu. Because of the terrain and the high incidence of flood rescue operations there, Station 65 had been provided with PFDs and other swift-water PPE for one engine. But as luck would have it, engine 265 was already committed to yet another water-related emergency, and they took the swift-water equipment with them. Consequently, engine 65's technical rescue gear was limited to one over-the-side rope pack and hardware. Dwyer knew his crew was about to be pushed to the maximum of the safety envelope.

Under these conditions and based on his limited training in shore-based swift-water rescue, Dwyer could see few desirable rescue options for this situation. He knew that under ideal conditions, a rescue boat attached to a rope system over the water might work, but this was out of the question because no boats were available, there was no time to set up such a system, and his personnel weren't trained to do it.

A helicopter hoist rescue would be out of the question in this case because of low clouds, heavy rain, darkness, heavy tree cover, and the steep, unforgiving terrain. A variety of rope systems over the water might work, but engine 65 had no line-throwing device. Without PFDs and other personal safety equipment, Dwyer knew that sending firefighters into the water to attempt a contact rescue could end in disaster and might make the situation worse for the original victims. Most important, the firefighter's wife was screaming that she could no longer hold onto the roll bar of the jeep in the cold, rushing water. In seconds, she and the boy would be swept away. There was no time. They had to do something now or two lives would be lost.

Captain Dwyer instructed his firefighters to begin tossing the rescue rope at the woman, hoping she might somehow grab it. If she could just hang onto the rope, Dwyer thought, his crew might be able to *pendulum* her and the boy back to shore. It was risky, but he could see no other alternatives at this point.

Dwyer radioed instructions for assisting units to find access points downstream to provide a backup to this rescue, and to search for the male victim already lost. Then he instructed his crew to begin a series of rescue attempts at the Jeep. The firefighters from engine 65 were having trouble getting the rope to the victims' location. The distance between shore and the Jeep was just far enough that tossing the rope far enough became difficult. Once the rope hit the water, it was immediately swept downstream by the fast current.

Repeatedly, the rope became tangled on snags beneath the water. Finally, they managed to toss the rope far enough, and the water swept the rope directly to the driver, but she could not grab it without releasing her grip on the roll bar or her boy (and even then she would only have one shot at the rope). When the firefighters tried pulling the rope back for another throw, they met with resistance, and the rope inexplicably became taut. They had no way of knowing that, in the turbulence of the current, the rope had somehow become wrapped around the woman.

The cold water was quickly taking its toll. The rescuers could no longer hear the boy screaming over the roar of the water, and the woman continued to be battered by the current. Finally, she screamed that she was about to let go. As the firefighters scrambled to get another rescue rope into service, the woman and boy were swept downstream and disappeared under the water.

Acting on their limited training, engine 65's crew moved downstream along the bank and pulled the line as hard as they could. There was resistance, but then it diminished, indicating the load had released. They continued to pull. Miraculously, the woman and the boy resurfaced and moved closer to the shore in a pendulum arc. When she got within 15 ft of shore, the firefighters formed a human chain and pulled both victims from the water alive.

A search for the male citizen swept away during the first rescue attempt found him along the shoreline. He had managed to get himself into a shore eddy that allowed him to escape the current. Drenched and exhausted, he was alive.

Later, Captain Dwyer learned that the woman never was able to grasp the rope; it had simply wrapped itself around her in the current. Fortunately, the rope did not unravel as the firefighters pulled her in. Conversely, it was good fortune that prevented the rest of the rope from tangling on snags. If this had occurred, Dwyer knew, she and the boy might have been drowned while tangled at the end of the taut rope.

Several days later, Dwyer was asked to recount the scene for personnel attending a first-responder swift-water rescue training session. Dwyer described the frustration of knowing that there were probably better and safer ways to rescue the victims, but neither he nor his crew had been trained or equipped to attempt them.

Dwyer talked about the impact of a lingering image in his mind. As he stood on the shoreline commanding the operation, he had direct eye contact with the woman, trapped just 60 ft away. The boy was screaming as he was repeatedly submerged. The woman was unable to keep her son's head above water. And she was looking directly at Dwyer, yelling that she couldn't hold onto her baby. Dwyer said he would never ever forget the look on her face, the direct eye contact between them, or the feeling of helplessness he had at that moment.

"These incidents are relatively rare," said Dwyer to the firefighters, "and few of us will see more than one or two situations like this in a career. But when you're the one who's standing there watching someone plead with you to save his or her baby, you want to know exactly what to do." He went on to say, "You don't want to stand there helpless, without a plan and without some way of helping that person. And if you're running the incident, you will be thankful if you've done your homework to make sure your people are ready."

Shore-based Rescue Techniques

As a general rule, conducting rescue from shore is preferred over going into the water with the victim. Other options are boat-based rescue (including the use of PWC and inflatable rescue boats) and helicopter-based rescue. All other things being equal, the more rescue options we can give firefighters and rescuers, the less likely they will be forced to resort to the most hazardous method (contact rescue in the water) from the very beginning.

The techniques discussed here will allow properly equipped rescuers to make (or at least attempt) effective rescues from shore. The setup time for these methods ranges from seconds to just a couple of minutes. If the victim is missed, a unit may have time to pick up and try again at the next downstream access point. These quick hit-and-miss techniques allow flexibility in rapidly changing situations.

Throwbags

Throwbags are just about the most versatile tool used for water rescue. They can be used for shore-based rescue from currents, to line streams, to shuttle equipment and/or victims across streams, and various other tasks (Fig. 3–38).

Fig. 3–38 Shore-based Rescue Options

Throw bags are similar in size and construction to the drop bags used for firefighting operations. One main difference is the foam flotation ring sewn into the bottom of the bag. Rope stuffed into the float bags is a blend of nylon and polypropylene, which allows it to float on the surface. This rope is strong enough to haul victims to shore. However, it is not generally rated as lifeline rope. Unless specifically approved, throw bag rope should not be used for vertical rescue or for tensioning systems.

Throwbag rescues from moving water. When a victim is being swept downstream, the quickest method of rescue is to toss some type of floating object attached to a rope. The throwbag gives us that capability. A single firefighter can make an effective rescue with a float bag if the victim is within throwing range and can grab the rope or the bag (Fig. 3–39).

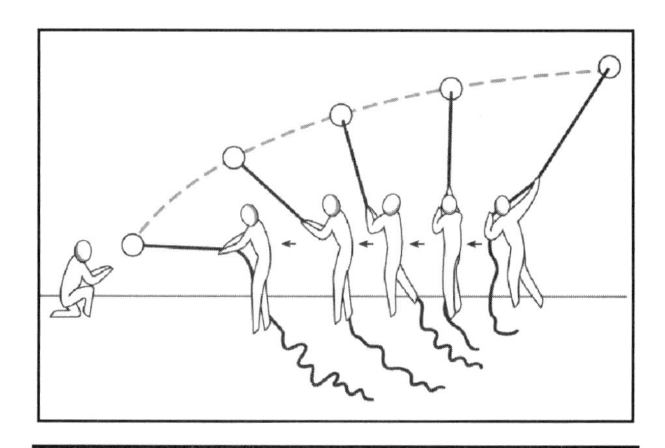

Fig. 3–39 Standard Throwbag Rescue Technique Utilizing the Pendulum Effect

The average rescuer can normally toss the bag 40 to 60 ft using an underhand, overhand, or side-arm throw. Once the victim grabs the throwbag, the rescuer begins to pull in the line. The victim will move toward shore in an arcing motion that resembles the swing of a pendulum. This is known as the *pendulum technique*, and it has a variety of other applications (See Fig. 3–29).

Line guns or similar devices

Line guns (where available) are excellent tools for *lining* a waterway (Fig. 3–40).

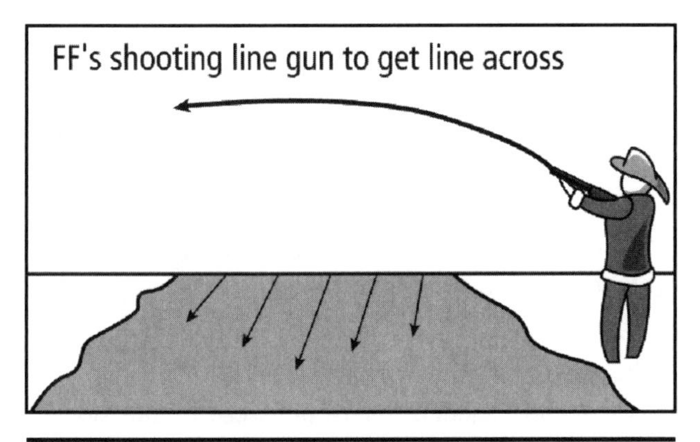

Fig. 3–40 Line Guns for Bridging Waterways

Lining the river using helicopters. In places where power lines or other obstructions might prevent helicopters from making a direct rescue, they still might be used to take a line across the waterway for shore-based rescuers. Safety dictates that the pilot will make the final decision to attempt any type of helicopter operation. If a helicopter is used, a crewman secured to the inside of the helicopter by nylon webbing should hold the rope while leaning from the side of the craft. If the rope snags, the crewman can easily drop the rope clear of the helicopter.

Rope tosses

If the waterway is reasonably narrow, rescuers may be able to throw the line across. Pull the desired length of rope from the rope pack (or float bag). Tie an in-line figure-8 at the desired length coming from the pack. Then attach the in-line figure-8 to the carabiner at the anchor point (if applicable). Make the desired number of small coils 1 ft in diameter. Make the desired number of large coils 2 ft in diameter. Separate the small coils from the large coils and bind the small coils with a bight. Then, throw the small coils, while allowing the large coils to pay off. Hang onto the rope.

Belaying

There are some situations in swift-water rescue that require rescuers to *belay* a line. In this context, belaying simply means to put tension on a line while using friction from either the rescuer's body or from a stationary object. Belaying can also be done by moving along the shoreline holding onto the rope, as in the pendulum technique. The purpose of belaying may be to haul a victim to shore, to secure a rescue line above the water from shore to shore in a mobile rope system, or for a holding action until the rope can be anchored to each shore for a fixed line system.

Body belay. For a *body belay*, the rescuer's body is used as a temporary or mobile anchor. The rope is passed around the rescuer's back to add friction, allowing the rescuer to control movement of the rope through his hands. The rope is *not* passed completely around the rescuer's body (to form a loop). The rescuer may not be able to get free of the line and may be injured or pulled into the water. With the rope across the rescuer's back, he can give slack or haul the line, using the friction of the rope to control rope movement.

Static belay. This technique uses trees, boulders, or other solid objects as frictioning devices to hold the rope. With a half turn or one full turn around a tree, the rescuer has plenty of friction for most needs, yet can still quickly release tension if necessary. By combining a static belay with a body belay, the rescuer increases holding power while maintaining quick-release capability.

Dynamic belay. A *dynamic belay* is used to pendulum a victim to shore. The rescuer runs downstream along the bank to prevent the victim from getting downstream and to decrease the load shock on the victim. While moving downstream, the rescuer hauls in the line.

Tensioned line systems for stranded victims

There are a number of rescue options available if a rope can be stretched across the waterway and tensioned sufficiently to keep it out of the water. These methods are normally used to rescue people from stationary positions (i.e., vehicles, trees, islands, etc.). Remember: Always assign a downstream safety in case a rescuer falls into the current or the victim is swept past the planned rescue point. If something goes wrong and the rescuer and/or victim becomes tangled in the line, be prepared to cut one end of the rope, allowing both of them to pendulum to shore.

Tensioned diagonal with anchors. This method is used for victims already in the water or in imminent danger of being swept off a perch. It often is used when there are no high anchor points to keep victims and rescuers out of the water. It requires the victim to enter the water using the force of the water to propel him to shore while holding onto a sling attached to a carabiner. The carabiner slides on a tensioned line set at an angle diagonal to the current. To set up a tensioned diagonal, perform the following steps:

1. Tension the line at a 45° angle to the current.
2. Clip a carabiner to the tensioned line. Attach control lines to the carabiner if possible.
3. Attach a sling to the carabiner. It should be long enough to reach just above or into the water.
4. If possible, attach a life float to the sling. In the absence of a life float, almost any other sturdy floating object will work. If there is no float available, the victim can hang onto the sling.
5. Attach a PFD and helmet to the carabiner/float and shuttle it to the victim to don if possible.
6. Have the victim get on the float or hold the sling while rescuers pull the victim to shore with the control line.
7. If no control lines are available, a strong current and a good angle on the tensioned line should move the float across the water to the victim. Change the angle of the line as necessary to propel the victim back to shore.

Tensioned diagonal—no anchors available. This version of the tensioned diagonal can be performed where no anchors are available or when time does not allow anchor setup. Rescuers on both shores belay the line at a 45° angle to shore and use the same techniques listed for the tensioned diagonal (see the previous section).

Tensioned line (no anchors) for victim pickoff. This version of the tensioned line rescue uses belayers and rope to *catch* victims moving downstream in the current. It is particularly suited for incidents in flood-control channels where units can be positioned downstream for rescue before the victim arrives. A float bag is ideal in most situations, although a standard drop bag or even a 250-ft lifeline can be used.

1. Line the channel and place sufficient rescuers on both shores.
2. It is not absolutely necessary to use a 45° angle for this evolution. A 90° angle may be desired, depending on conditions (i.e., straight channel vs. curved channel).
3. *Do not attach the line to fixed anchors.* Instead, use firefighters to belay the line until the victim arrives at the intended rescue site. Determine which end of the line will be dropped (depending on particular conditions at the rescue site).
4. Always provide downstream safety and establish a lookout for the victim approaching from upstream. As the victim approaches, keep the rope just above the surface of the water.
5. At the instant the victim hits the rope, release one end. Rescuers on the opposite side can then pendulum the victim to shore. It may be necessary to run downstream while pulling the victim to shore to avoid drag, which might tear the victim from the line. If possible, a pike pole or other long-handled object should be used to pull the victim from the water. This reduces the chances for the rescuer to be pulled in.

Case Study 2: Swift-water Rescue and Firefighter Safety

Until the swift-water rescue revolution of the early 1980s, few fire departments had official responsibility to manage flood and swift-water rescue incidents. Because of this ambiguity, standard firefighter training had for decades been directed at so-called traditional fire service roles (e.g., firefighting, pre-hospital emergency medical care, fire prevention, hazmat, vehicle extrication, etc.). Rarely was technical rescue (including swift-water rescue) part of the curriculum of fire department training academies, and departments rarely provided formal technical rescue training and continuing education to their firefighters.

This lack of training regularly placed firefighters at risk. The examples are many, but one stands out as emblematic of the willingness of firefighters and others to risk their lives to save another, even though they lacked adequate equipment and training.

In 1982, before the advent of the swift-water rescue program, a young boy was swept into a vicious flood-control channel known as Walnut Creek. The first-due engine company captain was not formally trained in swift-water rescue, but he had responded to enough rescues in Walnut Creek to know that units were needed downstream to *head the victim off* before he was lost. He requested a number of fire department units be dispatched downstream. The first-due captain became the IC, and he coordinated the assignments of units responding downstream. His goal was to get the boy out of the channel somehow before it converged with the larger and more deadly San Gabriel River.

One of the downstream units was a truck company, an engine company, and a paramedic squad assigned to attempt to rescue the victim from a bridge that crossed Walnut Creek. Without the advantage of swift-water rescue training, the firefighters did what made sense; they decided to lower a firefighter on a manila rope system from the upstream side of a bridge in an attempt to pluck the boy from the water as he went past. Firefighter David Theis, a rookie at the time, got the assignment. He tied a rescue knot on himself and soon found himself dangling inches over the water at the end of the manila rope, with firefighters holding the line as they stood on the bridge. The only problem was that Theis was on the upstream side of the bridge.

As the boy was swept past the bridge, Theis was able to snatch the boy from the water with his bare hands. He and the boy were immediately dragged beneath the bridge by the powerful current, where they remained suspended from the taut rope, skipping on the surface of the rushing water like a pebble.

Meanwhile, the other firefighters struggled mightily to raise their charges back to the roadbed of the bridge. Finally, the victim was ripped from the arms of Theis, who by then was exhausted and practically drowning. Further downstream, other firefighters managed to swim into the San Gabriel River, grab the boy, and ferry him to shore. But even they ran into trouble and found themselves in a life-threatening situation because they were attired in full turnouts.

Eventually, everyone was safely removed from the water. However, this incident crystallized the argument about the need for firefighters and other rescuers to be properly trained and equipped to conduct swift-water rescue operations. It also reinforced the need for a multi-layer response system that gets the right resources to the right places without delay.

Tether systems

When a line can be placed across the river, a floating object can be controlled from both shores to effect rescue of a victim stranded midstream or on a far shore. This is known as a *tether system*.

Two-point tethers. The *two-point tether* consists of two lines attached to a life float or inflatable rescue boat. In the absence of two lines, a single line with a figure-8 knot tied in the center can be used as the attachment point for the life float or inflatable rescue boat. A two-point tether can also be used to control a hose rescue device. When using tether systems, be mindful of the following important points:

- Tether systems are generally mobile. Lines should not be anchored to stationary objects on shore. Rescuers should be able to *walk* the entire system upstream or downstream to place the float where the victim is.
- Tether systems are quick and simple to set up and operate. Little technical rope knowledge is needed to use a tether system.
- Straight stretches of river are preferred. A tether system may be difficult to set up at a bend.
- Extremely wide channels may cause problems due to the drag of the tether lines in the current. Extra rescuers may be required.
- Even in relatively narrow channels, swift water may require additional rescuers due to the drag created by the tether lines.
- If the victim is physically and mentally able to assist, it is preferable to shuttle the float to him, instruct him to climb onto the float, and shuttle him to safety. This keeps all rescuers out of the water.
- If the victim's position allows him to don a PFD and helmet, attach these two items to the float for his use. At least one downstream safety position should be used in case the victim becomes separated from the float.
- If the victim is unable to help himself, it may be necessary to consider sending a properly equipped rescuer out on the float to assist. Again, measure the risk to the rescuer against the possibility of successful rescue.
- If it's deemed too dangerous for rescuers to conduct the operation with a reasonable chance of success, do not commit personnel to the water.

Hose rescue device

Firefighters from the Dayton, Ohio Fire Department have developed a rescue method that uses a fire hose inflated with air from SCBA bottles. The hose rescue device was designed specifically for rescues at lowhead dams (Fig. 3–41).

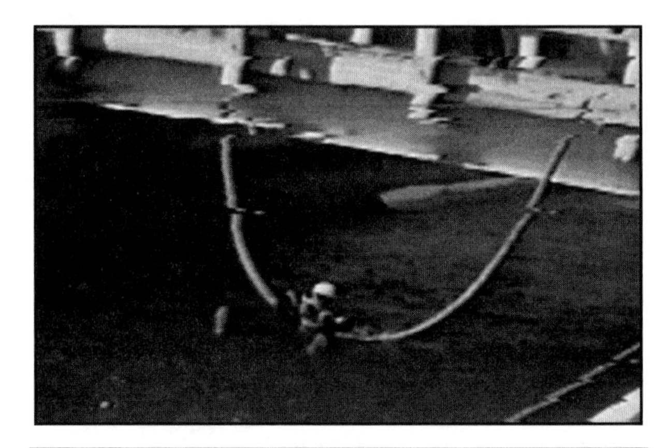

Fig. 3–41 One Use of Inflated Fire Hose to Capture Victims being Swept Downstream

The hose rescue device can also be used to perform bridge-based rescues of victims in moving water. This method is very effective on the steep, fast-moving rivers and flood-control channels found in L.A. County and surrounding areas.

Rescue device components. The hose rescue device consists of the following parts:

- A 2½-in. brass cap with a ring welded or brazed onto its exterior—The ring allows rescuers to attach rope to the cap.
- A 2½-in. brass plug tapped for quick-connect air valve fittings—A shield is welded or brazed on to protect the air valve fittings. Another hole is

tapped in the plug for a pop-off safety valve that relieves at 125 psi. A hole is cut into the shield; this allows rescuers to attach rope to the brass plug.

- A high-pressure, stainless steel pigtail air hose from the plug to the regulator
- An adjustable regulator preset at 100 psi—The regulator eliminates the possibility of hose burst if the pop-off valve fails. Rather than relying on one pressure safety device, the regulator and the pop-off valve add redundant safety.

Hose rescue device for lowhead dam rescue. The severe life hazard presented by lowhead dams cannot be overemphasized. When sufficient water is flowing to create a hydraulic, the only rational way to effect rescue is to work from shore, from bridges, from the air, or from a boat a safe distance downstream of the boil line. One of the most effective rescue methods is the hose rescue device. There are two variations of this method.

Method 1: Looped Hose. Inflate the hose, bend it back on itself to form a bight, and push the bight end out across the face of the dam. A control line can be connected to the rings on the hose rescue device for safety and control.

If the victim can grab the hose, pull him back (parallel to the face of the dam) toward shore. If a vertical wall prevents extrication at this point, pull the hose and victim through the boil line at the shore (here the boil should be weakest) and extricate him from the downstream current. A pike pole may be used to help pull the victim to shore. There is one major limitation: On a wide river, it may be difficult to control long sections of inflated hose. In this event, use method 2.

Method 2. The following equipment is needed:

- One 4,500-psi SCBA bottle—One bottle will fill approximately 200 to 300 ft of 2 ½-in. hose.
- One section of rope or a drop bag
- Two spanner wrenches
- Hose rescue device

Case Study 3: El Niño Effects on Flood and Swift-water Rescue Operations

The deadly storms that struck the west coast in winter 1992–93 resulted from a mild El Niño event. These storms and the havoc that sometimes accompanies them is gaining increased attention from scientists and emergency responders alike.

For example, when one of the most significant El Niño effects in recent history was predicted for the winter of 1997, the L.A. County multi-agency swift-water rescue system (MASRS) task force was convened in August to discuss plans for upgrading MASRS to help manage the onslaught of a powerful winter flood season. At the time of the meeting, held near downtown L.A., the sun was blazing down and smoke from a wildfire in the nearby San Gabriel Mountains towered in the sky like thunderheads. It hardly seemed like the time to be talking about floods. But everyone in the room knew it was necessary because in southern California: first comes fire and then comes heavy rain and floods.

The meeting's agenda included planning to train local firefighters in new swift-water rescue techniques and distributing additional rescue equipment to fire and rescue units before the winter hit. The task force would also get a progress report on a fire department project to develop WRPs. These were intended to streamline the emergency response to swift-water rescue incidents. The team would also discuss newly developed techniques for fire/rescue helicopters to pluck victims from the torrent of fast-moving flood-control channels. But the main topic of discussion was the expected return of El Niño and the attendant predictions of massive flooding and catastrophe in southern California.

By now, most readers have heard of El Niño, the natural phenomenon named for the Christ child by Peruvian fishermen two centuries ago because it reappears at Christmastime every two to seven years. El Niño is an event steeped in legend and frequently the subject of misinformation and unproven assumptions. Even today, scientists are trying to understand more about the root cause of El Niño events and how to predict effects. In basic terms, El Niño is marked by the reversal of the Pacific trade winds, dramatic warming of the equatorial Pacific waters to bath-water temperatures, and resultant changes in weather patterns in many parts of the world.

Historically, the effects of El Niño in southern California have included wild storms that stack up off the coast like the cars of a freight train. They lash the coast region with record rainfall and waves up to 30 ft high and wash away homes, bridges, roads, and people. The infamous El Niño event of 1982-83 saw homes and people swept away in flash floods, houses ripped from their foundations in Malibu by flooding and surf, and large sections of several municipal piers torn away by towering waves. Several L.A. County firefighters nearly lost their lives rescuing citizens from the roiling waters of overflowing streams and swollen flood-control channels. In some areas, the water can gain velocities of 35 mph or higher.

Yet even those storms paled in comparison to some of the others that have struck California during the past two centuries. Scientists from the U.S. Army Corps of Engineers recently determined that one storm in 1868 flooded an area roughly equal to one-third of L.A. County. This storm sent 300,000 cubic ft of water per second (cfs) roaring down Big Tujunga Canyon into the northeast corner of the San Fernando Valley. After flooding a huge swath of the Valley, it made its way to the Los Angeles River, which during the storm, carried more water than the Mississippi River. For comparison, consider the typical flow of the Colorado River through the Grand Canyon, which is currently somewhere around 20,000 cfs.

Now imagine a flow of water 15 times greater than that roaring out of the mouth of the narrow canyons of the San Gabriel Mountains above Los Angeles. This incredible flow of water headed across densely populated areas before joining the Los Angeles River in its headlong rush to the sea. In the 1860s, most of this region was sparsely populated, much of it dedicated to ranching and farming, with the rest in its natural state as undeveloped scrub or marsh.

Today, even with elaborate flood-control structures in place, it is entirely possible for such a flood to quickly inundate thousands of homes in the San Fernando Valley, washing many of them away. Evacuation and rescue operations under such conditions would be harrowing at best, and made worse if the flood were to occur at night. The devastating effects of a 300,000-cfs flow from a single mountain tributary would be multiplied when it met with the flow of the Los Angeles River. The river itself would already be swollen, and it can barely carry 100,000 cfs without the threat of catastrophic failure of its levee walls. Such a breach would have the effect of a dam break, sending walls of water into tens of thousands of homes in low-lying neighborhoods across the Los Angeles Basin, where more than a million people might be affected.

Emergency planners and fire/rescue officials are alarmed because the 1868 storm wasn't long in terms of the geologic scale. There is mounting evidence that the 1868 storm was the direct result of a major El Niño event. Officials are concerned that a future El Niño period might cause storms even larger than the one that devastated parts of southern California in 1868.

At the conclusion of the meeting of the MASRS, a group of firefighters waited by the elevator of the county's public works building and discussed the implications of what they heard about the El Niño predictions. Initially, some members were inclined to downplay the predictions, arguing that there was no guarantee that the storms would hit, but they were reminded of the other unusual events that had plagued California in recent years. Since 1992, the state had experienced two deadly earthquakes, including the Northridge quake, the most costly disaster in U.S. history to that time. It had also fallen victim to several avalanches in the high mountains that separate L.A. from the San Andreas Fault. The state had also experienced lethal flood events in five consecutive years, several damaging landslides, tornadoes, the deadly Los Angeles riots, the 1993 firestorms that burned hundreds of homes, and a level C volcano watch issued for the Mammoth Lakes ski resort area. (This particular event was followed by the implausible movie *Volcano* that depicted L.A. being swallowed whole by molten lava!) In addition to actual events, there were continued predictions that the San Andreas Fault was long overdue to rupture with an 8+ earthquake.

"About the only thing that we haven't had," said one firefighter as the elevator opened, "is a hurricane." At this comment, everyone laughed with relief because at least they could be confident that hurricanes are impossible in the cool Pacific waters of California. Or so they thought.

Hurricane watch

In truth, under current world climatic conditions, the likelihood of a full-fledged hurricane coming ashore in California is extremely small. There is no known instance of a fully formed hurricane-making landfall anywhere in the state. However, California has occasionally been ravaged by the effects of violent tropical storms and disintegrating hurricanes that churned their way up the coast of Baja, Mexico before breaking up as they encountered cooler air near the Mexico/U.S. border.

In 1976, tropical storm Kathleen dropped nearly 13 in. of rain in parts of southern California, causing severe flash flooding and the inundation of vast tracts of land. Ten people died in the ensuing flash floods. In 1939, a hurricane-spawned tropical storm made landfall between San Diego and Los Angeles and left 46 people dead as ships and yachts sank, people were swept away in flash floods, and others were buried in mud and debris flows.

The periodic return of El Niño—a phenomenon that is still not fully understood by scientists—creates conditions that are favorable for hurricanes to travel much farther north than they would in non-El Niño years. Because accurate records have only been kept in California for the past 200 years or so, there is little direct evidence that a full-blown hurricane has struck land in the recent past. However, the absence of historical data also makes it difficult to discount the possibility that past El Niños heated the Pacific waters sufficiently to allow fully formed hurricanes to venture as far north as L.A. Based on recent experience, some residents insist that they are no longer surprised at anything that happens in their state, so why not a hurricane?

Sure enough, almost as if to prove the point, meteorologists began reporting on September 10, 1997 that a hurricane had spawned in the Pacific hundreds of miles south of Cabo San Lucas. It was growing stronger and tracking northward at a rate that might be of some concern to Californians. It was exactly one month after the El Niño planning session. By September 12, Linda had developed into a category 5 hurricane, packing winds over 180 mph, with gusts approaching 220 mph. It was the most powerful hurricane ever observed in the eastern Pacific Ocean, and by far the most forceful to approach the west coast of Mexico in recent history.

Meteorologists warned that the size and strength of hurricane Linda was of a magnitude to cause alarm for residents and emergency services in California. The storm was tracking directly along the west coast of Baja and threatening to cause devastation to populated areas. Since much of the Baja coast is uninhabited except for occasional enclaves and fishing villages, initial reports of damage were sketchy. But the storm remained about 250 miles away from Baja, roughly parallel to its coast.

Eventually, however, Linda began losing power as it ventured north of Cabo. As the winds slowed, Linda was downgraded to a category 4, then quickly to a category 3. By September 14 the once-powerful hurricane was disintegrating into a major tropical storm, further weakened by weather patterns that drew it further into the open sea. Still, the U.S. Weather Service was warning that rainfall in San Diego and Los Angeles might range between 6 and 13 in. in one day's time. Rainfall of that magnitude by itself would have caused tremendous devastation and probable loss of life in a region that frequently floods when more than a few inches of rain fall in a 24-hr period.

Fire and rescue agencies from the Mexican border to Santa Barbara braced for the possible onslaught of record floods. Fortunately, by the time the outer remnants of tropical storm Linda swept over southern California, the rain had subsided considerably, falling mostly in the open deserts and mountains of Mexico and southern California. The most noticeable effect came in the form of 10- to 15-ft surf on south-facing beaches. For those expecting disaster once again, southern California had dodged the proverbial bullet—at least for now.

Upgrading storm disaster readiness

Fortunately, at the conclusion of the MASRS meeting, LACoFD Fire Chief P. Michael Freeman had ordered the technical operations section to expedite planning and preparation for the predicted El Niño event. Several members of the department's USAR unit and the water rescue committee were assigned to work full time under technical operations to upgrade planning, training, instructional materials, swift-water rescue team development, mud and debris flow training, and other preparations for the storms.

These members set about upgrading the LACoFD's existing system, which consisted of four helicopter-borne swift-water rescue teams and eight ground-based teams. These teams are intended to augment the department's normal three-tiered system of USAR units. This system is one of the most comprehensive to be found and had its roots in past tragedies. Members were digitizing and distributing WRPs for every major waterway in the county, as well as revising the department's dispatch matrix. Their goal was to enhance the ability of dispatchers to respond quickly with fire units in a linear pattern along flood channels, rivers, and streams. In part, the intent of this project was to ensure that fire units had sufficient time to set up for rescue before victims in fast-moving flood channels were swept past their positions. Enhanced first-responder swift-water rescue training was being planned for the fall, and continuing education courses for swift-water rescue were in full swing.

Anticipating the potential for major El Niño storms to overwhelm even these provisions, the LACoFD administration was reviewing a proposal to staff up to 20 reserve engines and other apparatus with swift-water rescue and USAR-trained firefighters. This would establish additional swift-water rescue teams that could be deployed across L.A. County or dispatched to mutual aid requests, including requests for multiple swift-water rescue strike teams and task forces.

At the county level, MASRS had been actively preparing for El Niño. At the state level, the LACoFD's water rescue committee chairman was attending meetings held by the California OES, and other agencies. Their goal was to enhance the statewide swift-water rescue mutual aid system, as well the swift-water rescue components of the state's eight OES/FEMA USAR task forces.

All of these disparate groups were expecting at least two months of lead time before the first effects of the El Niño were expected to be felt in California. Normally, this time of year is marked by hot, dry Santa Ana winds that whip wildland fires across the rugged hills of southern California. In part, this held true: Through most of August and into September, the region suffered through an intense heat wave that seemed to indicate that winter was still a long way off. Several fires blackened more than 40,000 acres across the southern part of the state during the late summer. Then, on September 13, 11 homes were lost in a fast-moving wildland interface fire in the nearby San Bernardino Mountains.

Already however, there were indications that the El Niño was about to make its presence felt. The August/September heat wave was accompanied by abnormally strong monsoon conditions as moisture drifted from the Gulf of Mexico across the Arizona deserts to California. The monsoon flow unleashed a series of thunderstorm events that caused dramatic flash flooding in the mountains and deserts of southern California. One flash flood sent a 12-ft wall of water, mud, and rocks sweeping through Red Rock Canyon State Park. The flood wiped out Highway 14 nearby, stranded many motorists, and buried park facilities—including the rangers' homes and offices—in 15 ft of mud and rock. The next day, a thundering mud and debris flow buried fully one-third of the town of Forest Falls, at the base of the San Bernardino Mountains.

Hurricane Linda drastically altered some of the planning and preparation by speeding up the time clock. Linda was soon followed by an even worse hurricane that plowed through the Mexican resort city of Acapulco, causing hundreds of deaths in the flooding that resulted from the storm's heavy downpours. In the greater Acapulco area—a region with terrain similar to many parts of L.A. County—the destructive power of fast-moving water funneled into populated areas by canyons and streams was demonstrated on a scale previously not seen there.

Viewing what happened in Acapulco, one might be tempted to extrapolate what might happen if similar rainfall amounts were to fall in the greater Los Angeles area. One might also be tempted to contrast that view with the results of recent flood events in 1992, 1993, 1994, and 1995, which occurred as a result of far lesser rainfall amounts. Hundreds of people were in need of rescue assistance during these storms, and thousands of homes were in danger of being inundated or swept away altogether. During the 1993 storms alone, firefighters and lifeguards from LACoFD rescued at least 111 citizens during more than 51 swift-water rescue operations, as well as saving millions of dollars of damage to homes and public property in Malibu and other areas.

The year 1995 was another year of heavy flood rescue activity. On one day, January 4, flood-control channels in Carson overflowed into surrounding neighborhoods, quickly swamping homes and automobiles. Lifeguards, rescuers from the county and city fire departments, and the Torrance fire and sheriff's departments rescued 200 people from the fast-rising water.

On January 10, county firefighters and lifeguards used boats, ropes, contact rescue techniques, and helicopters to rescue 35 people in the Malibu area alone. At least one dozen people were rescued by fire/rescue helicopters that employed one-skid landings, hoist cables, and newly devised short-haul techniques (with specially trained firefighters dangling from ropes beneath the helicopters to pluck victims from the currents).

On the same day, quick action by swift-water rescue teams saved at least 30 homes from severe damage or destruction in the Big Rock area. As fast-rising floodwaters and mudflows struck a long line of homes on the south side of the Pacific Coast Highway, the homes acted as a dam. The water rose rapidly, trapping dozens of citizens and threatening to push their homes off their foundations into the surf. Firefighters and lifeguards used boats, ladders, and skip loader tractors from the department of public works to pluck them to safety. They began a dangerous operation to breach private fences between homes and unplugged drains choked with debris. Properly equipped firefighters and lifeguards had to wade or swim into position with axes and other extrication tools. They then had to chop holes in the fences and block walls to allow floodwater and hundreds of tons of mud to drain through backyards and under homes to the ocean.

The effect was not unlike the breaching of a conventional dam: Tons of water rushed though the newly-created holes, causing a torrent that threatened to pull any loose objects, debris, and even people through them to the other side. It would have meant instant death for any rescuers who found themselves in the wrong place. These actions alone may have prevented millions of dollars in damage to private property, utilities, and the highway.

The decision to breach dozens of private fences was quickly made by swift-water rescue team members who witnessed similar destruction during the mud/debris flows of 1993 and 1994 in the Big Rock area. This time, recognizing the potential for a similar flood of mud and water along the Coast Highway, they planned ahead of time and were ready to breach fences and walls quickly. Their actions prevented the devastating buildup of water and the ensuing pressure that might have forced homes off their foundations and into the ocean.

These were just some examples of the dangers that could be presented by future El Niño events in rugged, densely populated areas like southern California. Although several deaths occurred in flash floods and the mud and debris flows that accompanied tropical storm Linda, it could have been far worse. As the Pacific continues to warm periodically under the influence of future El Niño events, it remains to be seen if the worst is yet to come.

Bridge-based rescue using a life float

This evolution is particularly suited for three-man engine companies (where they are found) and for units that do not carry fire hose (i.e., truck companies, paramedic squads, hazmat squads, USAR companies, etc.).

A life float is tied to a single rope, which acts as both a belay line and a pendulum line. The line is manned by one belay position stationed on the bridge, and one pendulum position on shore downstream of the bridge. A spotter is required on the upstream side of the bridge. Although this rescue can be performed with only three rescuers, four to six are desirable.

Equipment needed. For this method, you will need the following:

- Adequate rope. A 300-ft section of hose is desirable for extra flexibility. More or less may be required based on conditions such as the river width, water speed, etc. (Remember that drop bag lines can be clipped together if necessary.)
- One life float

Strategy. The process for this rescue method is as follows:

1. The belay ties the line to the life float and lowers the float into the water. The life float will *plane up* onto the surface of the water.
2. Meanwhile, the rest of the rope is stretched to the shoreline and tended by the pendulum. Once

the life float is dropped in, the pendulum has the same duties as the hose rescue evolution.

3. The belay can let the life float drift a good distance downstream of the bridge and still retain control. This is advantageous because it provides more time to position the float when the victim appears from underneath the bridge.
4. The belay still keys on a spotter on the upstream side of the bridge.
5. As the victim appears from the downstream side of the bridge, the belay maneuvers the float into the victim's path. When the victim reaches the float, the belay drops the line, allowing the float and rope to drift in the current with the victim.
6. Following their work on the bridge, the spotter and belay assist on the shoreline. If no other personnel are available, the spotter should take up the downstream safety position.

Bridge-based rescue with life float—when the shoreline is not accessible

In some cases, the banks of a river or flood-control channel may be inaccessible because of high water levels or obstructions such as fencing, cliffs, vegetation, etc. If possible, use another site that offers better access. (This is one of the advantages of preplanning potential rescue sites.) If there is no alternative site, use the following evolution:

1. Set up the evolution as outlined.
2. Eliminate the pendulum position along the shore. Instead, place the pendulum position on the bridge, at one end. The pendulum will attempt to swing the victim to shore from the end of the bridge. This will require careful technique. It may also require additional rope. The pendulum must attempt to get the victim to a spot along the shore where rescuers can effect rescue with ropes, roof ladders, etc.
3. There is a good possibility that the pendulum may run out of rope, having lost the ability to move downstream with the victim. In this case, the pendulum simply tries to *hold* the victim at shore until a rescue can be completed.
4. Use the downstream safety position if at all possible. The downstream safety position may be required to move a long distance downstream to find an access point.
5. Remember that these conditions constitute an added danger to any rescuer who falls in with the victim. If our ability to rescue a victim is hampered, we definitely want to keep firefighters out of the water.

Vertical Rescue for Victims Stranded on Immobile Objects

Victims often find themselves stranded on bridge supports, islands, trees, automobiles, and other midstream objects. Nearby bridges, trees, or elevated shorelines may offer the possibility of a vertical rescue. Aerial ladders, aerial platforms, and even ground ladders can be used to effect rescue from above the victim. For the fire department, the one overriding concern should be the safety of any rescuer placed in a vertical rescue situation. If there is any possibility of being knocked into the current, firefighters should not be tied to ropes.

The sensible alternative is to place foot loops and hand loops in the rope. This allows a quick escape if something goes wrong and a firefighter ends up in swift-moving water. He can hang onto the rope while other personnel attempt rescue, but he is not attached to the end of a static line in moving water, which would be a life-threatening situation.

There are two main methods of effecting such a vertical rescue. The first requires a static line, and the second is a raising and lowering system. (Note: Only basic, non-mechanical advantage rope rescue techniques will be discussed in this instruction.)

Vertical rescue over moving water—static line method

This method consists of the following steps:

1. Communicate your intentions to the victim if possible.
2. Measure the length of rope needed to reach the victim.
3. Make a rope ladder by tying loops along the length of the rescue line, or use an etrier if available.
4. Anchor the line to a bridge, aerial ladder, or other support. Double-check the anchor.
5. Lower the rope to the victim. Check to make sure the line is stretched taut from the anchor to the victim. This way, the rescuer doesn't experience a sudden loss of elevation when stepping into the first loop.
6. If the victim is able-bodied, find out if he can climb the rope ladder. If the victim agrees and it is safe to do so, lower a helmet and PFD and have the victim don them.
7. If there is doubt about the victim's ability to climb out, it may be necessary to send a firefighter down to assist. PFDs and helmets should be provided for both rescuer and victim.

If a rope ladder is not available, a roof ladder can be used in the static line method.

Vertical rescue over moving water—manpower lower and raise

This method consists of the following steps:

1. Communicate your intentions to the victim if possible.
2. Find a suitable anchor point.
3. Set up a lowering system.
4. Just as in other vertical rescue situations, provide a backup safety line when possible. Do not tie the safety line to the rescuer; place loops in the line or attach an etrier to provide handgrips and footholds for the rescuer and victim.
5. Place the rescuer in a PFD and helmet.
6. Lower the rescuer to the victim with an extra PFD and helmet attached to the line.
7. If possible, the rescuer should place the PFD and helmet on the victim. Then he should place the victim in a harness.
8. Personnel on top convert to a manpower raising system.
9. The rescuer attaches the victim to the raising system.
10. Haul the victim and rescuer to safety.

Note: Make sure the rescuer has a safety knife or rescue scissors attached to the PFD (where provided) to use if either person becomes tangled in the rope under the water.

Self-rescue of Stranded Victims

A conscious victim stranded in a river (or on the other side of a river) can present a dilemma for rescuers. The same conditions that left the victim trapped may also endanger the lives of rescuers.

Under such conditions, we need to ask ourselves two important questions:

- *Is the victim safe at the current location, or in immediate danger of being swept into the current?* If in a safe spot, leave the victim there until conditions improve (i.e., the water level drops) or a safe rescue evolution can be established. If conditions deteriorate to the point where the victim is in imminent danger of being washed away (i.e., the water level is rising or the victim cannot hang on to the object), an immediate rescue attempt is warranted.

- *Is the victim fit to assist in the rescue?* If the victim is injured, very young, or very old, self-rescue may not be possible. Under such conditions, rescuers at the scene will be forced to decide whether lives should be risked to attempt rescue.

If there is no way to get to the victim with a reasonable degree of safety, assign units and/or personnel to set up downstream safety positions in case the victim is swept off the perch. Support the victim in any possible way until conditions improve (i.e., attempt to provide a rope, PFD, and helmet, etc.) In the meantime, alternate rescue plans should be made. If the victim appears able to assist in the rescue, there are several options available.

Remember: The following methods are to be used only if the victim is in danger of being swept away and there is no safe way to assist the victim in any other way.

Single-line self-rescue system

This method is very basic:

1. Get a line to the victim
2. Shuttle a PFD and helmet for the victim to wear.
3. Shuttle a life float for the victim to use.
4. Pendulum the victim to shore on the float.

The main advantage of this system is that it keeps rescuers out of the water. The main disadvantage is that the victim will be in the water and may be unable to hang onto the float or rescue line. However, this may be the victim's one chance of being rescued. A downstream safety position should be established. Other units should be placed farther downstream to attempt rescue if necessary.

V-line self-rescue system

This method is used when it is imperative to get the victim to shore as quickly as possible with the least downstream movement. It is most useful when a downstream hazard must be avoided (i.e., lowhead dam, strainer, etc.). The V-line system uses the same basic principle as the single-line self-rescue system with one major difference: A V is created by keeping both ends of the line on the near shore (the victim gets the middle part). If one line is too short, a second line can be attached to it. The normal pendulum technique is augmented by the downstream line, which is used to pull the victim almost directly to shore. Rescuers tend both ends of the rope (one upstream and one downstream of the victim).

Once the victim enters the water, the downstream line tender pulls as hard as possible while remaining downstream of the victim. The upstream line tender remains upstream of the victim, giving just enough slack to keep the victim coming toward shore without forcing the victim under the water. If one of the tenders falls or loses the line, the second line tender can still complete the rescue.

Entering the Water

Thus far, this instruction has emphasized rescue techniques that keep rescuers out of the water. Shore-based rescue techniques remain the most desirable course of action, followed by boat-based and helicopter-based rescues in some cases. However, some rescues may not be possible from shore, by boat, or by helicopter.

Take the example of a small child being swept down a flood-control channel with vertical cement sides. This is a particularly difficult rescue to complete from shore because of the speed and power of the water, the vertical sides, and the possibility that the child may not be able to grab anything we can throw.

In this case, personnel at the scene will be forced to make a quick decision: Should a firefighter jump in with the victim to assist, or should personnel hope that the victim can be rescued by other units further downstream? This is a tough call to make if conditions are obviously life-threatening for rescuers. Under these conditions, no written rules or policies in the world can take the place of proper training and good judgment.

Some situations are simply *no-win*: to go in with the victim is likely to result in death or serious injury to the rescuer, with little hope of a successful rescue (i.e., victims in the hydraulic of a lowhead dam). In these cases, use only methods that keep all rescuers out of the water.

Of all situations, the marginal ones hold perhaps the greatest danger for firefighters. When conditions are obviously hazardous, yet there is a chance of rescue, there is a strong temptation to go after the victim. If things go just right, the result is a successful rescue. But if something goes wrong, the result could be disastrous. This is where proper training, experience, equipment, and alternative rescue methods can make a crucial difference in the outcome. It must be reemphasized that shore-based methods should be totally exhausted before even thinking of going into the water.

Shallow-water crossing techniques

Under limited conditions, firefighters can effect rescue by wading through shallow water to victims. Several methods of shallow-water crossing have been developed by the military and rescue professionals. These methods are limited by deep or swift-moving water. Generally, water above waist-deep makes them ineffective. It is better to rely on tools to create a sturdy base (tripod) and/or by additional rescuers crossing as a group to overcome the force of the current.

Reminder: Never tie ropes around yourself when entering moving water. Be aware of foot entrapment hazards.

Case Study 4: L.A. County MASRS

In 1992, a significant development placed a new tool at the disposal of emergency officials in L.A. County. Known as MASRS, this powerful resource resulted from a mandate by the L.A. County board of supervisors to develop a system to coordinate the vast county-wide, multi-agency network of activities that is necessary for preparation and response to flood-related emergencies.

The catalyst for the board's decree was the deadly El Niño storm season of 1992-93, which caused the death of an 11-year-old boy in the Los Angeles River. In addition, dozens of victims were stranded—including firefighters—when the Sepulveda Basin flooded. Several deaths also occurred during this same storm season when the Ventura River rampaged through a trailer park. These and dozens of other dramatic water rescues occurred across southern California. The 1992-93 floods crystallized the need to improve coordination among local fire departments and other pubic safety agencies that manage storm rescue operations. MASRS was designed to meet that need.

A group known as the Los Angeles county multi-agency swift-water rescue task force guides the MASRS. The mission of the task force is to improve county-wide coordination of flood and swift-water rescue activities. The task force represents fire departments from the City and County of Los Angeles, Long Beach,

Glendale, Pasadena, the San Gabriel Valley, and various other public safety agencies. This group reports directly to the county administrative officer and the board of supervisors through Daryl Osby, deputy chief of special operations of the LACoFD.

The multi-agency task force consists of firefighters, lifeguards, and law enforcement officials who specialize in technical rescue, supported by specialists from the county department of public works and other associated agencies. The members meet periodically to evaluate the current status of MASRS, to consider and implement improvements, and to provide overall coordination of the county's storm emergency response system. The result is a unique blend of public safety and public works agencies that has created a county-wide tiered response to swift-water rescue incidents and flood disasters.

For victims trapped on automobiles, on submerged homes, and for children being swept away in flood-control channels, MASRS provides rescue assistance on a scale never before seen. First responders have a reasonable assurance that they will be reinforced by a county-wide system of ground-based and helicopter-based swift-water rescue teams of the MASRS system. Therefore, if they happen to be swept away while attempting a rescue, they have a reasonable chance of being rescued. Furthermore, they are given technical support by the special resources provided by the L.A. County and city departments of public works, the National Weather Service, the California OES, and other public agencies that have roles in planning and response to flood-related emergencies. Today, partly as a result of rapid implementation and *trial by fire* under the demanding conditions of flooding and rescues in the intervening years, MASRS is arguably the most comprehensive—and the most frequently tested—swift-water rescue system of its kind in the nation.

Although the county's designated swift-water rescue teams are among the most advanced, the effectiveness of MASRS remains heavily dependent on the first responders, whose fire and rescue officials have long recognized their crucial role. Simply put, it is often during the first minutes of a rescue that first-arriving firefighters have the best chance of spotting victims, providing them with emergency flotation to prevent them from disappearing below the surface, and plucking them from the water.

Another dramatic rescue

An example of the importance of the MASRS was demonstrated in March 1992. Sean McAffee, a firefighter assigned to LACoFD engine 69 in Topanga Canyon, employed techniques he learned that very day in a training course from swift-water rescue team members. Using his new skills, he rescued a man who was washed off a private footbridge and became trapped on a tree in the raging Topanga Creek, in the Malibu area of L.A. County.

The Topanga incident occurred on a night when airborne swift-water rescue teams were prevented from reaching the scene because of heavy cloud cover, driving rain, and darkness. The closest ground-based swift-water rescue team was 40 minutes away, having just completed another rescue. As a result, McAffee was the most highly trained firefighter at the scene for the first hour. The captain of engine 69 recognized this, and assigned McAffee to supervise other firefighters to establish a

Tyrolean rope rescue system over the raging floodwaters. McAffee volunteered to be placed in a rescue harness, suspended from the tensioned rope, and maneuvered across the flood to perform the rescue.

Through the use of a system of ropes and pulleys, firefighters on the shore were able to maneuver McAffee across the river on the highline system until he reached the man, who was clinging desperately to the tree. On McAffee's command, the firefighters manipulated the pulley system to lower him to the victim. There, he secured the man in a harness, attached him to the rope system, and signaled to be hauled to the safety of shore. The ultimate outcome of this rescue would have been in doubt before the advent of MASRS and the county fire department's first-responder swift-water rescue program, yet the victim and rescuers alike walked away unscathed.

It is in these situations that the more advanced swift-water rescue teams of MASRS prove their worth because of their proven ability to conduct more elaborate operations. These include high-risk options, like swimming into deadly currents after victims, grabbing them, and dragging them to shore without being drowned in the process. In cases where victims have somehow managed to climb atop automobiles or rocks, or are found clinging to trees or other objects, it is the job of first responders to begin stabilizing the situation until the arrival of a swift-water rescue team. Team members can then establish more advanced rescue systems using ropes, inflatable boats, and helicopters, or they can actually swim across raging floodwaters to reach victims.

In cases where first-responding firefighters are compelled to enter the water to save a victim's life (generally after shore-based rescue methods have failed), the MASRS tiered response system provides the necessary backup capabilities to save them if they cannot extricate themselves and their victims from the water. This type of redundancy is considered crucial to the success of all high-risk emergency operations, and it is a cornerstone of the county's MASRS.

Fixed-line rescue. This method uses rope as a guide for firefighters as they make their way across rivers and streams. The first step is to line the waterway (Fig. 3–42). The line is anchored to sturdy points on both sides of the channel. Then, it is tensioned (similar to the carabiner slide) and used by the rescuer to cross the river.

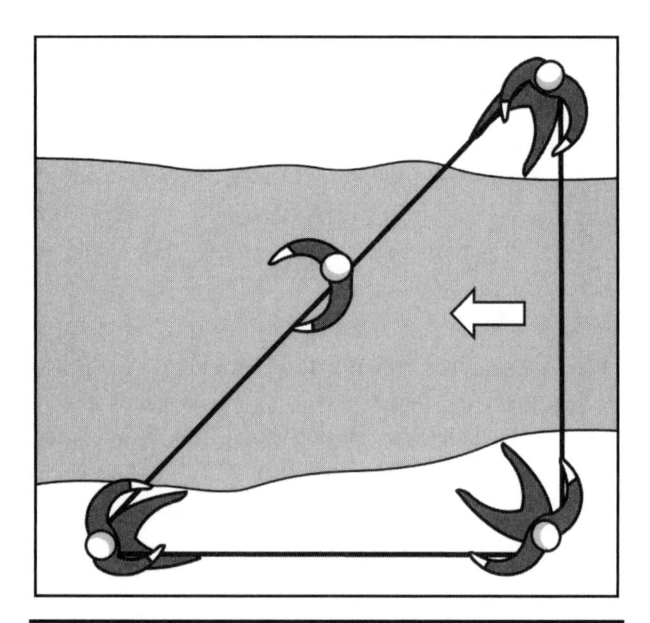

Fig. 3–42 Continuous Loop Rescue System

Continuous-loop rescue system. This rescue method is particularly useful when only one side of a river is accessible, and a victim is stranded midstream or on the other side. This method requires a minimum of four rescuers. Because it requires a shallow-water crossing, this method is limited to the same basic conditions of water speed and depth.

A rescue line approximately three times the distance from the victim to shore is required. The ends of the rope are tied or clipped together forming a loop (thus the name continuous-loop rescue system).

1. Rescuer 1 and rescuer 2 act as belayers on shore. Rescuer 1 is positioned upstream of the rescue site, and rescuer 2 sets up downstream of the site. This creates the most efficient pendulum/ belay angles for the rescue.
2. Rescuer 3 makes a hand loop and crosses to the victim, becoming the third belay point.
3. Rescuer 4 makes another hand loop and crosses to rescuer 3's position.
4. A PFD and helmet are attached to the line and shuttled over for the victim.
5. Rescuer 4 accompanies the victim to safety while the other three rescuers belay.
6. When the rescue is complete, rescuer 3 returns to shore.

The continuous loop of rope provides redundant safety factors. If the victim or any of the rescuers are swept off by the current, the belayers can pendulum them to shore. Therefore, even if two belayers were to drop the rope, a third belayer is available to take over.

Even so, a downstream safety should still be positioned downstream of the entire operation when manpower permits. In addition, as with any swift-water operation, additional units should be assigned to set up rescue points downstream in the event the downstream safety misses a firefighter or victim. This duplication of safety factors ensures maximum safety for rescuers and the victim (Fig. 3–18).

Contact rescue in swift-water situations

Some situations will defy every attempt to rescue victims from shore. If victims are unconscious, trapped, injured, very young, or very old, they simply may not be able to be rescued unless someone gets in the water and pulls them out. As a general rule, contact rescue in swift-water situations should be considered only when it's apparent that other lower-risk methods won't be effective. These rescues are especially dangerous for three main reasons:

- Rescuers are placing themselves in the same environment that is threatening the victim's life.
- The danger is multiplied because rescuers are attempting to remove the victim and themselves from the water.
- Conscious victims may attempt to use the rescuer for flotation. This places the rescuer in a true life-threatening situation.

The decision to make a swimming rescue is often made in a split second. It is difficult to set hard and fast rules, but the following general guidelines should be followed:

- If the rescue can be completed from shore, from a boat (or PWC with trained rescuers) or from a helicopter, attempt those methods first.
- If conditions in the water are likely to overwhelm the rescuer, attempt contact rescue only as a last resort, and with the knowledge that you may not escape safely and you may become the object of a rapid-intervention operation (RIO). Weigh the danger to yourself, your teammates, and other rescuers before committing yourself.
- If there is any reasonable doubt about your ability to complete the rescue safely, ensure that downstream safety and/or other rapid intervention means are established.
- In all cases, rescuers should use appropriate PPE when it is available. If the decision is made to attempt contact rescue, the following instruction will be helpful.

Entering the water for contact rescue

It is important to know what is downstream of the entry point. Rescuers should never enter the water if there are severe hazards downstream (i.e., lowhead dams, underground waterways, etc.). This emphasizes the need for accurate WRPs.

Ensure that at least one downstream safety position and/or helo/swift-water (HSW) capabilities are functional. If possible, assign multiple downstream rescue points to ensure maximum safety for rescuer and victim. Again, this should be included in WRPs.

Pick an entry point that will give you maximum advantage in the water. Enter the water and approach from upstream of the victim (if possible).

Personnel assigned to search functions should anticipate the possibility that they may discover the victim, and may have only one chance to attempt immediate rescue in the water. This emphasizes the need for all personnel to don their equipment before they need to use it.

Ensure that the IC and all responding units are notified if a rescuer enters the water. This situation qualifies as an *emergency traffic* radio message. Rescuers should bring some type of flotation device into the water for the victim whenever possible. This will allow the victim to hang onto the float while the rescuer tows the victim to shore. Flotation devices may include lifeguard rescue tubes, rescue boards, life floats, or PFDs.

Approaching the victim. In moving water, always approach the victim from upstream. If possible, communicate with the victim and explain what you are going to do. Make it clear that you will help *only* if the victim cooperates and follows your instructions. Caution the victim against grabbing you.

Making contact with the victim. Tell the victim to roll over on his back with feet downstream in the self-rescue position. Tell the victim to relax and let you come around from behind to take control. Ensure that the victim understands and will cooperate before coming within his reach.

If the victim tries to climb on top of you to stay afloat, drop under the surface, kick away from the victim, and move upstream. A tired victim will rarely fight you if it means swimming against the current to get to you. If the victim continues to come at you, kick water in the victim's face or submerge and move upstream. Surface away from the victim and begin your approach again. By now, the victim should be very cooperative.

If the victim grabs you, use your arms, fists, and legs to break the hold. As a last resort, force the victim under the water until the victim lets go of you. *Never* let the victim jeopardize your life in the water. If the victim presents a danger, simply swim upstream until the victim cooperates or becomes too tired to fight.

If you have a float, hold it at arm's length or push it to the victim and instruct him to grab it. If the float has a tether line, you can tow the victim to safety.

Getting out of the water. Swift-moving water is almost always a bigger danger than the victim. Both the victim and rescuer will have their hands full just trying to avoid obstacles along the way. This is particularly true in flood-control channels, where the water may be moving between 20 and 30 mph.

Protecting the victim from further injury is important, but the most important thing is to get the rescuer and victim out of the water as soon as possible.

Under life-and-death conditions, the rescuer may be forced to use the victim as a buffer against obstacles. As harsh as this may sound, it is an accepted fact in rescue that an injured rescuer is of little help to a victim in the water.

There are several techniques used to remove injured swimmers from the water. Some use rescue boards or other floating objects to stabilize the spine. Backboards can also be used to stabilize a patient prior to lifting from the water. These methods are effective in calm water.

However, it is neither safe nor rational to expect cervical spine care and airway management to be delivered effectively when both rescuer and victim are trying to escape from swift-moving water. Under these conditions, the victim should be pulled out of the water in the safest and quickest manner possible.

Cervical spine precautions and other treatment should be initiated after the patient and rescuers are out of danger.

Under ideal conditions, rescuers on shore will be able to assist. Under less than ideal conditions, the rescuer will have to pull himself and the victim without any help.

River and Flood Search Operations

Search operations for victims reported in rivers, flood-control channels, aqueducts, or flash floods can be dynamic incidents, extending for miles along affected waterways. During storms, water (and victims) may be moving up to 30 mph in some channels. Therefore, it is important for the IC to request sufficient resources in a timely manner.

Positioning of downstream SAR units is important not only for the victim, but to ensure rescue of firefighters or other rescuers who may end up in the water. Units should be assigned downstream according to WRPs if such preplans exist. WRPs will relieve the IC of many of these decisions and allow for more effective incident management. For channels without preplans, the IC must assign units to downstream positions based on knowledge of the channel, access points, current conditions, and the equipment and capabilities of responding units. The logistics of swift-water rescue operations may be very complex. The IC should expand his organization in a pro-active manner to ensure the greatest life-saving capability. He should be prepared to divide the incident into divisions along the channel.

No rescues can be made until the victim is found. It is important to remember that the first unit to spot the victim may have the only opportunity to make a successful rescue. Therefore, even search units should be prepared to attempt immediate rescue if necessary. It is also important to remember that there may be more victims than originally reported, particularly in cases where citizens, firefighters, and others may have entered the water to attempt rescue. Continue search/rescue operations until assured that all victims and rescuers are accounted for.

In addition to search units, the IC should ensure that other units are strategically placed along the waterway to attempt rescue if the victim is spotted. Until the victim is found, the IC should ensure that the following objectives are met.

Determine the point last seen (PLS) and place marker floats

Quickly determine from witnesses the point where the victim entered the water and where the victim was last seen. The first unit to arrive at the PLS should toss a marker float into the water. Marker floats seen in the water are an indication that the victim *may have* come past already. Any units spotting a marker float in the water should transmit this information to the IC. The IC should use these reports to assist in the development of a search pattern.

Because marker floats (and victims) may become trapped in eddies or tangled in debris, ICs should consider ordering the first three or four units to place maker floats in the water when they reach the waterway. While not as accurate as the original marker float, additional floats still give an indication of how far the victim may have traveled. *Do not use marker floats as your only means of determining possible victim travel distance.* They are intended only to be used as an adjunct to the search effort.

Request sufficient resources

LACoFD units and automatic aid units are dispatched according to the appropriate area response form, or as requested by the IC. These units are given assignments according to the appropriate WRP (if one exists for the involved waterway).

WRPs have been developed for most major channels in L.A. County. Incidents occurring on smaller channels without preplans must be handled without the benefit of a preplan. However, most small channels will eventually drain into a major channel that has been preplanned. In these cases, the IC provides appropriate coverage for the original channel and uses the WRP to set up downstream SAR functions.

Outside agencies that may be requested include the following:

- Additional downstream fire/rescue agencies
- Appropriate law enforcement
- Lifeguards where appropriate
- Harbor patrol where appropriate
- U.S. Coast Guard where appropriate
- Dive rescue teams where appropriate
- Flood-control agencies
- Other appropriate authorities

These agencies should be listed on the appropriate area response form and WRP.

Anticipate the victim's route and make plans to intercept at every possible rescue point. WRPs for the involved waterway should be of assistance to the IC. Observation of reference floats and reports from downstream units, citizens, and other witnesses will also affect decisions regarding the placement of units. Be aware of waterways that drain into other channels and ensure timely coverage of the secondary waterways. Where the affected waterway drains to the ocean or a lake, ensure the response of lifeguards, Coast Guard, harbor patrol, etc. Units conducting searches should begin at the downstream end of their assigned areas and work their way upstream. This will maximize the possibility of spotting the victim.

Ensure that units are prepared to attempt rescue if the victim is spotted. Divide assignments to include a search component as well as strategically placed rescue units.

Strategy and tactics

ICs will find that a different approach is required to manage each of these situations. While it is not possible to plan for all the specific problems that may be encountered, some guidelines will help the IC use resources with maximum efficiency and safety.

ICS. Some modifications make the ICS particularly effective for the management of these incidents (i.e., linear divisions along waterways, passing command, etc.). Many water rescue situations will dictate the use of unified command.

Dividing the incident. Incidents in waterways can be effectively divided in a linear pattern along the involved waterway(s). Some ICs prefer to identify divisions or branches along battalion and field division boundaries.

Passing command. Some swift-water rescue incidents may require SAR operations along miles of waterways. Some waterways in L.A. County pass through several battalions and divisions. In some isolated situations, it may be advisable to pass command to an officer located downstream. However, in most cases the original company officer, battalion chief, or division assistant chief should maintain command throughout the incident.

Command post location. The location of the command post will depend on the situation. If the rescue involves a stranded victim, it should be easy to find a suitable location near the site. However, if long stretches of river are involved, the command post location may have to be moved several times as operations progress. Possible command post locations should be listed in all WRPs.

Radio frequency assignment. Mutual aid frequencies should be considered when multiple agencies are assigned to a swift-water incident.

Rescue of stranded victims. When a victim is found clinging to debris in the middle of a river, or stranded on an island, the IC should strive to achieve two main objectives:

- Ensure that personnel are stationed downstream to rescue the victim if swept from position (and to protect rescuers who may be swept downstream during the rescue operation).
- Rescue the victim with the least danger to firefighters and rescuers. Two questions must be answered to determine the best course of action:
  - Is the victim in imminent danger of being swept away? If the victim is stranded in a safe position and not in immediate danger of being swept away, leave the victim there until a reasonably safe rescue effort can be mounted (or until the water level drops).
  - Can the victim assist in the rescue? If the victim is able-bodied and able to follow directions, consider using a self-rescue technique that will keep all firefighters out of the water. If the victim is injured, trapped, very young, or very old, self-rescue may be impossible. Under such conditions, the IC must weigh the chances of successful rescue against the danger to personnel.

If rescue cannot be attempted with reasonable safety, personnel should support the victim in any way possible until more help arrives, or until conditions improve. Downstream units should be prepared to attempt rescue if the victim is washed into the current.

Rescue of victims from moving water. Swift-water rescues are dynamic events, and incident commanders may experience all manner of logistical problems. This is where WRPs may prove to be crucial to the success of the operation. All units should have clear assignments. Personnel should be familiar with the rescue site. It is important for rescuers to be aware of danger areas associated with particular rivers, streams, and flood-control channels.

When rescuers have spotted a victim being swept downstream in a channel but are unable to effect immediate rescue, the IC should concentrate on the following four main objectives.

Objective 1: *Ensure that units maintain visual contact with the victim and constantly update downstream units of the victim's present location.* Personnel who make visual contact with the victim should immediately notify the IC and all responding units. This message qualifies as emergency traffic. When possible, they should also throw a reference float into the water while attempting rescue. This may assist downstream units in locating the victim's approximate position if the victim goes under the water or becomes trapped upstream.

As downstream personnel spot reference floats, they should notify their superior, who will forward that information to the IC. In addition to dropping floats, the unit that spots a victim should make all efforts to follow the victim if immediate rescue is not possible. This may require units to drive along bike paths adjacent to rivers, or to drive from bridge to bridge ahead of the victim. Helicopters, law enforcement, and other resources should be used to assist in the search when possible.

Objective 2: *Anticipate the victim's route and make plans to intercept at every possible rescue point.* If there is a WRP for the involved waterway, responding units should automatically perform the tasks assigned to them in the preplan. The preplan will establish critical rescue points and/or search perimeters for each responding unit. If necessary, the IC can change any assignment via radio to cover actual conditions and needs. As in any other emergency, the IC cannot rely exclusively on the preplan. Adverse conditions can force drastic changes to keep up with changing conditions. During major storms, many units may be tied up on other incidents. During severe flooding, even the course of the waterway may change. Both these conditions would definitely force the IC to change strategy.

Objective 3: *Assign sufficient resources for search and rescue operations at identified rescue points.* The IC should ensure that units are positioned far enough downstream to intercept the victim. This may be especially difficult in rural/foothill/mountainous areas. Rivers, creeks, and flash-flood zones in these areas may be inaccessible for miles. Rescuers may only get one shot to rescue a victim.

Firefighters may be on their own without help for extended periods. These situations demand extra caution. In all cases, anticipate delays in the arrival of assisting units. Inclement weather and the flurry of activity that often follows can hamper plans. Roads and bridges may be washed out, and helicopters may be grounded by weather.

Finally, move-up companies may be covering local fire stations. These personnel may not be familiar with the rescue area. The same problem may be encountered when personnel are working overtime from other areas. These effects can be buffered with good preplans.

Objective 4: Rescue the victim with the least danger to rescuers. Again, WRPs can be used as a basic action plan in the initial stages of the incident. The plan spells out what types of rescue attempts will be made at critical rescue points along the particular waterway.

Marine Rescue Operations

Marine rescue includes emergencies where victims are in need of rescue on the open sea or in some cases, lakes. These emergencies may result from airplane crashes, capsized boats, boat collisions, boat fires, etc. Generally speaking, marine rescue/disaster plans should include the automatic response of multiple agencies that can provide the most timely rescue resources. This would include the coordinated response of marine-based, air-based, and land-based fire/rescue/EMS resources, to effect timely rescue and recovery, provide rapid transportation to shore, and process (as necessary) through a landside Multi-Casualty Incident (MCI) system.

Minor and Major Marine Disasters Off the L.A. County Coast

Fire/rescue agencies in L.A. County have banded together to establish an air/sea disaster plan (A/SDP) for the waters off the coast. The system includes the automatic response of land/air/sea rescue resources simultaneous to the report that a plane has crashed, a ship has wrecked or sunk, a vessel is burning, a helicopter has gone down, or some other marine mishap has occurred. This helps ensure a coordinated response of marine-based, air-based, and land-based fire/rescue/EMS resources, to effect timely rescue and recovery, provide rapid transportation to shore, and processing (as necessary) through a landside MCI system.

In the waters off L.A. County, ultimate responsibility for offshore SAR operations is assigned to the commander, 11th Coast Guard District as the federal search and rescue coordinator. As stated earlier, the plan calls for the first on-scene unit from any agency to establish command that will ultimately be transferred to the Coast Guard, which may use unified command at their discretion.

Dispatch

A marine emergency with fewer than 10 reported victims should be dispatched as a *minor marine disaster*. A marine emergency with more than 10 reported victims should be dispatched as a *major marine disaster*. If there's any question as to the nature of the marine incident, it is appropriate to err on the side of caution and dispatch a major marine disaster. Incidents with victims reported in the water should *not* be dispatched as public assist.

Special resources

The following special resources should be available.

- Personnel regularly assigned to LACoFD USAR task forces are trained as rescue swimmers. These personnel can be deployed into the ocean or lakes from helicopters or boats.
- The LACoFD lifeguard division's rescue swimmers and divers may be deployed from baywatch boats and inflatable rescue boats.
- The U.S. Coast Guard will dispatch rescue helicopters and cutters to the scene.
- Other local fire departments may dispatch ground units to assist with MCI operations.
- Agreements exist between the LACoFD and the U.S. Navy for protection of Catalina Island and response to marine disasters. Therefore, the IC may request the response of large hovercraft(s) from the U.S. Navy, with an estimated time of arrival (ETA) of approximately 1 hr. The hovercraft may be used to transport engine companies, USAR companies, hazmat units, and other apparatus to support emergency operations at sea or on Catalina Island.

Emergency operations

Weather, helicopter availability, and other conditions permitting, USAR task forces respond directly to the scene via LACoFD helicopters with rescue swimmer teams. These teams are equipped with 10-person inflatable rescue platforms to locate and remove victims from the water. If conditions preclude helicopter deployment, USAR task forces will notify the IC and respond on the ground to an appropriate location for boat pick-up or other assignments.

The first boat or helicopter on the scene will establish ocean incident command (OIC). Upon arrival on the scene, all boats activate their blue lights, and all assisting boats and helicopters report to OIC for offshore operations. All land-based companies report to the landside IC. The personnel from these units will be assigned to assist with boat-based victim rescue/extraction or landside MCI operations.

Ultimate responsibility for offshore SAR operations along the coast of the United States is assigned to the local U.S. Coast Guard District as the SMC. However, the first on-scene agency should establish initial incident command until the Coast Guard arrives on the scene. Upon arrival of the Coast Guard and other agencies, unified command may be established, with the Coast Guard as the ultimate authority for incident command and final disposition of the emergency.

Submerged victim rescues

When a victim is missing and presumed to be submerged in the ocean, a lake, a pond, or other body of water, the IC should immediately request a dive rescue team in addition to other units that may be responding. If conditions are reasonably safe, first responders may initiate a preliminary search. The following instruction provides some guidelines to begin search operations until the arrival of a dive rescue team.

Searches. Fire/rescue agencies are frequently called upon to search for swimmers, boaters, fishermen, and others who have disappeared in deep water. Searches under these conditions are usually hampered by lack of visibility due to murky water or the depth of the water.

Even swimming pools can present significant search problems. Chalky-looking, cloudy water may conceal the bottom of a pool. In 2002, a missing boy was eventually found at the bottom of a backyard pool, where he'd been for days. He was hidden from view by an optical illusion created by chalky water. If there is any doubt whatsoever, even pools must be physically searched.

In all cases where a victim cannot be immediately spotted in deep water, the IC should ensure that the following objectives are met:

- Request one or more dive rescue teams and get their ETA.
- Establish the PLS and mark it.

When responding to a search rescue in calm water, your only source of information regarding the victim's PLS may be witnesses on the shore. In some cases, firefighters or other rescuers approaching the victim may actually see the victim go under the water. To create a PLS marker, make sure the rope is longer than the water depth and tie a life float, reference float, or other floating object to a weighted line. This marker serves two purposes:

- It pinpoints your immediate search area.
- It will be useful in plotting the possible location of the victim for the dive rescue teams when currents, winds, and other factors are taken into consideration.

Consider a safety officer and other measures required when personnel are working in extremely hazardous environments. When dive rescue teams arrive, give as much information as possible to assist them in locating the victim.

Beginning the initial search. When you get to the location where the victim was last seen, begin a quick visual surface search to make sure you are in the proper location and the victim has not *popped up* near you. Before attempting an underwater search, get your bearings by identifying prominent geographical structures such as cliffs, points, or mountains. Other points of reference include objects near the shore, such as trees, buildings, and other landmarks. Try to line up a couple of these points to triangulate your position in the water. Once you have properly identified the search area, take another look to verify that the victim has not surfaced. If there is no sign of the victim, an underwater search can begin.

Beginning search operations. If they can swim and have appropriate safety equipment, firefighters should begin search operations. In some cases, they may locate the victim before the dive rescue teams arrive. To conduct a search, perform the following:

1. Before submerging, take a few quick deep breaths, being careful not to hyperventilate.
2. Make a surface dive, looking for in the immediate area where victim was last seen.
3. When swimming underwater, it is better to make short duration dives rather than trying to stay underwater for long periods. You will be less fatigued; you can cover more area and continue to check your location.
4. If you're unsuccessful in locating the victim but you're able to remain in the area, continue to search as long as it's safe. If you are becoming tired or cold, exit the water.
5. If you are able to locate the victim, grab any available part of the victim's clothes, hair, or body and swim to the surface.
6. Place the victim on a float or rescue tube, if available. If possible, perform artificial ventilations while moving toward the shore. If this is not possible, simply get the victim to shore as soon as possible.
7. The IC should ensure that sufficient resources are waiting on shore to begin definitive medical care.

It may take several surface dives to get a victim to the surface. This author and a USAR company colleague once located a victim while conducting surface dives, and brought him up from a depth of about 15 to 20 ft. It took three separate dives to get the victim to the surface, because every time we got down to the victim and began towing him upward, we ran out of air and were forced to surface. It was a frustrating reminder of the need for USAR companies and rescue companies to have dive rescue capabilities.

Submerged vehicles

Rescuers responding to vehicles submerged in water are faced with a number of special problems ranging from the need for scuba or snorkeling gear, to powerful currents, to poor visibility in murky water. In addition to vehicle damage from collisions, many automobiles tend to turn upside down as they submerge, and become buried in mud. In addition to this *turtle* effect, air/water pressure differentials, failure of electrical systems, and other complications make vehicle rescue a special challenge.

Fortunately, if there isn't severe damage, many passenger vehicles tend to float for several minutes before submerging, and electrical windows and door unlocking systems tend to keep working long enough for conscious victims to make an escape before the auto goes below the surface. Buses, however, tend to sink almost immediately, often going upside down as they descend.

Stories of firefighters attempting dramatic rescues abound. Often, they arrive without scuba and other necessary capabilities, but make valiant attempts to rescue victims. At least four of the author's personal acquaintances have rescued people in just such conditions, and the author and other colleagues have unsuccessfully attempted to rescue victims under similar circumstances.

In 1979, Ventura County firefighter Gordon Daybell, a rookie at the time, was dispatched on engine 13 to *a vehicle in the water* at Lake Piru, a small farming community 40 miles northwest of downtown Los Angeles. Arriving in darkness, the firefighters saw headlights shining up dimly from beneath the murky lake. No scuba or snorkel gear was available.

Without hesitation, Daybell dove into the water wearing only his work uniform (sans boots). He dove several times in the cold water, finally reaching the motorist and yanking him through the driver-side window after using a spring-loaded center punch to shatter it. Daybell swam the man—now pulseless and non-breathing—to shore, where firefighters initiated CPR. Paramedics feverishly worked to regain his pulse, and by the time they arrived at a hospital, the victim was actually attempting to breathe on his own. The man survived practically without any deficit.[3]

In 1983, LACoFD firefighter/paramedic Marty Romero was working Squad 36 when he was dispatched to a traffic collision in the industrial city of Carson, on the San Diego Freeway. When he arrived on the scene at an elevated freeway interchange, Romero and his partner discovered one damaged vehicle on the interchange and one upside down, barely afloat in 20 ft of water, in a storm channel adjacent to the freeway. The car had plunged some 40 ft to the surface of the water, crushing the roof. Stripping off his raingear, and without the advantage of scuba or snorkel equipment, Romero entered the water and swam to the car, where a woman was trapped inside. Amazingly, he was able to muscle the victim from the car just as it started to go under the petroleum-contaminated water. His partner was waiting to pull her from the water as Romero swam her to shore.

In 1987, Ventura County fire department captain Curt Brown received a still alarm that a vehicle had driven into Lake Sherwood, a private lake near the world-renowned country club. Operating without scuba or snorkel equipment, Brown made several surface dives, finally pulling the driver from a van in 15 ft of muddy water and swimming him to shore, where he was successfully resuscitated.

On July 4, 1998, a man was attempting to pass other cars at a high rate of speed on California Highway 138, which is notorious for its deadly head-on collisions. In L.A. County's high-desert region, he ran an oncoming car off the road. The second car, containing two women and a nine-year-old girl, veered off the road, plowed through a chain-link fence, and plunged into the 20-ft deep California aqueduct. The aqueduct, which transports water to the L.A. Basin, moves at a steady 4 to 8 mph, a forceful current in which to conduct an underwater rescue under the best conditions.

Engineer Ron McFadden, responding from L.A. County Fire Station 79 some 13 miles away, arrived nearly 15 minutes after the accident. An LACoFD lifeguard division dive team had been dispatched and was to be flown by helicopter from the coast, but that would take at least 45 minutes.

After positioning engine 79, McFadden and the other firefighters were told that two women had self-rescued before their car had submerged, and bystanders had assisted them from the water. But no one had been able to get to the 9-year-old girl who was seat-belted in the back seat. The car now rested on the bottom in 14 ft of moving water, and the firefighters had no scuba or snorkel equipment.

Making one of those pressure decisions that would be debated by others later, McFadden informed his captain that he was going to attempt a rescue. A life literally hung in the balance. Nobody objected, although this was clearly a risky maneuver. McFadden donned an SCBA from engine 79 and entered the water. As he dove, he was initially startled by the violence of the blow-by pounding his SCBA mask; the pressure regulator had begun free-flowing due to the pressure change. He surfaced briefly, rechecked his mask, and dove once again.

Under a free-flow condition, the breathing time is (obviously) drastically reduced, so McFadden knew he had precious little time. He managed to reach the car in zero-visibility conditions, and was able to find the frame of the blown-out rear window. Squeezing himself partially through the opening, McFadden was able to feel a limp body still strapped into the back seat. Fighting the current, he was able, after several attempts, to unbuckle the seat belts and pull the girl from the car. Surfacing, he swam her to the waiting arms of firefighters along the shore. Within minutes, she was airborne in Air Squad 17, a Bell 412 fire/rescue helicopter, with firefighter/paramedics running a *full code* on her as CPR continued. By the time they reached Antelope Valley Hospital, they had restored a pulse. The girl survived the ordeal, was able to regain her speech and ability to walk, and continues to recover.

McFadden's decision to use a firefighting SCBA to perform a dive rescue was questioned in some circles. Certainly it was a risky move; if McFadden had become incapacitated, some said, it might have prompted others to enter the water, starting a chain reaction that could have resulted in the loss of several would-be rescuers. But others noted that, without any equipment whatsoever, diving for the victim might still have resulted in some sort of entrapment for McFadden. Undoubtedly, the remoteness of the accident site left the firefighters in a catch-22 situation.

The manufacturer of the SCBA naturally reiterated that the equipment wasn't designed for dive operations and shouldn't be used for that purpose. In addition, the fire department was essentially forced to issue a memo reminding firefighters that SCBA isn't intended as dive rescue equipment, setting strict policy against such use of SCBA in the future. Eventually McFadden was awarded medals of valor from the LACoFD and the California State Firefighters Association (CSFA) for his heroic work. Most importantly, a little girl is alive today because of his efforts.

Five years later, another car crashed into the same aqueduct during the morning commute, this time with five people inside. Once again, McFadden was on duty on engine 92, and once again, he and another firefighter took turns making rescue dives wearing their SCBA while the LACoFD dive teams and a USAR task force station responded via helicopter and on the ground. Meanwhile, McFadden and his partner managed to pull four children out of the car, all in full cardiac arrest, and also managed to extract the woman driver. Two LACoFD fire/rescue helicopters landed to perform MedEvacs, flying all patients to waiting hospitals. The final toll was four fatalities, but the incident once again demonstrated the need for dive capabilities wherever there is significant danger of people becoming trapped or incapacitated beneath the surface. It also demonstrated an extraordinary level of valor and ability to adapt and overcome, on the part of McFadden and his colleagues.

In terms of rescue systems development and planning, these incidents raise interesting points. First, the sheer number of firefighters faced with victims trapped in submerged vehicles across the United States is somewhat staggering. Second, the number of firefighters who find themselves in this dilemma without the availability of (or the training for) SCBA or snorkeling equipment is equally staggering. Clearly, submerged vehicles are a significant hazard anywhere there are roads near water, which means that relatively few fire/rescue agencies are immune from it.

Even in calm water, these are challenging rescues. If there is a moving current, the danger is compounded by the possibility of a vehicle shifting and pinning a rescuer.

The effect of crash impact on submerged vehicles. One of the most important factors is the amount of impact the vehicle experienced before and after it entered the water. Regardless of the type and size of the vehicle, the amount of air trapped within it is related to the amount of damage sustained as it hit the water—and whether or not the windows were down. Low-speed entry (often occurring around marinas and boat launch ramps) offers a greater chance for rescue than a high-speed crash off a bridge or cliff.

Vehicles on the surface. Because vehicles may float for several minutes before sinking, it may be possible to make a rescue at this time. The vehicle will tend to tilt somewhat depending on where the center of gravity is. A front-engine car will tend to tilt forward, exposing the rear window. This is ideal if the windows are rolled up, because the tempered glass of the rear window will shatter with a blow from a sharp object like an axe. A blow to one corner of the glass is most effective. Victims can then be pulled out through the rear window before the car goes down.

Of course, rescue can be made from the side windows if the windows were already open or the occupants can roll them down. Electric windows may still operate for a time even when a vehicle submerges. If the windows are inoperable, or the victims are unconscious, the side windows (which are also tempered) can be shattered just like the rear window.

Windshields are notoriously difficult to break because of the lamination process that protects them from projectiles on the highway. *Don't* waste valuable time trying to break a windshield unless there are no other options.

Be aware that the vehicle can suddenly tilt vertically and plunge to the bottom. You don't want to be climbing through a window (or in the car) when this happens. In fact, rescuers should move far away from the car if it appears to be ready to submerge. Rescuers do not want to be near such a large moving object in the water. Rescue efforts can continue once the auto has come to rest below the surface.

If it appears that the auto will submerge before victims can be removed, it may be possible to tie a rope to the bumper. The rope will not keep the car afloat, but it can guide rescuers to the auto after it goes to the bottom.

Locating the vehicle below the surface. In deep or murky water, rescuers may have difficulty finding the vehicle. However, a submerged vehicle will leave clues to its location. These clues may include a continuing trail of escaping air bubbles, an oil or gasoline slick, or articles floating to the surface. At night, headlights may be seen shining up from below.

Marking the vehicle's location. In any submersion rescue, it is important to mark the location of the victim(s). In the case of a vehicle, the first rescuer in the water should bring a safety line and tie it to a bumper or other structural member (if this was not already done while the vehicle was still floating). A reference float, life float, or other floating object can then be tied to the line to mark the location at the surface. This is important for three reasons:

- It marks the location for other rescuers arriving at the scene.
- If a dive team or other rescuers arrive in boats, they must know where the original rescuers are operating to avoid endangering them with propellers.
- Rescuers can use the guide rope to *pull* themselves down to the vehicle. This conserves precious time and energy.

Stabilizing the vehicle. In some cases, it may be necessary to stabilize a vehicle before rescue can proceed. The guide rope can be used for this purpose in calm water. Tying it to a bridge, tree, or other strong anchor will at least lessen the chance of the vehicle moving and pinning a rescuer. A winch, tow truck cable, or crane can follow the rope to provide additional stabilization. Other measures may be necessary under some conditions.

Moving a submerged vehicle. In some cases, the position of the vehicle can hinder rescue operations (i.e., a vehicle on its side with inoperable doors due to impact damage). Firefighters may need to move the vehicle to proceed with rescue operations.

A tow truck, A-frame with a winch, or crane should be used when possible. If these devices are available, it may be possible to lift the vehicle to the surface. A heavy-duty crane may be able to pick the vehicle out of the water and set it on the shore. Obviously, this would create a much safer work environment for rescuers.

Under limited conditions, a rope hauling system could be used to move a vehicle in a life-and-death emergency. The manpower pull raising system with a brake is one alternative. Mechanical advantage systems may also be used to move limited loads.

Note: Rope systems should be used only for life safety situations. If a current is present, rope may not hold the vehicle in place. If the vehicle is moving in the water, do not attempt to tie off to a stationary object; the rope is likely to snap. A cable that can withstand the stresses placed upon it will be a better choice.

Opening doors. Doors may be difficult to open due to the pressure differential between the inside of the vehicle and the water. In some cases, they may be opened when the vehicle fills with water, equalizing the pressure.

Doors may be difficult to open even when the vehicle is filled. Consider the possibility that the vehicle may have struck an object before going into the water. Previous impact may have jammed the doors.

Searching a submerged vehicle. If possible, rescuers should thoroughly search the interior. Victims may end up in unusual places, such as beneath dashboards. Visibility is likely to be poor, so repeated dives may be necessary to find victims. Air pockets may remain in portions of the interior; it is possible for victims to find these pockets and survive long periods of submersion.

Avoid tying ropes around the bodies of rescuers. Instead, place hand loops in the rescue line, allowing personnel to hold the line for support. If available, equipment such as swim fins, wetsuits, and dive masks should be used. Dive masks are especially useful by providing clear vision. Lighting is also useful at night and during inclement weather. Lighting up the surface of the water will increase visibility, limited only by the clarity of the water.

Submerged aircraft rescue

On first appraisal, the chances of this type of incident seem remote. However, the sheer volume of flights over L.A. County makes it possible (if not probable) for aircraft to go down in the ocean or in local lakes and reservoirs.

Aircraft into the water present similar problems found in a submerged vehicle rescue. Obviously, the impact of an airplane into the water may kill many victims outright. Others trapped in the fuselage may drown before help arrives. However, there have been many successful rescues of people from downed aircraft in the water. One reason for this may be the lack of fire, which often occurs when a plane crashes on land.

Generally, the people who survive these incidents were either thrown from the aircraft on impact or escaped on their own. However, there have been exceptions. It is possible for portions of the fuselage to stay afloat long enough to effect rescue. This is a dangerous undertaking and may require a dive rescue team to actually extract the victims.

The first minutes after a crash are critical. The rescue priority should be as follows:

1. Provide flotation for large groups of survivors found floating in the water. Move on to the next group. Pulling each victim out of the water should normally wait until the next two priorities are covered.
2. Rescue survivors trapped in the plane. These victims may drown without prompt rescue. Pull them from the plane and provide them with flotation.
3. Provide flotation for individual survivors floating in scattered patterns around the crash site.
4. Pull large groups of survivors from the water.
5. Pull scattered survivors from the water.
6. Treat and transport according to normal triage protocol and available resources.

The use of SCBA under water. It's important to comment on the use of SCBA under water. SCBAs are designed for fire and other hazardous atmosphere environments. They are not designed to be used as scuba gear for search/rescue under water. However, firefighters need not panic if they accidentally fall into water while wearing SCBA. Many SCBA will function under water to a certain depth. For example, the technical design group of U.S. divers states that Mark V and Mark VI SCBA will supply a free-flow effect under water, thus providing more than enough air to get to the edge of a pool (or to shore in most cases).

Actual tests conducted by members of LACoFD indicate that this is indeed true. They have also found that other SCBA, like Panthers, will function under water in emergency conditions. The wearer must be aware of the potential for free flow. To reiterate, however, firefighting SCBA of any brand is not intended for underwater operations.

Endnotes

1 Conversely, some Arizona residents refer to them as *California crossings*.

2 Typically known as *Arizona crossings*, roads built across dry or shallow riverbeds dot the landscape of the western United States, and they are the source of dozens of swift-water rescue operations each year. The road across Triunfo Creek is typical of many Arizona crossings: There are safe ways to cross during good weather, but they can become potentially deadly whenever a heavy rain falls.

3 Unfortunately, in a cruel twist of fate, Firefighter Daybell noticed tingling in his arms several weeks later, and was diagnosed with an inoperable brain tumor. Gordon Daybell died less than two years later, at the age of 21.

4

Mud and Debris Flow Rescue Operations

Mud and debris flows are among the most destructive forces on the planet. The power of these events was demonstrated in 1985 when an erupting Columbian volcano melted the snowcap. The sudden heating of the thick snowpack unleashed a huge mud and debris flow that rampaged down the mountain, where it wiped one town off the face of the earth and buried a large city in 20 ft of mud. More than 23,000 people perished in the Columbian disaster. The mud and debris flow shown in Figure 4–1 swept away apartment complexes 6–8 stories high and wiped out a huge swath of a Venezuelan city in 2002.

Fig. 4–1 Venezuela 2002

Today, similar conditions are a concern for people living in a number of towns and cities in northern California. Typical is the Mammoth Lakes area of California, where earthquake swarms and geothermic changes are signaling a possible eruption in the coming years. The same danger confronts other places where populated areas lie downstream of snowcapped mountains, especially in Washington, Oregon, and Alaska. Other nations are home to equal or worse threats. In parts of Japan that are vulnerable to mud and debris flows, the government builds concrete channels to divert the heaviest flows away from towns and cities. Other nations throughout the world aren't so fortunate. South American nations like Columbia, Peru, and Chile remain vulnerable to these deadly effects.

On the west coast, L.A. County is one place with decades of experience with deadly mud and debris flows. Recent events forced the LACoFD and other fire departments to establish formal training and response programs to deal with the lethal hazards posed by mud and debris flows. In 1993, a series of firestorms leveled hundreds of homes across L.A. County and left tens of thousands of acres of rugged mountains and hillsides scorched. When winter came, there was yet another danger facing local residents and firefighters alike: devastating mud and debris flows. Geologists, weather experts, and fire/rescue personnel warned that the danger was worst for people in populated areas located downhill (and downstream) from mountainous areas incinerated by the conflagrations in Malibu, Altadena, and Laguna Beach (in neighboring Orange County). As firefighters soon discovered, the hazards would be amplified for them and other rescuers.

To prepare, affected fire departments developed new awareness, training, and response programs to minimize the threat of life-loss from mud and debris flows. The LACoFD already had long-standing swift-water and USAR systems in place, and they formally adopted mud and debris flow emergency procedures. They also implemented a new series of mud and debris flow rescue first responder training programs. In Figure 4–2, a victim of one of the mud and debris flows that struck Southern California after the 1993 fire storms awaits treatment after being rescued from Laguna Canyon.

Fig. 4–2 Southern California Mud and Debris Flow

The training came not a moment too soon. December heralded a series of heavy storms that sent large mud and debris flows into fire-damaged areas in coastal Malibu and the inland mountain community of Altadena. In at least one case, the new training came just hours before it was put to use saving lives. Three hours after training, LACoFD firefighters from Fire Station 70 joined other Malibu units, a USAR company, and two LACoFD swift-water rescue teams to put their skills to work. Together, they saved the lives of more than a dozen people trapped by a massive mud and debris flow that swept through a canyon and roared across the four-lane Pacific Coast Highway.

The deadly wildland fire storms that struck Southern California in October 2003 burned more than 375,000 acres and left huge areas prone to mud and debris flows. In the fires, 21 people died, but fire officials knew from experience that secondary consequences of the wildfires might be even more lethal. Fire/rescue agencies across the southern half of California began preparing for the inevitable mud and debris flows that were certain to come as winter set in (Fig. 4–3). On Christmas Day 2003, a man was pinned by trees and debris following a mud and debris flow that swept through a mountain community burned by the Southern California firestorms of 2003. It took several hours for firefighters to locate and rescue this man, who lost two daughters when they were ripped from his hands by the flow.

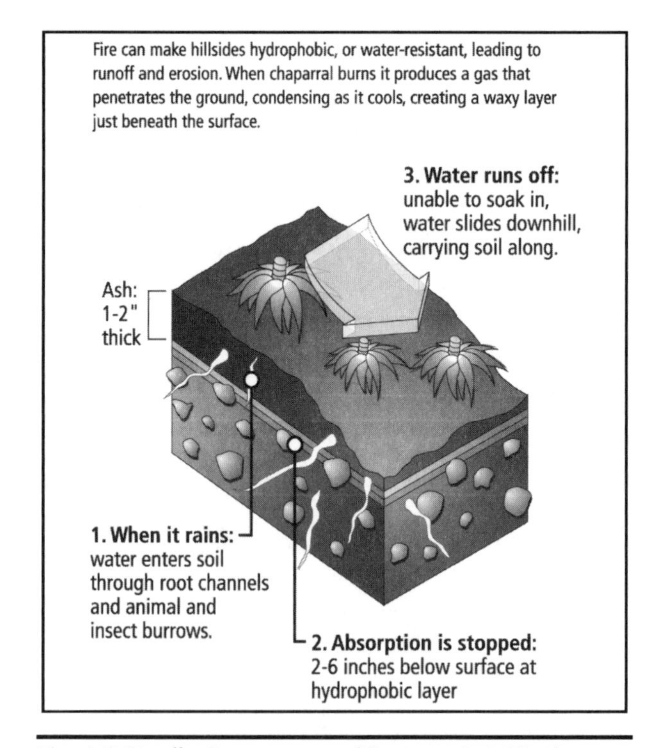

Fig. 4–3 Deadly Consequences (Courtesy *L.A. Times*)

Sure enough, 18 people were killed on Christmas Day when huge mud and debris flows tore through several canyons that had burned in October. Hundreds of people were stranded, and local fire/rescue agencies spent the night rescuing survivors. After more than three weeks of searching, it was finally concluded that some victims never would be found because they might be buried beneath dozens of feet of mud and forest debris that was swept four miles down one of the canyons. Such is the nature of the fire-flood sequence in Southern California and other steep mountainous areas prone to wildfires.

What is a Mud and Debris Flow?

A simple but eloquent definition of mud and debris flow is given in John McPhee's book *The Control of Nature.*[1] McPhee describes a mud and debris flow that buried part of a neighborhood in LaCanada, an upscale suburb of Los Angeles nestled below the southern face of the 10,000-ft San Gabriel Mountains. The flow occurred late one night in 1978 during a driving winter rainstorm, in an area below slopes that had recently been burned by a 60,000-acre wildland fire:

> *It was not a landslide, not a mudslide, not a rock avalanche; nor by any means was it the front of a conventional flood.... In geology, it would be known as a (mud and) debris flow. (Mud and) debris flows amass in stream valleys and more or less resemble fresh concrete. They consist of water mixed with a good deal of solid material, most of which is above sand size. Some of it is Chevrolet size. Boulders bigger than cars ride long distances in debris flows. Boulders grouped like fish eggs pour downhill in debris flows.... It was not only full of boulders; it was so full of automobiles it was like bread dough mixed with raisins.*

The 1978 LaCanada mud and debris flow buried several homes to their roofs. In one of the homes, an entire family—and most of their furniture—literally floated to the ceilings atop the invading mud, which poured from the cracks from the pressure of the flow. The couple and their two children were trapped for several hours, their faces pressed to the ceiling, with precious little breathing room.

LACoFD personnel from Stations 82 and 19 used plywood sheeting to create a path over deep mud to reach the roof of the home, protruding from the muck. Carefully using chain saws and axes to avoid injuring the victims who lay trapped just beneath the ceiling, the firefighters burrowed through the roof and ceiling to effect a dramatic rescue several hours after the flow ceased.

Mud and debris flows are endemic to places where steep mountains and foothills rise above valleys and flood plains. They are common where the mountains are cracked and fractured by earthquakes and tectonic forces, making them more vulnerable to erosion. If the mountains are covered with highly flammable vegetation, the probability of mud and debris flows increases exponentially. Vegetation is sometimes the only thing keeping boulders and soil clinging to the slopes. When fire denudes the vegetation, the rock and soil begin sliding and falling into the canyon bottoms. When intense rain occurs, tremendous amounts of debris can be quickly turned to slurry and mobilized into a huge flood.

Case Study 1: Two Hikers Killed Below Site of L.A. Firestorm

Following the 1993 firestorms in southern California, one of the worst mud and debris flows occurred on March 6, 1994. A surprise thunderstorm unleashed a 15-ft high wall of mud and debris that scoured several miles of a steep canyon of trees and boulders. Dozens of weekend hikers were exploring the canyon when the mud and debris flow occurred. Two of them, a father and his young son, were never again seen alive.

The morning broke with sunshine one day after a brief but violent Pacific storm swept across southern California. In Bailey Canyon, a popular hiking destination in the rugged San Gabriel Mountains, it was sunny and warming. However, by mid-morning, dark thunderheads were seen brewing over the peaks of the San Gabriels.

By noon, a fully formed thunderstorm was in progress, flashing lightning on the peaks. Unknown to hikers on the lower slopes, a severe downpour was pounding the upper reaches of the mountains. One of the areas that received the heaviest drenching was the Kinneloa burn area, where dozens of homes burned in a conflagration the previous year. Torrents of water and ash rushed into the deep canyons, building into a flood.

Eight miles to the west in Altadena, a strange roaring noise shook the windows of LACoFD Fire Station 66. This old rock-walled station is located at the base of the mountains immediately below the heart of the Kinneloa fire area. Captain Mike Minore and Firefighter Specialist Todd Mitcham, both members of the department's swift-water rescue teams and the FEMA USAR task force, went out to investigate.

The sky was clear to the south, but the peaks of the San Gabriels were hidden by dark thunderheads. Although no rain had fallen at Station 66, Minore and Mitcham looked out across a nearby ravine to see a torrent of muddy water roaring down a stream that is bone dry most of the year.

They knew this was a sign of trouble. Minore climbed into the cab of engine 66, grabbed the radio, and reported to command and control that flash floods were rolling through the area. Then he ordered the firefighters aboard, and engine 66 commenced to survey their rugged and widely burned first-in district for potential rescue situations. They found none. However, unbeknownst to Minore and Mitcham, a life-and-death drama was playing itself out in a steep canyon several miles to the east.

At the same time engine 66 was scouting its district, water, ash, and boulders were pouring into Bailey Canyon, a popular hiking area in the steep mountains above Sierra Madre, just east and downhill of the Kinneloa burn area. Quickly, the flow overran brooks and creeks and grew into a flash flood whose leading wall was filled with mud, trees, and boulders.

As the newly formed mud and debris flow thundered through Bailey Canyon, individual hikers strung out along the length of Bailey Canyon stopped what they were doing. Instinctively, they tried to pinpoint the source of the strange rumbling that emanated from the mountain above them. When the ground began to vibrate, most of the people knew it was no earthquake. Yells of *Flash flood!* were heard, and the hikers scrambled for high ground up and down the canyon. Most were able to reach high ground before the flood swept past. Several were later rescued by helicopter and ground-based rescue teams.

Unfortunately, a father and his young son were hiking in a steep draw within the canyon, surrounded by rock walls nearly vertical in places. They tried to head for high ground, but they had no way to climb above the flood. It took several days of intensive searching by the Sierra Madre rescue team, an L.A. County Sheriff's helicopter, USAR-1, the Department of Public Works, the coroner's office, and several wilderness search dog teams to find the remains of the boy and his father beneath 30 ft of mud and ash.

In an event that demonstrated the potential danger to rescuers, the effort to recover the bodies nearly became a tragedy in itself when searchers came within seconds of being trapped by a large secondary mud and debris flow. This near-tragedy occurred on the third day of the search. It followed a sudden downpour that struck the upper elevations of the San Gabriel Mountains, conditions similar to those that caused the original disaster. Television news cameras, perched on a hillside to film the search and recovery operations, caught the drama on film.

A dozen searchers and public works personnel used boats and heavy equipment to muck through the layers of mud that clogged a debris basin built at the mouth of Bailey Canyon for flood control. By design, most of the mud and debris from the first flow had been deposited in the 2-acre, 30-ft deep debris basin, and most of the water had been separated and siphoned off by a large, screened pipe that protruded from the center of the basin. The pipe was designed to carry water through an underground shaft, where it is dumped into a series of flood-control channels that carry it across the densely populated Los Angeles Basin, finally flowing into the Pacific Ocean.

Typically, the L.A. County Department of Public Works clears the basin of mud and debris with heavy equipment after exceptionally large storms or at the end of winter. However, because it was assumed that the bodies had also been deposited in the basin, a round-the-clock emergency clearing and recovery operation was ordered.

Suddenly, as the news cameras rolled from above, most of the workers stopped what they were doing and fixed their sites up at the mountain. Something was coming. At first, the only sign of trouble was a strange rumbling noise that quickly grew louder. This was followed shortly by vibrations that shook the ground. As the workers began to abandon their positions in the basin, news cameras panned up the canyon just in time to film a huge flash flood of mud and debris plunging from the mouth of Bailey Canyon. Trees as large as 40 ft were carried along like matchsticks in the leading edge of the flow. The roiling mud was full of boulders and other debris.

The workers scrambled to get out of the debris basin, and the last one barely escaped before the flood roared into the basin. For several minutes, an immense black flow continued, depositing thousands of tons of additional mud and debris into the basin, filling it nearly to capacity again.

It was a chilling reminder of the scenario that confronted the hikers several days earlier in Bailey Canyon and left little doubt that the father and son were now buried beneath nearly 30 ft of mud. It was also a sobering reminder of how vulnerable firefighters and SAR personnel are to secondary mud and debris flows. The county personnel in this case were separated from disaster by only a few seconds and an uphill climb of some 40 ft. If not for the warning provided by the audible roar of moving debris, some of them might not have escaped in time.

The mud and debris flows in Malibu and Bailey Canyon clearly demonstrate that the concerns of geologists were credible and accurate. They were followed by dozens of other damaging flows near the three major burn areas. Today, it is widely recognized that effective mud and debris flow rescue training for firefighters and other emergency responders in threatened areas can greatly enhance personnel safety and operational effectiveness.

Mud and Debris Flows: The Landslide Connection

Mud and debris flows can also be associated with the occurrence of landslides and mudslides. This process sometimes occurs where steep slopes are underlain by siltstone bedrock. An analogy can be made to a layer cake, with the frosting representing topsoil. If the layers are tilted (as in the case of geologic upthrust related to earthquake faulting), gravity begins to act upon them, pulling them toward the canyon bottom. If the slope is covered with vegetation that has deep root systems, it can resist slippage. However, as the slope becomes saturated, water percolates through the topsoil, eventually reaching the siltstone. The effect is intensified where recent fires have damaged vegetation (Fig. 4–4).

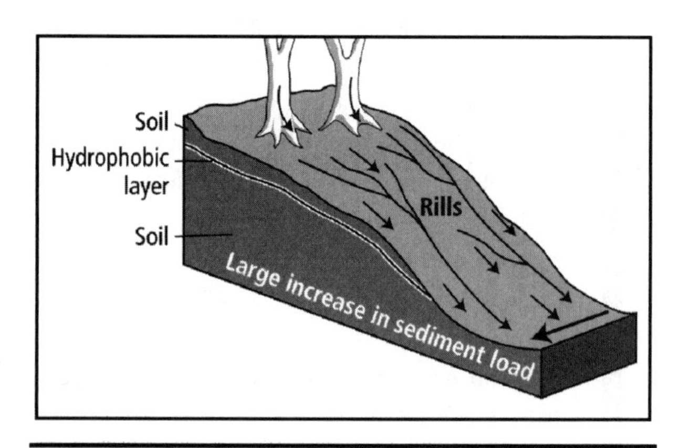

Fig. 4–4 How the Hydrophobic Layer Sets the Stage for Large and Deadly Mud and Debris Flows

The moisture begins to act as a lubricant between the layers, increasing the tendency of the soil layers to slide away. Eventually, the bedrock softens and begins to break up.[2] When the weight of the water-laden soil reaches a critical point, gravity can overcome the holding power of the root systems. Portions of the slope begin to break free, creating a cascade of mud and rock that can carry homes away and bury entire neighborhoods.

In some cases, mudslides and landslides can block the flow of streams, creating natural earthen dams that might not be detected downstream (especially if the slide occurs at night, far upstream from populated areas). The danger lies in the potential for the stream to breach the dam, loosing a flood of mud and debris upon populated areas downstream. Figure 4–5 shows the remains of a mountain community devastated by large mud and debris flows that struck after the 2003 fire storms in Southern California.

Fig. 4–5 Mountain Community Devastated

In February 1998, just such an event occurred in the city of Ventura, 50 miles north of Los Angeles. In a place called Hell Canyon, a large landslide blocked a flooded creek during a major storm. By the time the slide was discovered, millions of gallons of floodwater had backed up behind the unstable blockage, and the whole thing threatened to collapse. Downstream from Hell Canyon, hundreds of residents were in immediate peril from a massive mud and debris flow. Fortunately, the blockage was discovered in time, and Ventura city and county officials were able to keep the dam from giving way while they created a safe outlet for the stream and pumped out the water. Nature cooperated by lessening the storm, which gave officials precious time to lower the water and avert a disaster.

L.A. County: Mud and Debris Flow Country

As pointed out by McPhee, the mountainous terrain of L.A. County has all the features of a classic mud and debris flow zone, creating a worst-case scenario for tens of thousands of people living in the area. The San Gabriel mountain range is known to be one of the most dangerous and prolific producers of mud and debris flows in North America.

The San Gabriels are a young mountain range, owing their fast growth to the tectonic pressure exerted by the only sharp bend of the entire 1,000-mile long San Andreas Fault. This bend, or dogleg, has caused the fault to lock, creating tremendous forces on the land directly east of Los Angeles. The active faulting that results from this tectonic struggle is still visibly at work. There are now hundreds of earthquake faults underlying Los Angeles County, creating a patchwork of broken land. The 1994 Northridge earthquake raised the western end of the San Gabriel range a full 12 in. Hundreds of aftershocks in the first 24 hrs raised the mountains another 3 in. In this manner, the mountains are growing faster than they can be eroded by weather.

The San Gabriel Mountains are young, but the materials that form them are extremely old. Much of the mass of the San Gabriels consists of ancient bedrock that has been shattered, lifted, tilted, and conveyed from places hundreds of miles to the south. Today, the American plate and the Pacific plate continue to grind against each other along the San Andreas fault, which marks the borders of the plates as well as the northern margin of the San Gabriels. As these plates grind together, new parts of the earth's crust are smashed together to raise the mountains higher, even as they are being broken apart.

Rock breaks and falls off the steep slopes continuously. This is especially true when water trapped within the rock repeatedly freezes and thaws, causing the rock to expand, contract, and eventually break. Rain carries the broken rock down the slopes and ravines, and eventually into streams. Heavier rain begins to move larger rocks and boulders. As the hillsides become saturated, mud and rockslides fill the canyons with yet more debris. The heaviest rock and debris help define the streams. Occasionally, natural dams of rock and piles of debris are formed as part of this process. This material may accumulate for years, preloading the canyons for mud and debris flows (Fig. 4-6).

Fig. 4–6 A Malibu Creek Shows Signs of Mud and Debris Flows in Storms of 1994

Case Study 2: New Mud and Debris Flow Training Pays Off

The LACoFD's first mud and debris flow training program was presented to a selected group of personnel assigned to USAR truck companies in 1993. These members, in turn, acted as field trainers for these specialized truck companies. The USAR truck companies took turns presenting the new course to all fire stations in high-hazard areas across L.A. County through the winter of 1993/1994.

In early December 1993, a series of threatening storms approached southern California. Following newly implemented standard procedures for approaching storms, the LACoFD deployed eight swift-water rescue teams to strategic fire stations, including two teams to Malibu.

On the morning of December 7,[3] Captain Scott Smith and Firefighter Specialist Troy Flath, attended the mud and debris flow safety and awareness course conducted by personnel from USAR Truck Company 125. Less than 2 hrs later, severe thunderstorm conditions occurred directly over the Old Topanga burn area in Malibu. More than $^3/_4$ in. of rain fell in 30 minutes. This perhaps was not a great amount of rain in some places, but the 35,000-acre Old Topanga fire had left many of the canyons preloaded for mud and debris flows. Almost immediately, Malibu experienced serious trouble as a number of life-threatening situations occurred simultaneously while the heavy rain continued.

At 1130 hrs, an immense mud and debris flow roared down Big Rock Creek, immediately overflowing the banks of the creek where many homes had burned a month earlier. Big Rock Creek is normally a trickling stream that crosses under the Pacific Coast Highway and drains into the Pacific Ocean. Suddenly, it became a river of mud and boulders, scouring the bottom of the canyon just minutes after the downpour struck the upper peaks of the Old Topanga burn area. The flow swirled around pockets of homes in the canyon and headed directly for the million-dollar homes that line the Pacific Coast Highway—the same homes that firefighters from all over southern California managed to save from the Old Topanga conflagration.

As the downpour continued, mud and debris poured off the hillsides, clogging Big Rock Creek in several places. As these smaller debris dams burst, they sent waves of mud and debris down the canyon, transforming it into a torrent of mud and rock. The flood immediately clogged the flood-control system that normally directs storm runoff under the Pacific Coast Highway to the beach. With its normal path beneath the highway blocked, the flow quickly swept up and across the road, trapping people in cars and homes.

LACoFD's command-and-control facility was suddenly flooded with 911 calls reporting people trapped along the Coast Highway. Battalion Chief Cliff Dyshart, engine 70, the lifeguard rescue team, Squad 88, truck 125, USAR-2, Air Squad 8, and other units hurried to the scene. Dozens of people were trapped in a variety of dangerous predicaments as the mud and debris continued to flow like a river, accentuated by periodic waves of secondary flows. Extra units were requested and full-scale rescue operations began.

The most life-threatening situation appeared to be a couple whose Jeep Cherokee had been swept into the middle of the debris flow. Mud was up to the windows and rushing by at great speed. The Jeep was foundering adjacent to the point where the creek normally passes beneath the highway in the flood-control channel, which was now overflowing with mud. If pushed into the dropout point, the vehicle and its occupants would be buried in 15 ft of slurry

After sizing up the situation and assigning a firefighter as upstream lookout, USAR-2 Captain Scott Smith assigned Firefighter Flath (USAR-2) and Firefighter Steve Linnel (engine 70) to attempt rescue of the couple. A front-end loader from the L.A. County Department of Public Works had already arrived at the scene, and the driver bravely volunteered to drive his vehicle into the flow if necessary to rescue victims.

After a rescue plan was hastily developed, Flath and Linnel, wearing swift-water safety gear, climbed into the bucket of the front-end loader carrying ropes and other rescue equipment. They gave the signal, and the public works driver plowed the big tractor into the flowing mud. A tremendous amount of mud and water was still rushing out of the canyon, and additional walls of debris were a constant threat. Fortunately, they were able to remove the couple from the roof of the Cherokee and into the safety of the bucket without mishap.

Later, several front-end loaders were used to rescue citizens from rooftops and second-story balconies of homes along the Coast Highway. After the rescues were completed, Chief Dyshart ordered the entire hazard area evacuated to prevent people from being trapped by large secondary flows. This was the first of several debris flow incidents to occur in Malibu during the winter of 1993–94. It provided a good illustration of the value of the recent training, which came not a moment too soon.

Rainfall: The Trigger

Intense rainfall can act as the trigger to mobilize mud and other material in the preloaded canyons. Typically, this happens when powerful winter storms march in from the Pacific Ocean, stalling over the San Gabriel Mountains to drop millions of tons of water into steep, narrow canyons. The San Gabriel Mountains have seen some of the most intense rainfall recorded. In 1978, 12 in. fell in 24 hrs; during that time, 1½ in. poured down in one 5-min period, causing a debris flow 25 ft high that swept at least 13 people to their deaths. In 1933, more than 30 people were killed by a large debris flow from the San Gabriels near Glendale. In 1943, 26 in. fell in 24 hours.

Conditions that can mobilize debris flows often cause other simultaneous phenomena that increase the danger for rescue personnel and victims. If one canyon is flooding, chances are good that adjacent canyons are being hit hard by heavy rainfall at the same time. Major mud and debris flow events have occurred in California during at least 24 rainy seasons since 1905.

One mud and debris flow sequence in the Wrightwood area of the San Gabriel Mountains dislodged an estimated 18 million yards of material from the mountain and transported it 15 miles into the desert. Debris flows in steep terrain commonly move faster than 20 mph, and have been known to exceed 100 mph. Debris flows have moved boulders as large as five-story buildings. A mud and debris flow in the Tujunga area of Los Angeles County carried a 15-ton boulder into a residential area nearly two miles downhill from the edge of the San Gabriel Mountains.

In the nearby Santa Monica Mountains, which hug the Pacific Coast in the Malibu area, conditions are also ripe for mud and debris flows. These mountains, while less steep and of lower elevation than the San Gabriels, are equally broken and fractured. In Malibu and Pacific Palisades, ancient landslides cause mudslides nearly every time it rains. Debris flows and mudslides have caused property damage and deaths several times in the past decades.

The Role of Wildfires

When major wildfires burned more than 150,000 acres in the Santa Monica and San Gabriel Mountains in the fall of 1993, the stage was set for mud and debris flows to become a major hazard. Even before firefighters could rest from the fires, geologists and soil engineers began warning of severe life threats from mud and debris flow in the burned areas. In Figure 4–7, firefighters search for victims of a mud and debris flow in San Bernadino after the 2003 California fire storms.

Fig. 4–7 Searching for Survivors

Case Study 3: Malibu Slide and Laguna Beach Mud and Debris Flow Disaster

On Monday, February 23, 1998, fire departments and rescue agencies braced as the worst El Niño-related storm of the season bore down on southern California. The region was already water-logged after a week of pounding downpours that pushed annual rainfall totals to nearly double that of the average year. The ground was saturated, unable to absorb much more water. As a result, nearly all precipitation was running off the steep slopes, where it was funneled into swollen creeks and rivers. Steep hillsides, already unstable and prone to sliding in dry weather, became water-laden and heavy; at any time, gravity could take over and force the slopes to seek their angles of repose.

Based on past history and warnings from geologists, local officials knew that massive mudslides and mud and debris flows were probably imminent in some areas. Yet, they were caught in a dilemma. Current geological capabilities were not yet honed to the point where accurate predictions could be made about which hillsides might slip or where mud and debris flows would actually form. Ironically, one group of geologists was fulfilling the terms of a grant from the National Science Foundation to use real-time aerial surveys to study the movements of L.A. area hillsides during the storms. Their hope was to determine how to predict when and where certain slides would likely occur. For fire and other municipal officials, there were too many unanswered questions: Should evacuations be ordered in areas prone to mud and debris flows and mudslides? If so, who would decide exactly which streets and homes should be evacuated? Would it then be necessary to evacuate the same areas every time rain threatened? At what level of precipitation should evacuations commence?

Without solid answers to these daunting questions, officials could do little more than pre-deploy specialized resources, maintain a high state of vigilance, prepare to order evacuations when actual slope slippage or mud and debris flows were reported, and respond quickly to actual slides and flows.

In Los Angeles and Orange Counties, nearly two dozen swift-water rescue teams were pre-deployed in strategic locations to respond to swift-water incidents, as well as, mudslides and debris flows. Bulldozers, skip loaders, and other heavy equipment were moved into high-risk zones. Camp crews, normally used to cut line during wildland fires, were mobilized to assist firefighters with personnel-intensive emergency operations like placing sandbags, stabilizing levees, shoring collapse areas, and digging through mud to find victims.

The storm struck with full force, causing flooding and deaths from San Luis Obispo, CA to Tijuana (Mexico). In a single day, a number of events occurred. In Santa Maria, 100 miles north of Los Angeles, two highway patrol officers were killed when their vehicle was swept into a raging river that washed out a major highway. Homes were flooded in Santa Barbara and Santa Ynez. The Ventura County Fire Department rescued several people stranded by overflowing rivers. The LAFD performed a number of rescues, including the helicopter rescue of a man and his two dogs from a flood-control channel.

On the same day, the LACoFD performed 35 swift-water rescues in its 3,000 square-mile jurisdiction. In a 24-hr period, a single LACoFD unit, Air Squad 8 (pre-deployed with a helicopter-based/swift-water rescue team aboard), performed 18 swift-water rescues with support from other fire and rescue units. The pilot of Air Squad 8 logged nearly 6 hrs of flight time performing the rescues. An L.A. County Sheriff's helicopter rescued six other people. A man was swept to his death while attempting to assist a neighbor whose car was trapped in an Arizona crossing in San Dimas Canyon. L.A. County Fire and Sheriff's units spent an entire day searching for the victim, to no avail.

In Orange County, all hell broke loose once again. Laguna Beach, which was the site of a major conflagration during the firestorms of 1993, was among the hardest-hit areas. As often happens, the storm reached a crescendo just as dozens of slides, accidents, and debris flows were occurring simultaneously. This is a familiar scenario to firefighters who work in places like Laguna Beach and Malibu. Because the slopes are preloaded to slide by previous storms, and the runoff tends to reach maximum force at many places at once, the number of incidents seems to blossom into a burst of activity that can quickly stretch fire and rescue resources.

The situation was exacerbated by heavy downpours and darkness that impaired visibility and by slides and washouts that caused large areas to become isolated. Moving emergency resources under such conditions can be difficult and time-consuming, even if pre-staged. Adding to the trouble was the fact that many affected areas were mountainous neighborhoods where the only access is winding, single-lane roads on which passenger vehicles are forced to squeeze past one another. Additionally, there was the problem of communications, which can suddenly become overloaded (or be lost altogether) during the height of a disaster.

Attempting to maneuver fire engines and other rescue units to the scene of a mudslide or swift-water rescue under such conditions can prove difficult and dangerous. The situation becomes even worse as residents try to escape in their own vehicles, adding to the traffic. Rescuers must also remain mindful that additional mudslides and debris flows could block escape routes during the course of emergency operations. If it's dark or the weather is stormy, helicopter flight may not be possible, so ground teams are often left to their own devices.

Such was the situation that faced firefighters and lifeguards in Laguna Canyon when the storm reached its crescendo in the darkness of February 23. Just before midnight, residents of Laguna Canyon were awakened by a deafening downpour as the skies opened up with the heaviest rainfall of the day. Then the electricity went dead, plunging them into darkness. In an instant, the mountain above the homes seemed to give way, sending rivers of mud crashing through the walls of apartments and houses.

People were swept from their beds, from their rooms, and in some cases completely out of their homes. Others were trapped inside their homes by mud that rose nearly to the ceilings. As previously described, the effect of a mud and debris flow is not altogether unlike an avalanche or a flood; people are subjected to tremendous forces that are completely beyond their control, sometimes physically ripping loved ones apart.

One woman reported hearing the mountain moving. Instinctively, she and her husband moved to their baby. The mother grabbed the infant from her crib. In an instant, a flow of mud and debris literally came through the walls of their home, flushing mother and daughter through the building. The baby was ripped from her arms and disappeared in the darkness. Still caught in the avalanche-like wave of debris, the mother could not even scream because her mouth was filled with mud. She literally had to scoop mud from her mouth with her hands to breathe and yell for help.

When the flow subsided, the parents were buried in mud, trapped several feet apart, pinned in place by debris and remnants of the building that disintegrated around them. They could hear the baby crying, but neither was able to move toward the sound. They cried for help. Soon, the baby's crying stopped. Fortunately, neighbors rushed in from adjoining apartments to help. One man came across the silent baby completely by accident as he fumbled in the mud looking for survivors; he later described her as a *ball of mud*. Firefighters were able to clear the baby's airway, and the infant survived the ordeal. The parents were eventually freed, both suffering from hypothermia and other injuries.

Elsewhere, the situation was not so good. Two people were missing, and resources were scarce in the mountainous little neighborhood. Mudslides and debris flows were occurring in other sections of Laguna, and dispatchers were unable to keep up with the 911 calls. Firefighters were unable to get through to the main debris flow, and mutual aid was requested from the Orange County Fire Authority. Meanwhile, Laguna Beach lifeguard personnel committed themselves to physical rescue operations at the scene of the biggest debris flow. They were able to access the site in Jeeps. Working in waist-deep mud, in a pouring rain, and equipped with wetsuits and other water rescue gear, six lifeguards began a house-to-house search.

Reports indicate that communications were interrupted or overloaded, and the multitude of emergency calls strained the ability of dispatchers to accurately identify the extent of the disaster. Lifeguards radioed for help, advising *a very large slide* and added that it was growing fast. Throughout the incident, secondary slides and debris flows occurred, threatening the rescuers. Eventually, more than a dozen residents were rescued. Two occupants died in the mud, and their bodies were recovered the next day.

The next day, a local citizen knocked on the door of L.A. County Fire Station 70 to report that the *whole mountain* was beginning to slide above a neighborhood of homes on a hillside decimated by the 1993 firestorms. Engine 70 and Patrol 70 responded to find a 12-ft high, 140-ft long, reinforced concrete retaining wall buckling under the weight of a quarter-mile square slope that was indeed beginning to move. It was nighttime and difficult to immediately determine the boundaries of the potential slide. However, to firefighters who saw concrete spalling from the wall as it creaked and separated, it was evident that homes and roads were in imminent danger. Battalion 5, USAR-2 (a company staffed by this author and other firefighters), and Swift-water 70 (an LACoFD lifeguard division swift-water team) were called in. Soon, a local soil engineer arrived and confirmed the hazard.

Several of the homes at the topmost boundary of the slide had already been yellow-tagged for limited occupancy due to hazardous conditions by inspectors from the Department of Building and Safety. However, others appeared in danger of a massive slide. Battalion 5 established an IAP, including plans for evacuation, alternate access (in case emergency access was required before and after a slide) rescue, and medical treatment. Roadblocks were established by sheriff's units, residents were warned, and fire units began a surveillance of the area as public works personnel assessed the situation.

After significant movement throughout the night, the slide seemed to stabilize without a massive release. For months afterward, dozens of homes were in various states of danger from the slide, and contingency plans were put in place for additional storms that might cause the slide to accelerate.

The Effects of Wildfires on Mud and Debris Flow

Exactly what role did the wildfires play in creating such heightened concern over the possibility of mud and debris flows? The answer lies in the ability of southern California wildfires to coalesce plant life and soil into a coating that repels water. In southern California, chaparral contains high levels of long-chain hydrocarbons (oils) that help maintain internal moisture in the arid climate. Chaparral litter gives up these complexes to the soil as piles of it thicken in the years between wildfires.[4] When intense fires sweep the area, fed explosively by the oil-laden plants, the waxy compounds are vaporized, condensing in a layer a few inches below the ground. The layer of soil just below the surface essentially becomes hard and waterproof during this reaction.

After the fire, dry unconsolidated particles of soil, rock, ash, and other materials are left in thick layers on steep canyon walls. Without plants and roots to hold this material to the slopes, it begins cascading down hillsides in a steady, dry stream. High winds under these conditions have been known to create dust storms comparable to a bad day in the Sahara. Thus, the canyons are *preloaded* for debris flows.

Rain is the trigger that can literally move large portions of mountains. The topsoil quickly becomes saturated with water, increasing the pore pressure just above the hydrophobic (waterproof) layer. The soil liquefies and begins moving downhill. The liquefied soil seeks the path of least resistance (as in a stream), forming small canals called *rills* across the hydrophobic layer. The rills speed the water downslope, sometimes increasing the velocity three-fold (and increasing the transport capacity one thousand-fold). As rills develop across the face of the burned slopes, they connect, forming small tributaries leading to bottom drainages. Massive loads of debris can now be mobilized by relatively small amounts of rainfall. With the onset of intense rainfall, the trigger is set, the trap is sprung, and anyone down canyon is vulnerable to the unannounced arrival of huge walls of mud, rock, water, trees, and sometimes homes and cars.

For reasons still not entirely understood, wildfire burn areas seem to attract tremendously intense rain. The burned areas appear to act as separate microclimates, attracting (or creating) storm cells whose force is concentrated directly over old fire zones. In L.A. County and surrounding areas, this phenomenon has been witnessed on many occasions. The combination of extremely heavy rainfall on fire-scarred terrain creates one type of worst-case scenario (denuded soil) overlaid by another worst-case scenario (intense rain in steep topography).

As noted by McPhee, 1 in. of rain on an area of mountain 10 x 10 miles in area equals approximately 7,232,000 tons of water. Mix that amount of water with millions of tons of rock, soil, and other debris, add the runoff factor caused by the hydrophobic layer, and the stage is set for disaster.

Fig. 4–8 Mud and Debris Rescue Ops in Italy

Managing Rescue Operations at Mud and Debris Flows

Effective emergency response to rescue situations caused by mud and debris flows requires the combination of several related disciplines, including rope rescue, collapse rescue, shoring, roof operations, extrication, swift-water rescue, and other special skills (see Fig. 4–8).

The following is a compilation of basic recommendations for managing these incidents:

- Recognize the conditions that create mud and debris flows, and take appropriate protective actions.
- Understand that the first wave of mud and debris may not signal the end of the event, but merely the beginning. Frequently, multiple waves of mud and debris will follow the first. The ensuing or secondary waves may be larger than the initial flow.
- Do not commit personnel to hazard areas without establishing an upstream spotter to warn of additional waves, a warning system, proper safety equipment, and an escape plan.

- Remember that mud and debris flows (like water) generally follow the path of least resistance. Although they tend to follow streambeds, the large amount of solid material within them may block channels, divert flows elsewhere, and completely change the hydrology of a large area of land in a matter of seconds.
- Realize that a solid object is only 22% of its normal weight when being carried by mud and debris flow, and take appropriate precautions to keep personnel out of danger. Rocks and huge boulders may roll or even float in a mud and debris flow.
- In some areas below major burns, entire neighborhoods are in danger of being buried. If conditions are ripe for mud and debris flows, strongly consider emergency evacuations before these areas are heavily impacted.
- Severe flash flooding, independent of debris flows, is possible in adjacent areas that have not been burned. This may cause other problems for rescuers.
- Flooding and mudslides may completely block access to many areas. Bulldozers, helicopters, four-wheel drive vehicles, and other special transportation will be required. In some cases, all these modes may be ineffective, and hiking in by foot may be required.
- Some events may create life-threatening conditions to rescuers approaching the vicinity. In these cases, all SAR efforts may be delayed until water and mud stops flowing, until hillsides can be secured, until daylight breaks, or until weather conditions abate.
- In some cases, the most prudent action is total evacuation of entire canyons and other endangered areas as a precaution during the approach of major storms.
- Special equipment, tools, and supplies will be required to conduct rescue operations when mud and debris flows occur. Required items may include plywood, screw jacks, pipe, 4 x 4-in. and 6 x 6-in. wood struts, air bags, umbilical air SCBA systems, shovels, buckets, hammers, nails, avalanche poles, swift-water rescue gear, etc.
- Search dog teams may locate live victims if promptly dispatched. Specially trained cadaver dogs can also expedite the discovery of deceased victims.
- There are effective methods to rescue victims buried in void spaces created by mudslides. All personnel should become familiar with the basic game plan for these rescues.
- Fire crews and other sources of personnel will be especially important to rescue efforts.
- Nighttime operations may be especially dangerous. Anyone who has operated at a nighttime rescue operation in the pouring rain knows that visibility and communication may be extremely poor. Personnel efficiency is impeded by wet, cold, muddy (or icy) conditions. Upstream safety personnel may not hear or see a wall of mud and debris until it is too late. Personnel committed to a rescue operation in the danger zone will be even less likely to escape in time if something goes wrong. The most extreme caution should be used for mud and debris flow SAR operations at night.
- Unified command is likely to be required because multiple agencies may be involved.
- A systematic approach should be used when homes and autos are buried. Search thoroughly, document the results, and mark the sites using the standard USAR collapse site marking system.
- A systematic approach should be used when homes and autos are buried. Search thoroughly, document the results, and mark the sites using the standard USAR collapse site marking system.
- For structures buried in mud and debris, remember that victims may be alive inside, even if the structure is completely covered. Access via the roof may be necessary. This may require traditional truck company ventilation/access techniques (cutting through the roof with axes and chain saws). Victims may be trapped in void spaces in the house.
- Remember that live victims may have *floated* to the ceilings as mud entered the structure. Therefore, use only shallow chain saw swipes when cutting through the structure into the living spaces to prevent injuring victims.

- Trench collapse rescue materials and techniques may be required to make access and conduct rescue.
- Access may be made via helicopter or aerial ladder. Place multiple ladders over the mud to spread the weight of rescuers and prevent sinking or plywood for the same purpose. Heavy bulldozers, front-end loaders, and other similar vehicles have also been successful.

Endnotes

1 John McPhee, *The Control of Nature: Los Angeles Against the Mountains*.

2 See W. Wells III, "The Effects of Fire on the Generation of Debris Flows in Southern California," *Geological Society of America Reviews in Engineering Geology* (1987) Vol. VII.

3 All LACoFD USAR companies and USAR task forces are certified swift-water rescue teams at all times. They are trained and equipped to conduct shore-, boat-, and helicopter-based swift-water rescue operations, and they are prepared to make contact rescues in swift-water and flood conditions. They are also equipped and trained in mud and debris flow rescue operations.

4 See DMG Note 33, "Hazards from Mudslides: Debris Avalanches and Debris Flows in Hillside and Wildfire Areas," California Department of Conservation, Division of Mines and Geology.

5

Helicopter Rescue Operations

Helicopters are among the most useful tools ever invented for rescue, firefighting, and EMS operations (Fig. 5–1). Considering the adverse weather, terrain, and unusual emergency situations under which public safety helicopters are called upon to rescue victims, they have an impressive safety record. Because of their proven life-saving capabilities, helicopters are increasingly in demand in the fire and rescue services, and helicopter rescue methods are becoming increasingly sophisticated and effective. In terms of cost-effectiveness for agencies with combined fire, EMS, and technical rescue, helicopters are unsurpassed as aerial resources because of their multi-mission capabilities.

Fig. 5–1 USAR FF and LACoFD Firehawk Crew Practicing Marine Rescue over a Lake

As shown in Figure 5–1, the integration of helicopters and rescue/urban search and rescue has produced very effective hybrids for technical rescue operations. This integrated, systematic approach to rescue is part of the new frontier of technical rescue in the modern fire service. Here, a new version of the military Blackhawk helicopter, called the Firehawk, lands to pick up the crew of an L.A. County Fire Department USAR company, who are trained and equipped as rescue swimmers to be helo-deployed into the ocean during marine disaster operations.

In places with diverse demographics and terrain, helicopters may be called upon to perform a variety of different tasks, all in the course of a single shift. Helicopters are used to drop water or foam on wildland fires, control fire spread at large urban fires, and deliver firefighters to the roofs of burning high-rises. They are also used to rescue trapped victims from buildings, water, cliffs, and mountains. Helicopter crews also perform on-scene EMS treatments of patients in remote areas and fly them to trauma centers. Few other tools can claim such versatility in actual application.

Advantages of Helicopters in the Fire/Rescue Environment

Every day, fire/rescue helicopter teams extract citizens from predicaments in which emergency access, treatment, packaging, and evacuation might otherwise be extremely time-consuming or unsuccessful. Most of these high-risk operations are completed safely, especially when the crews are well-trained for the job at hand and when the evolutions are practiced on a regular basis.

There is an added, difficult-to-define safety benefit to the appropriate use of fire/rescue helicopters: rescues are often accomplished faster and with fewer personnel than might otherwise be required. While not immediately evident, this benefit should be considered when deciding whether to conduct rescues from the ground or from the air.

Hazards Associated with Helicopter Rescue Operations

Even though most fire/rescue agencies operating helicopters have excellent safety records, helicopter rescue teams are exposed to inherent dangers. These should be fairly obvious to anyone who has observed helicopters hoisting victims from the sides of cliffs, performing short-haul rescues (victim and rescuer suspended on a fixed line beneath a helicopter) at floods, or extracting victims from other predicaments. While the overall level of danger is consistent with typical flight hazards, fire/rescue helicopter crews experience periodic hazard spikes. These spikes occur when they are attempting certain maneuvers over rugged terrain, in inclement weather, with poor light conditions, in fire and smoke, or at the scene of chaotic emergencies and disasters.

In Figure 5–2, L.A. County USAR and Air Operations firefighters practice helicopter-based rescues at a specialized training facility at the department's Special Operations Headquarters. In the foreground is a memorial, sculpted and donated by the artist John Cody, in tribute to LACoFD pilot Tom Brady (killed during a mid-air collision during a wildfire) and Air Ops firefighter/paramedic Jeff Langley (for whom the tower is dedicated), a pioneer in helicopter rescue who lost his life during a cliff rescue operation in a rugged canyon in Malibu. For personnel training at this facility, the memorial is a potent reminder of the risks inherent in helicopter firefighting and rescue operations, and a reminder to follow established personnel safety protocols.

Fig. 5–2 The Integration of Helicopters and USAR/Rescue Units Can Greatly Enhance Copter Rescue Capabilities And Create Additional Options (Photo Troy Case)

Although not often understood or discussed by the general public, these dangers are familiar to those who actually perform the rescues. Experienced rescuers understand that the helicopter is a machine that relies on thousands of moving parts working in unison to defy gravity and stay in the air, often in hazardous conditions.

This generic danger is accentuated when rescuers are dangling below the helicopter on a hoist cable or rope and when the helicopter is at a hover, making it vulnerable to adverse events. Consequently, the ever-present potential for mechanical problems cannot be ignored; neither can the danger of rotor strikes, flying debris, heavy wind and rain, darkness, fog, impaired communication due to rotor wash, and other inherent dangers. Obviously, these hazards can be magnified during unusual operations such as helicopter high-rise fire/rescue operations, where strong thermal columns and smoke columns can result in reduced lift and impaired visibility.

Although the danger is slight, helicopter rescue teams also know about the potential for hoist cable failure, rope separation, collision with objects, entanglement in power lines or vegetation, and other equally unpleasant hazards. When suspended below a helicopter, experienced helicopter rescuers know that they and the victims being rescued are in essence a *load* that could be jettisoned. Under certain conditions, to prevent a crash, it's generally accepted that the pilot and crew may be forced to jettison the load by punching a button that will explosively cut the cable, or to take a knife to the rope.

For a rescuer suspended below a helicopter, the decision made by the pilot and crew about whether to release the load is somewhat academic, because either option may have equally unpleasant personal consequences. Obviously, the sudden severing of a cable or rope is likely to result in catastrophe for the rescuer, who will free-fall to the earth. The rescuer's chance for reprieve is limited unless dangling close to the ground or above some feature of the terrain (e.g., trees, snow drifts, etc.) that might dampen the force of impact.

In Figure 5–3, the rescuer (and victim) are legitimately a "jettisonable load" whenever suspended from a helicopter in flight. This is a key point for any firefighter or rescue team member who may be called upon to conduct hoist, short-haul, or rappelling operations beneath a helicopter.

Fig. 5–3 Jettisonable Load: Rescuers Must Understand They May Be "Cut Loose" If the Copter Experiences an In-Flight Emergency

Whenever possible, pilots will attempt to get the rescuer close to the ground before severing the cable, but that may not always be an option—more than likely, it won't be a choice. There have been times in which cutting the rescuer (and in some cases, the victim) loose actually saved their lives. This is exactly what occurred during a rescue on Mt. Hood in May 2002. During a rescue, a Pavehawk helicopter crashed into the mountain seconds after the flight engineer severed the hoist cable, allowing the packaged victim and rescuer to fall harmlessly to the snow. The helicopter rolled 1,000 ft down the mountainside, ejecting the crewmen and rolling over them in the process (see Case Study 1).

In 2001, many observers saw the consequences of rescuers being released in another mishap that occurred during a mountain rescue training session in Europe. Two firefighter/rescuers were being hoisted into a helicopter during a cliff rescue simulation. Without warning, they were unceremoniously released onto the cliff face, apparently the result of the cable being severed from within the craft. The rescuers, still connected to one another via their harnesses, tumbled several hundred feet over rocks and crevices until they mercifully reached a flat spot that stopped their headlong fall. Both firefighters were severely injured and were lucky to have survived at all. This mishap—one of the few ever filmed in such graphic detail—serves today as a case study for some helicopter rescue operations courses (Fig. 5–4).

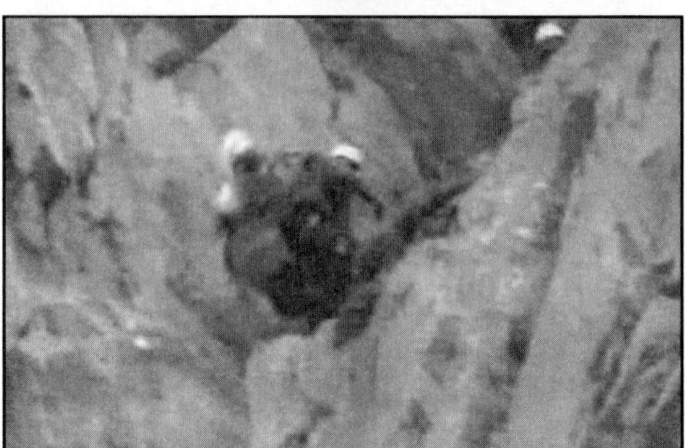

Fig. 5–4 Spanish Firefighters Fall Down Cliff After Being Released by Accident during a Rescue Exercise

This is not to say that rescuers should not be released if the helicopter gets into trouble. To the contrary, the choice between losing one life and losing many lives is simple. The loss of one person is the lesser of two evils.

In addition, there may be unhealthy ramifications for the rescuer if the pilot and crew decide *not* to cut the line when a helicopter gets into trouble. For example, the helicopter becomes entangled in power lines and suddenly free-falls, or an engine is lost and the pilot is forced to make an escape move. If these or some other serious mishap occurs, the rescuer could be dragged through trees, over rocks, or into water (and perhaps even dragged below the surface of the water as the helicopter sinks like a stone). Even if the rescuer reaches the ground safely, he may be tethered to a hoist cable or rope directly below a falling helicopter or exposed to flying debris if the rotor system disintegrates on contact with the terrain.

The bottom line is this: any rescuer suspended beneath a helicopter in an emergency or in training must assume the risk that he and any victims may be sacrificed to save the helicopter, the crew, and anyone on the ground.

Helicopter crewmembers also know that unsecured equipment within a helicopter may find its way out of open cabin doors during hoisting and short-haul operations, creating yet another hazard. If such objects are heavy (e.g., carabiners, medical boxes, etc.), they create a falling-debris hazard for personnel on the ground. Lightweight objects (e.g., rescue blankets, articles of clothing, etc.) may be pulled into the main rotors, creating an obvious problem for those in the helicopter (and for people on the ground if a crash results).

This is not to say that helicopter rescue is excessively dangerous. To the contrary, helicopters are frequently the safest method available to rescue some victims. But the risks associated with helicopters must be considered when developing emergency policies and procedures, when making decisions about methods of rescue, and during the training process. Fire/rescue officials must be familiar with modern helicopter rescue capabilities and limitations.

High expectations bred of success

With such versatility and success come expectations. Movies and other media representations of helicopters fuel the notion that helicopters can perform practically any maneuver, any time, under any conditions. The public has been conditioned to expect that the day is saved when helicopters show up. Increasingly, this expectation is becoming a demand that helicopters be used when things are at their worst.

Sometimes these expectations lead to conflict. Practically every year, there are disputes about whether helicopters should have been employed to rescue some stranded person in the United States. It's become rather commonplace for public officials and the news media to call into question the tactics employed by helicopter rescue crews and other rescuers, particularly after prominent events are televised live.

Such disputes aren't limited to the United States. Consider a situation in the Taiwanese province of Chiayi, where in July 2000, four citizens perished in a raging flood. The situation involved a horribly botched, hours-long rescue attempt in a raging flood on the Pachang River. As of this writing, a police helicopter pilot faces a 10-year prison term, and several firefighters are threatened with 5- to 10-year prison times. The firefighters were indicted on charges ranging from breach of duty to negligence and manslaughter for their roles in the ill-fated, swift-water operation. This situation clearly called for the use of rescue helicopters.

Case Study 1: Pavehawk Helicopter Crash in Rescue on Mt. Hood

On May 30, 2002, the risks associated with helicopter hoisting operations were demonstrated graphically when an Air Force Reserve Pavehawk helicopter[1] augered into the side of Oregon's Mt. Hood. Rescuers were attempting to hoist an injured climber trapped in a deep crevasse. The spectacular crash—and urgent RIO that followed—were filmed by a local news helicopter and aired live on national news throughout the day.

The incident began with several teams of climbers making a bid for the 11,235-ft summit in sunny, calm weather. Each of the climbing teams, which included several off-duty firefighters, was roped together separately on the 65° slope. As they reached a point about 800 ft short of the top, one member of the topmost team apparently lost his footing and dragged his teammates down the slope before they could arrest their respective falls with ice axes.

The lower teams heard yelling and had just enough time to look uphill to see three climbers—still roped together in a tangled ball of bodies and backpacks—hurtling at them. The teams attempted to brace and arrest their falls. One team further to the side was able to arrest, but two teams were knocked down the slope by the impact. Nine climbers were bound into a knot of arms and legs, rocketing down the slope at terminal velocity, heading directly toward a deep yawning crevasse. This crevasse characteristically splits the face of the slope in the same location year after year. In seconds, they shot into the 40-yard long crevasse and slammed into the bottom of the 40-ft deep split in the ice.

One climber struck an ice ledge face-first and was instantly killed, and two other climbers died beneath a pile of bodies at the bottom of the crevasse. The survivors had various levels of injuries. The climbers who had managed to arrest their falls and avoid being struck by the hurtling tumble of bodies now hustled down to the crevasse and were confronted with a horrific scene. Nine climbers—some injured and others dead—were trapped in a deep crack in the face near the top of the highest mountain in Oregon, with a protracted rescue operation required. They were relatively fortunate in that they had good weather, it was early in the day, and one member of the party was able to get a signal on his mobile phone to call 911.[2] Fortunately, several firefighters and a paramedic were in their group, as well as a nearby climber who happened to be a physician.

One of the uninjured firefighters was able to retrieve his rescue pack, and the rescue party began to set up rope systems to get into the crevasse and begin triaging, treating, and extracting the victims from its depths. Meanwhile, local mountain rescue teams were mobilized, as well as helicopters from the Air Force Reserve's 935th Rescue Wing and the Oregon Army Air National Guard.

The air national guard arrived and dropped off mountain rescue team members, who assessed the situation and began working with the climbers to extract the live victims from the crevasse and to prepare them for helicopter evacuation. Once the first victim was packaged and ready, the air guard helicopter came in and flew him back down the mountain for treatment.

The 935 Rescue Wing's Pavehawk was signaled to come in to extract the next victim, who had been packaged in a rescue litter and readied for a hoist operation. As the Pavehawk hovered overhead, the litter was attached to the hoist cable. It was right about then, as the helicopter hovered 20 ft over a group of rescuers, that there was an in-flight emergency and the Pavehawk began wobbling and losing altitude. It practically slid backward down the mountain on a diminishing cushion of air. Fortunately for the victim, an alert flight engineer severed the hoist cable as the pilot struggled to regain control, and the basket dropped to the ground in front of surprised rescuers instead of being pulled down the mountain with the helicopter.

In a scene aired live nationwide, with commentary by the distraught pilot of a local television news helicopter, the Pavehawk's nose struck the slope, its main rotors disintegrating as it began a 1,000-ft tumble down the mountain. The ejected flight crewmembers were flung out ahead of the helicopter, which immediately rolled over them as it continued tumbling down the slope. From all appearances, the pararescuers seemed to be crushed by their own 20,000-lb ship. After agonizingly long seconds, the Pavehawk finally came to rest upside down on a brief upslope that finally interrupted its down-slope trajectory.

The rescuers and surviving victims were fortunate on a number of counts. Despite the potential for a catastrophic fire, there was none. Despite being steamrolled by a 20,000-lb helicopter, both ejected pararescuers survived. Instead of being flung down the mountain at the end of the hoist cable like a thimble tied to the end of a spool of string, the victim in the rescue litter had been cut loose. The fast-thinking flight engineer recognized that the helicopter was losing control and altitude in a place where there was no room to lose either. After being cut loose from the Pavehawk, the victim didn't hurtle down the mountain like a toboggan while secured in the litter. The flight crew wasn't physically trapped inside the helicopter. There was a group of highly-trained rescuers right on the scene when the Pavehawk went down, and they immediately mounted an RIO to extract, treat, and package the injured crewmembers. Additional rescue helicopters were also available to extract the crew from the crash site, as well as, evacuate the original victims who had fallen into the crevasse.

The crash did, however, present a dilemma for the rescuers and the IC, who now had twice as many injured victims, at two separate rescue sites, on a steep mountain slope, with one less helicopter. One result of the crash was a delayed rescue for the original victims, whose rescuers were now contending with two emergencies and the ever-present potential for yet another mishap during the course of the operation. Another effect was that additional rescue crews had to be scrambled to the scene, under conditions that were less controlled, with some of their own colleagues awaiting help on the mountain. Yet another uncertainty was the cause of the crash. Was it related to altitude, too strong wind gusts, or the effect of the wall of snow on the pilot's vision?

In a remarkable demonstration of organization, professionalism, and adaptability (not to mention moment of valor), the rescuers were able to recover from the setback. They reorganized, requested additional resources, and extracted all the Pavehawk crewpersons and the original victims from the mountain before sundown.

This mishap emphasizes a number of lessons that can be applied to all rescue emergencies. Not the least of these is the need for ICs to have a constant backup plan (and a backup plan to that) for those times when things go wrong. In addition, a sufficient number of rescuers should be maintained on the ground and in the air to ensure an immediate RIO if there is a mishap. Finally, ICs must carefully assess conditions when committing firefighters and other rescuers.

In a news story carried by Reuters, a Taiwanese chief prosecutor was quoted as saying, "It was (the firefighters' and pilot's) duty to provide disaster relief, but they were unwilling to carry it out and passed it on to other people." In Taiwan, as in other nations, there is an implied responsibility for firefighters and other recognized rescuers to take calculated risks to save the lives of people in trouble. In some places, the fire department or other public safety organization is specifically designated with primary responsibility for conducting swift-water rescue and other rescue operations. Could American firefighters and rescuers be criminally prosecuted for choosing not to initiate rescue operations under similar conditions? Many experts agree that it's only a matter of time before a criminal case of this nature is brought before U.S. courts.

Clearly, some expectations about helicopters go beyond reason. Helicopters aren't infallible, nor are their crews. Things sometimes go wrong, and when they do, it's often catastrophic for everyone inside or below the helicopter.

However, if conditions are within acceptable limits, a well-trained helicopter crew can perform a number of highly technical rescues. Many rescue helicopter crews are set up to perform hoist operations. Others prefer short-haul systems and fixed-line flyaways. These are both effective for stranded victims.

Reducing Hazards of Helicopter Rescue Training

To reduce the risks associated with helicopter rescue, public safety helicopter providers devote significant resources to ensure the best equipment, best mechanics and pilots, and best-trained crewmembers. Not surprisingly, training is critical for safe helicopter rescue operations (Fig. 5–5).

Fig. 5–5 Helo High-Rise Operations: A USAR Firefighter Preparing to Rappel to the Roof of a High-Rise Building

In Figure 5–5, LACoFD USAR Company members practicing "live flight" helo high-rise team operations beneath a Bell 412 helicopter. Although basic precautions are being taken (e.g., a bottom belay, the copter hovering within 50–60 ft off the ground), there are inherent risks in live-flight training of this nature.

- The firefighters are rappelling on a single rope with no belay (a belay is considered impractical for putting firefighters on the roof of a burning high-rise building in a timely manner with the least complications).
- They are wearing full battle gear for working above the fire in a high-rise structure.
- They are under time constraints due to the nature of the evolution, which emphasizes timely deployment to reduce the "hover time" above a dangerous high-rise fire environment and other factors.

It is for these reasons that the LACoFD conducts some of its helicopter-based training using a three-story steel-frame tower with a stripped-out Bell 412 helicopter (with a working rescue hoist system and other realistic features) mounted atop it, which reduces the need for higher-risk "live flight" training where practical.

Naturally, the process of training within the helicopter environment has inherent risks that mirror many of those found in emergency operations. Training personnel to conduct hoist, short-haul, helo/swift-water, and other rescue evolutions is (to a great extent) reliant on the use of helicopters in flight. More specifically, it often requires helicopters to hover with personnel standing on the skids and/or suspended below the helicopter. In addition, other personnel below often act as victims, in sometimes hostile terrain. Even in a training situation, these may be very unforgiving environments that involve a high degree of exposure to possible flight or fall hazards.

Another less critical but important concern is the cost of training with helicopters. There is a significant hourly cost (including fuel and maintenance) associated with helicopter flight, regardless of whether it is for emergency duty or training. There is also an unavoidable period of downtime for routine maintenance for a given number of flight hours, and a cost associated with that as well. In short, it makes economic and operational sense to limit helicopter flight time where it is feasible to do so without jeopardizing the mission of the helicopter rescue team.

In what ways may helicopter hoist and short-haul training hazards (and the associated costs) be reduced to a reasonable level? One way is to increase classroom training time for three basic tasks: (1) communicating policies and procedures (2) conducting preplanning and blackboard simulations, and (3) critiquing past incidents.

However, classroom sessions are no substitute for effective manipulative training using actual helicopters in the air. While acknowledging the truth of this statement, one might ask: Must helicopters *always* hover in the air under engine power for manipulative training to be effective? The answer, fortunately, is *no*.

Some agencies have found that a combination of in-flight training and the use of *mock-up* helicopters mounted aboveground, provides effective training while solving the problem of excess flight time for routine training and practice sessions.

Helicopter Training Towers

One example of the use of mock-up helicopter rescue props is found at the LACoFD. This department operates a fleet of four Bell 412 and three Blackhawk/Firehawk helicopters as air squads[3] and a Long Ranger for aerial command and control. During a typical week, these helicopters conduct hoist rescue operations in terrain that may include mountains up to 11,000 ft high, coastal cliffs and waters, urban areas (i.e., high-rise buildings) and Catalina Island. During the winter, they conduct short-haul and hoist rescues at swift-water and flood emergencies. Short hauls may also be used in cases where a hoist is not available, or as a backup if the hoist motor fails (Fig. 5–6).

Fig. 5–6 USAR Firefighter Rappelling from Air Ops Training Tower.

To accomplish the goal of effective, repetitive, and safe hoist and short-haul rescue training at reduced in-flight costs and risk, the LACoFD built a tower-mounted helicopter training prop with an operating hoist. In the early 1990s, a group of firefighter/paramedics,[4] pilots, and mechanics assigned to the LACoFD's air operations section wanted more time to practice manipulative skills without *burning up blade time*. They proposed the development of a tower-mounted helicopter prop. The concept went through several iterations until 1994 when the project was green-lighted.

This project began with approval of the site at the department's special operations division headquarters, which also houses air operations at Barton Heliport. It was an extremely cost-effective project. Through the federal surplus equipment program, the LACoFD acquired a surplus Bell 205 helicopter, which was later stripped out by the air operations mechanical staff and painted in the standard yellow/black/white scheme of LACoFD's air operations. Later, Breeze Eastern would donate one of their working rescue hoists. During basic air operations crewperson training, LACoFD firefighter/paramedics practice helicopter-based "pickoffs" on the Air Ops training tower (Fig. 5–7).

Fig. 5–7 Air Ops Training Tower

The tower itself is actually part of a petroleum cracking tower being disassembled and removed from Shell Oil's Carson refinery. Shell Oil, a good neighbor of the LACoFD for many years, was good enough to donate the tower for this project. The tower and its mountings were designed by a battalion chief who is also a licensed contractor. A private engineer was hired by the department to assist with the project. With the engineering completed and the building permit issued, the foundation was laid with embedded connections of the tower, which was later lifted into place with a 60-ton crane.

The hoist rescue-training tower provides an effective means of conducting certain types of repetitive manipulative training sessions while reducing wear and tear on departmental helicopters. It also improves some training safety conditions by eliminating the need to have an actual helicopter hovering in the air at all times (Fig. 5–8).

Fig. 5–8 A Rescue Company Firefighter Training on the Breeze-Eastern Hoist Donated by the Manufacturer and Mounted in the LACoFD Air Ops Training Tower Helicopter

Considering the number of personnel who require primary and continuing hoist rescue training and the number of hours required to support this training, it is clear that the new tower will dramatically reduce the flight hours required of the LACoFD's helicopters. Currently, all air ops-qualified personnel and those permanently assigned to USAR-1 must perform a hoist rescue during training or emergencies at least once every 90 days to maintain their hoist rescue qualification. Although some real-time training and certification sessions require actual hovering helicopters, the training tower can be used to simulate a variety of rescue conditions, including night operations, pickoff rescues from cliffs and burning buildings, and rescue litter operations. Firefighters can practice basic hoist rescue operations in the tower indefinitely without using a minute of "blade time" (actual helicopter flight) before graduating to "live flight" rescue practice, which comes later. This is an extremely cost-effective way to develop and maintain the basic mechanical skills required to conduct these operations, saving the more expensive (and higher risk) helicopter flight time for more advanced training and skills maintenance (Fig. 5–9).

Fig. 5–9 The Risk of Serious Injury or Death during Air Ops Training is Reduced Using the Air Ops Tower

The Use of Helicopter Hoist Rescue Operations for Static Swift-water Rescue Situations

Helicopter hoist rescue operations are a time-honored tool for extracting victims from *static* swift-water rescue situations (where the victim is on an immobile object like a car, rock, tree, high ground, etc.). Fig. 5–10 provides an example of a *static* swiftwater rescue, where the victim is stranded atop an SUV that is not moving, and where she is not being subjected to the forces of moving water.

Fig. 5–10 Hoist Ops in Static Water Situation

However, if the force of water is exerted directly on the victim, standard hoist rescue operations may be more dangerous to the victim, rescuers, helicopter crew, and bystanders. This requires good judgment on the part of the IC and rescue officers. It begs the judicious use of helicopters during swift-water rescue emergencies, matching the actual needs to the available resources without going to more high-risk operations before exhausting lower-risk options.

Load limits

There is a rated load limit for every hoist mechanism. Some common helicopter hoist systems are rated to lift between 460 and 600 lb, with the cable lifting *vertically* (no side-loading). The tremendous forces exerted by moving water pushing on a person's body can easily exceed the rated capacity of helicopter hoist systems by several magnitudes. This raises the potential for immediate (and life-threatening) damage to the hoist cable, motor, mechanism, or airframe support. Failure of any of these components may be catastrophic for the victim/rescuer, and possibly to the crewmember operating the hoist while standing on the skid of the helicopter.

Side-loading

Victims being swept downstream in a river, stream, or flood-control channel are in the grip of the water. This can create dramatic side-loading forces that can adversely affect the helicopter hoist system. If excessive side-loading is exerted on the hoist arm, it may upset the delicate balance of the helicopter and cause a catastrophic phenomenon known as *dynamic rollover*. Because the helicopter is generally hovering close to the ground during swift-water rescues (with a cable and victim/rescuer attached to it, no less) there is little or no room for recovery if something goes wrong. Yet another danger is related to the potential for the hoist cable to snag on heavy debris in the water, or on trees, bridges, or other objects.

Helicopters routinely conduct standard hoist rescue operations to pluck victims from houses, rocks, cars, other another predicaments where the force of moving water is not pushing them). This is perfectly acceptable and is often preferred because the rescuer and victim can be immediately hoisted back into the cabin of the helicopter. If there are multiple victims, the rescuer can be lowered again to pluck others from additional static swift-water predicaments.

Short-haul systems for swift-water conditions

Helicopter *short-haul* rescue generally works like this: A rope or cable is attached to an electrically-operated hook mounted to the underbelly of the helicopter, or to a "belly band" that encircles the floor of the cabin and the belly of the helicopter. A rescuer (and/or a rescue device for the victim) is attached to the end of the cable or rope. The pilot flies the helicopter to a point at which the rescuer can snag the victim, or where the victim can climb into the rescue device. (Fig. 5–11).

Fig. 5–11 Rescuer Being Lowered on Hoist

In Figure 5–12, a helicopter short-haul operation is adapted to the swift-water and marine rescue environment. In this particular evolution, the rescuer had been lowered out of the hovering helicopter into the water, where he connected the victim to his system using a Cearly Rescue Strap®. Rescuer and victim were then lifted from the water by the helicopter and were short-hauled to the shoreline. In a marine disaster, the victim could be short-hauled to a rescue boat, allowing the rescuer to return and retrieve more victims without disconnecting from the short-haul system.

Fig. 5–12 Short-haul Operation in Swift-water

Helo/swift-water short-haul rescue operations. If a victim is being swept downstream in fast-moving water, helicopter rescue becomes more complicated and dangerous. In the early 1990s, the LACoFD pioneered a new variation of the short-haul system for pulling victims from moving flood channels. This method is now being used by a number of fire/rescue agencies.[5] Using a rope-lowering system rigged inside the helicopter, a crewman standing on the skid lowers a rescuer below the helicopter while the pilot matches the victim's water speed. The rescuer is "dipped" into the water, and secures the victim with a nylon strap known as a Cearly Rescue Strap®, developed by an LACoFD helicopter pilot in 1993. The pilot then lifts them out of the water while still matching the water speed.

This is a sort of "fixed-line flyaway" performed in the water. In a variation of this evolution, a rescuer enters the water, contacts and connects the victim to the same rescue harness, and then floats with the victim as the helicopter flies overhead. A crewman on the skid lowers a fixed rescue rope to the rescuer in the water as the pilot matches the water speed. The rescuer connects himself and the victim to the line with a large locking carabiner. Then the pilot ferries the rescuer and/or victim (dangling beneath the helicopter) to a safe location, where they are lowered to the ground beneath the still-hovering helicopter.

Either of these evolutions may also be performed in the ocean or lakes, with adjustments made for the obvious difference in environments. For example, although ocean currents can be fairly strong, they are not moving at the speed of a flood control channel. Further, most lake and marine rescues involve victims in waves or calm water, or possibly ice.

In cases where a rescuer cannot immediately be placed in the water with the victim, another device called the Cearly Cinch Harness® may be lowered from the helicopter to provide immediate flotation for the victim. The device is designed so that victims will automatically understand it is intended for them to slide the looped device over their head and arms. The helicopter crew can also signal for victims to do just that. Once the device is over the head and arms, the crewperson snaps the line up, releasing a velco strap that allows the harness to cinch securely around the victim's torso. Once the victim is suspended in such a manner, it is virtually impossible for him or her to slip out.

In Figure 5–13, an Air Operations crewperson demonstrates how the Cearly Rescue Strap can secure a victim within seconds for a short haul or hoist lifting operation. In Figure 5–14, an LACoFD Air Ops firefighter/paramedic demonstrates the use of the Cearly Cinch Harness for securing and lifting a victim from the water, from the side of a cliff, or some other stationary location. Figure 5–15 demonstrates how tugging the attached rope causes a velco strap to release, thereby allowing the harness to cinch around the victim's torso and preventing accidental release until the victim is placed back on solid ground (or the deck of a boat).

Fig. 5–13 Securing a Victim

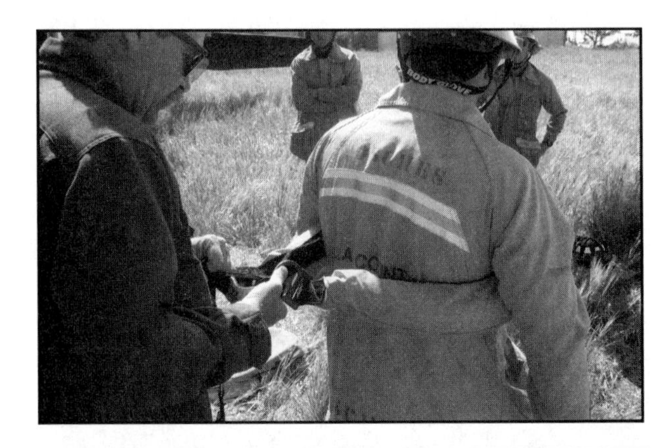

Fig. 5–14 Securing a Victim in a Stationary Location

Fig. 5–15 PPE for Helo/Swift-water Rescue Ops

All personnel assigned to LACoFD rescue task forces and swift-water rescue teams are certified in these methods. These swiftwater-helo trained personnel are also assigned to the LACoFD's FEMA/OFDA USAR Task Force for domestic and international response (which adds another level of options when they respond to water and storm-related disasters like hurricanes, cyclones, etc.). Annual recertification is required, and the maneuvers are practiced often in a fast-water aqueduct channel. Because this operation carries a higher level of risk for the rescuer and the helicopter crew, it is generally attempted only when shore-based rescue methods have proven unsuccessful or when the victim's condition or predicament is such that an immediate helicopter rescue attempt is warranted. For an IC, the availability of these capabilities is an important tool and may also be used to rescue a firefighter in trouble.

One requirement for short-haul operations is maintaining sufficient altitude to avoid dragging the rescuer/victim through treetops or other objects. This skill requires a certain amount of experience and training to maintain a safe margin, not to mention excellent depth perception. It can be difficult for the pilot to ensure adequate clearance when poor weather and darkness reduce visibility and depth perception or when wind and rain are buffeting the helicopter. A crewmember is usually assigned the job of acting as a second set of eyes to observe the rescuer/victim at the end of the rope and to communicate with the pilot to ensure adequate clearance over obstacles. At night or in poor weather, even with night-vision goggles, a crewmember's depth perception may also be impaired, adding to the potential danger of short hauls.

Dynamic swimmer free evolution. For this method, the rescue swimmer enters the water independent of the helicopter. After the swimmer contacts and secures the victim using a Cearley strap, the helicopter, which has been tracking the rescuer, lowers the short-haul rope. The rescuer and victim are clipped onto the rope, and the helicopter (still matching their speed in the current) lifts them from the water and deposits them on shore. Figures 5–16 through 5–19 show this evolution.

Fig. 5–16 Dynamic Swimmer Free Evolution—Rescuer Captures Victim While Copter Sets Up

Fig. 5–17 Rescuer Secures Himself and Victim to Capture Device at End of Rope, While Copter Continues to Move Downstream

Fig. 5–18 Ready for Liftout

Fig. 5–19 Rescuer and Victim Being Extracted

As shown in Figure 5–16, the pilot of this LACoFD Sikorsky Firehawk (the civilian version of the military Blackhawk) hovers downstream of the victim. This facilitates downstream flight as the victim is swept past the rescuer, who will quickly follow to secure the victim and make the rescue. In this evolution, the rescuer (usually assigned to a ground-based swift-water team or USAR company) enters the water from the shoreline instead of from the helicopter. He will rendezvous with the helicopter via the short-haul line. The crewman is verbally guiding the pilot to move river-left or river-right to line up with the approaching victim and rescuer. The rescuer has captured the victim with the Cearley strap and is preparing to hook both himself and the victim to the short-haul line.

Figure 5–17 shows the crewman dangling the capture ball (a stainless steel cage) within reach of the rescuer. The rescuer connects the victim and himself to the short-haul line by snapping a carabiner (attached to the Cearley strap) to the capture ball. Simultaneously, he snaps himself to the capture ball using a carabiner attached to a rescue strap connected to the rescue harness beneath his PFD. The crewman directs the pilot to begin ascending while still moving downstream to take up slack until the rescuer signals he is ready to be lifted from the water (usually just an instant after he hooks in to the capture ball).

Once the rescuer signals he's ready for extraction (Fig. 5–18), the pilot lifts the helicopter until the crewman reports the rescuer and victim are free of the channel and any obstructions. Next, they may be deposited on the shoreline. If there are multiple victims, the rescuer may disconnect the victim and remain connected to the short-haul line. Now the helicopter can lift off and place him in position to attempt another rescue. The dynamic swimmer free evolution may also be used to rescue victims and rescuers who have passed shore-based rescue points and whose last resort may be helicopter-based rescue. In Figure 5–19, the rescuer and victim are short-hauled to dry land and slowly placed on dry ground (or, in the case of a boat, the deck).

These evolutions require pinpoint accuracy and a steady hand (and feet) on the controls. The victims directly beneath the aircraft are not readily visible, and therefore "real time" communication between the pilot and the crewperson standing on the skid is critically important to keep the copter in position to allow the rescuer to shag and secure the victim (who may be in water moving 20–40 miles per hour).

Dynamic tethered rescuer evolution. The rescuer is lowered from the helicopter on the short-haul line. When he is dangling about 30 ft below the helicopter, the crewman locks the rope off at the anchor. The pilot, guided by the crewman, matches the speed of the victim and lowers the rescuer into the water. The rescuer contacts the victim, wraps a Cearley strap around him, and signals to be raised. The helicopter, matching the water's speed, lifts the rescuer and victim to the shore. Figures 5–20 through 5–24 show the dynamic tethered rescue evolution. In Figure 5–20, a rescuer is attached to single rope (short-haul) system, which has been weaved through a brake bar rack that will be used to lower the rescuer when the copter reaches its hover position over the water.

Fig. 5–20 Short-haul System

Fig. 5–21 Rescuer Lowered from the Helicopter

Fig. 5–22 Rescuer Nearing the Water

Fig. 5–23 Rescuer in Rescue Position, Searching for a Victim

Fig. 5–24 Rescuer Snagging Victim and Securing Him with a Cearly Rescue Strap®

In Figure 5–21, the rescuer is lowered from the helicopter on the short-haul line by the crewman, who uses a brake bar rack as the friction device. To avoid shock-loading the rope, the rescuer tilts backward and maintains his feet on the helicopter's skid or step until he completely inverts (i.e., goes below the skid), and then he removes his feet. In Figure 5–22, the rescuer is lowered to the water level while the copter maintains its hover position downstream of the victim (who is fast approaching the rescue point. The pilot of this LACoFD Bell 412 helicopter prefers to face *upstream* (allowing him to visually line up on the victim until he passes beneath the nose). This is simply a judgment call based on the personal preference and experience of the individual pilot. This upstream-facing approach will require the pilot to drop and go. In other words, he will have to make a rapid downstream turn while simultaneously descending to provide just enough slack in the short-haul line for the rescuer to make contact with the victim. Once the rescuer has inverted and removed his feet from the skid, he is lowered to the surface of the water, at which time the crewman locks off the brake bar rack.

In Figure 5–23, at the instant the victim reaches the rescuer (or a moment before this point) the pilot "drops and turns" and begins following the victim downstream while the rescuer makes a contact rescue and secures him with the Cearley strap. Once the rescuer connects the Cearley strap to the short-haul line, he signals that they are ready for extraction. Now the pilot gently lifts the rescuer and victim from the water while moving downstream.

Static swimmer free evolution. In this evolution, the rescue swimmer (sometimes deployed from the helicopter, sometimes from shore or a boat) makes his way to a victim stranded on a stationary object or in a pond or lake (no current). He attaches a Cearley strap to the victim and signals for the helicopter. The helicopter hovers overhead and the crewman lowers the short-haul rope. The rescuer connects himself and the victim to the rope, and both are short-hauled to safety. Figures 5–25 through 5–27 show a static swimmer free evolution.

Fig. 5–25 Deploying a Rescuer into a Lake

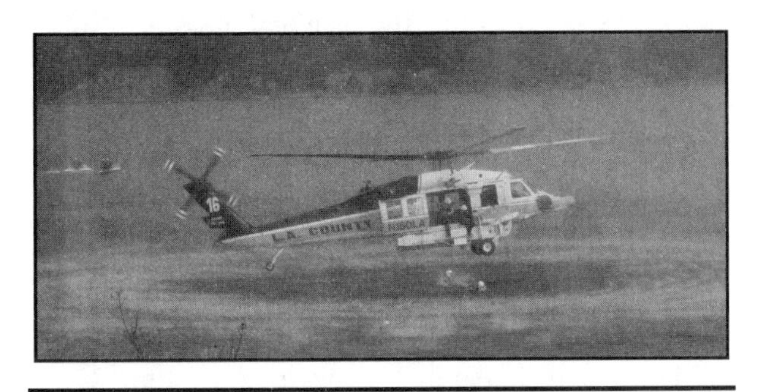

Fig. 5–26 Rescuer Connects with Victim and Signals for Extraction

Fig. 5–27 The Rescuer and Victim Short-hauled from Water

Case Study 2: HSW Extraction

Within one year of the 1992 Rubio Creek incident (Case Study 1), the new helo/swift-water rescue (HSWR) methods developed by the LACoFD would be tested twice on the same day. The events occurred when a heavy rain storm slammed into Los Angeles in March 1993, causing more than a dozen swift-water rescue incidents.

As it happened, this author was assigned to a HSWR team pre-deployed in the mountainous northwestern part of L.A. County that day. On the second or third swift-water rescue response of the day, we were dispatched to what became the first incident where the new LACoFD HSWR methods were used.[6] The victim was successfully rescued, proving that the methodology and equipment were in fact effective for conducting HSWR.

This rescue offers a case study not only of the use of helicopters to rescue trapped victims, but also a look at the unique ability of helicopters to locate and rescue victims submerged in water.

This successful swift-water extraction, and the many more that followed, also helps bury a long-held misconception among some swift-water rescue instructors that helicopter rescue should be considered only as a last resort. These instructors, who number many and are scattered across the nation, use an adage that explains their position: *Reach—throw—row—go—helo*, implying that helicopters should be considered for swift-water rescue only when reaching assists, throwing assists, boat-based rescues, and contact rescue have proven unsuccessful. Their view holds that the very nature of helicopters (one million moving parts, each reliant on the other to keep it in the air) makes them the most hazardous—and therefore the last—option. To support their position, they cite a long history of air crashes during swift-water rescue missions.

Unfortunately, the *helicopter as a last resort* school does not consider the fact that many air crashes during swift-water rescue incidents involved crews that did not specialize in HSWR operations. These crews had no formal training in understanding and evaluating the dynamics of moving water and had not trained with the fire departments and rescue teams on the scene. It also doesn't account for military, media, and public safety helicopters sometimes pressed into service without specialized training, experience, or equipment. Finally, the helicopter as a last resort view is often colored by the fact that the instructors themselves are not familiar with helicopter operations, do not have experience actually performing HSWR operations, and do not regularly work with agencies that do.

Beyond that, the incident about to be studied highlights some previous points about incident command and managing swift-water rescue operations. Typical of many rescue operations, this emergency occurred in a remote location away from lights and news cameras, and the only witnesses were the victims and the rescuers. In the next day's newspaper, mention of this incident would be limited to two paragraphs in a larger story about the other swift-water rescues that occurred that day.

The victim, a baby boy who was swept from his father's shoulders as the man and his wife attempted to wade across a flooded mountain creek, didn't survive. There were no heroes that day—just firefighters and other rescuers doing their job with their usual sense of purpose and innovation on a stormy day in March.

As the incident unfolded, the baby's parents had become stranded in a picnic area on the wrong side of a stream. With the water continuing to rise and darkness coming, they figured now was their only chance to avoid spending the night out in the open in a storm. The father placed the baby on his shoulders and led the way. As the water rose above his waist, the man stepped into a hole, lost his footing, and was swept beneath the water. In an instant, the boy was gone, swept into a rugged canyon inaccessible to ground vehicles except in a few points where a major mountain road passed close by. The mother was somehow able to get across the creek, and she went for help while the father began scrambling downstream in a panic to look for his missing son.

The 911 call was transferred to the LACoFD, which dispatched a standard first alarm water rescue response consisting of five engine companies, one truck company, one ground-based paramedic squad, one battalion chief, the closest USAR company, the closest ground-based swift-water rescue team, and two fire/rescue helicopters (each staffed with HSWR teams for this storm). The L.A. County sheriff's department was also notified to dispatch units, including the closest volunteer mountain SAR team (as per written understanding between the LACoFD and the sheriff's department).

The baby was missing somewhere in a two-mile stretch of rugged, cliff-lined stream flowing through the mountains to the upper end of Lake Castaic. Flying directly to the scene in a Bell 412 helicopter, we arrived in the area at virtually the same time that first-due Engine 149 got to the road that crossed the creek to the picnic area.

Engine 149 established incident command and named the emergency the *Creek Incident*. Their captain would serve as Creek IC. Next, they gave a size-up report to the fire command and control facility (FCCF) and assigned the airborne swift-water team leader (this author) as search group leader. Finally, they directed us to follow the stream toward the lake, where the baby's father had been heading. Engine 149 would serve as the command post at the river crossing.

We acknowledged our assignment on the radio, and pilot Rick Cearley began flying downstream through the steep canyon. He followed the twisting stream, brimming with whitewater from the rain draining from the steep mountains behind us. Both cabin doors were open, with air crewmembers and HSWR team members standing on the skids, scanning the stream for some sign of a victim or his father.

The rain had let up some and was basically a hard sprinkle that allowed fairly good visibility below the low-hanging clouds. After a couple of minutes, we spotted the father flagging us down from a sandbar where the creek drained into Lake Castaic. Cearley landed the helicopter on a sandy shoal and kept the blades turning while I exited the helicopter to talk with the distraught man. Meanwhile, the others readied their equipment and PPE for a potential HSWR operation.

"My baby's in the stream . . . up there!" exclaimed the man, pointing upstream into the steep canyon.

His look of inconsolable anguish expressed the gravity of the situation and eliminated any doubt that this was going to be a full-blown SAR operation. Seconds counted, and maximum effort was required. I asked the man what happened, where he last saw the baby (the PLS), when the accident happened (about 25 minutes ago), and what the baby was wearing (a white jumper).

I instructed the father to remain where he was and explained that another helicopter (from the sheriff's department) would soon land to get more information from him and reunite him with his wife. Normally, we might take the father with us to point out exactly where he lost his son, but we had the PLS. After considering the weight and room limitations inside the helicopter, we determined that it was better to leave the father on the ground for now and proceed with our rapid aerial search.

At this point, with half an hour gone since the mishap, the baby's best chance was for one of the HSWR teams to spot him quickly, extract him, and fly him directly to the closest trauma center. The baby would need a center capable of dealing with pediatric patients or the closest receiving hospital if found in full arrest.[7] The next best chance was for a ground-based team to locate and extract the baby and to transfer him to the closest helicopter for a direct flight to the hospital.

We lifted off, leaving the forlorn father standing alone on the sand bar to watch us depart into the canyon. I radioed Creek IC with a brief explanation of the status of the victim as we knew it and a request for one additional ground-based swift-water rescue team to help search the stream. I also requested that the closest sheriff's department helicopter be dispatched, preferably one with forward-looking infrared (FLIR). After removing the father to a safe zone, the extra copter could help us with the search. Finally, I requested that the firefighters and engineer from engine 149 begin searching the shoreline. We would mount an airborne search until the other units arrived. Creek IC copied the information and began making requests through FCCF for the additional swift-water team and a sheriff's department helicopter.

As Cearley began flying a search pattern up the canyon, the two firefighter/paramedic air crewmembers aboard the helicopter scanned the river beneath cloud-shrouded mountainsides.[8] We knew from years of experience that it's often possible to spot submerged victims from the air, even when shore-based and boat-based personnel can't see them.[9] So that was part of our search strategy: to detect that one telltale sign of the baby beneath the surface of the water that might lead to a rescue. My partner and I were already in our drysuits, PFDs, rescue harnesses, and other PPE, and we were setting up to conduct a tethered rescuer evolution if the baby was spotted. In this case, based purely on a regular rotation of duties, I was designated the primary rescuer, with my partner as the backup rescuer.

The plan was to lower me from the cabin of the helicopter on a single-line rope system, much like being lowered over a cliff. An air crewmember by the name of Bobby Fullove would control the brake bar rack that was our friction/lowering device for this evolution. The idea of this evolution is to *dip* the primary rescuer (me) into the water, allowing him to grab the victim with a special rescue tool known as a Cearley strap (designed by the very pilot who was at this time flying the helicopter).[10] If problems developed, the backup rescuer (my partner) could be deployed via the hoist cable or another rope system (or through the use of a one-skid or hover-step). However, for now, there was no sign of the baby. The search was on in earnest now, soon to be assisted by the ground units.

As we passed engine 149, located just upstream from the PLS at a well-known road crossing, Cearley turned the helicopter around and we began a downstream search. Cearley spoke with the pilot of the sheriff's department helicopter that had just arrived, requesting him to locate and interview the father for more detailed information while moving him to the command post. Meanwhile, I coordinated search operations with Creek IC and the various ground units. We also divided the incident into divisions and groups.[11]

Within minutes, the second LACoFD helicopter arrived (also staffed with an HSWR team) to help with the search. We were soon joined by the sheriff's department patrol helicopter, which had airlifted the boy's father to the command post. Over the aviation frequencies, the pilots coordinated their movements to establish a round-robin system whereby they would follow each other in a search from upstream to downstream, maintaining sufficient air space between them.

As each downstream-moving helicopter reached the lake, it would peel off to river-left (see chapter 3 for details on directional orientation in the swift-water rescue environment). Since the river-right side of the stream was bordered by a 400-ft high cliff, peeling off to river-left was the only viable option. Each helicopter would make its way back to the upstream boundary of the search area (the point where the father and baby were separated during the crossing attempt) and then begin working its way downstream again. We would continue this until someone found the baby or until it became necessary to break off to refuel, then return and continue as necessary.

By now, other first-alarm units, a sheriff's unit, and members of the local mountain rescue team were beginning to filter into the staging area designated by Creek IC. The battalion chief arrived, and Creek incident command was transferred to him. He immediately established unified command with the sheriff's department, which also had jurisdiction over the mountain rescue team. Engine 149's captain became operations. On our recommendation, Creek IC designated the first-arriving, ground-based swift-water rescue team as the search group leader, taking that responsibility off us and allowing us to concentrate on the airborne search operations.

By this time, firefighters from the engine and truck companies (equipped with sneakers, PFDs, and other PPE) were physically searching the shorelines and the eddies and pools, using pike poles as sounding/reaching tools, chain saws to cut off offending branches, and shovels as scooping tools. Firefighters from the ground-based LACoFD swift-water rescue teams were entering the stream to physically search every eddy, hydraulic, and strainer for the baby.

The search was nearing the first hour, and it seemed like the race against time was being lost. With the water temperature hovering in the 40s, the baby would have a chance of cold-water drowning resuscitation if he could be found soon. In the helicopters, we were becoming increasingly frustrated by our inability to spot the baby. We could see through the water to the creek bottom in many places. However, heavy vegetation overhung much of the creek's shoreline, making it difficult to see anything—including a baby's body—in the current.

The other fire helicopter crew reported to us that they were having some luck by bringing the helicopter down to a close hover over the water, which had the effect of pushing back the water and vegetation. (We might even use the term *parting the water* because that's what it was doing.) This revealed areas of the stream that weren't previously visible.

Cearley began employing the same technique, and sure enough, we found that the upper layers of moving water *peeled back* from the rotor wash. Not only that, but many of the tree branches and chaparral on the water were pushed back. The sheriff's department helicopter started doing the same thing. All of us were ready to try practically anything within reason at this point to locate the baby.

Standing on the skid with firefighter/paramedic Bobby Fullove, my harness connected to the anchor point in the helicopter with a pickoff strap, I was totally concentrating on the water below, praying to see some indication of the baby. I think we all were doing the same.

Suddenly the other LACoFD helicopter radioed that they had spotted something in the water. They stopped and hovered over the spot where firefighter/paramedic Bill Monahan (standing on the skid of the other helicopter) had seen a flash of white in the water. Cearley moved our helicopter into position behind them and hovered. From the tone in Monahan's voice on the radio, I knew he had the child's position in sight. We got ready to go into the water.

As the other helicopters hovered close over the stream, blowing back water in the process, Monahan had seen the flash of white T-shirt as the baby's arm flopped up in the rotor wash. Monahan thought the child might be wedged between a rock and a tree in the middle of the streambed, about 2 ft beneath the surface.

The primary rescuer on Monahan's crew was still getting his wetsuit on, so the pilot of his ship radioed that they would slide forward to allow us to attempt the rescue. Monahan and his crew kept their eyes on the point where they saw the baby as they slid forward, and they radioed instructions to us as we crept up over the spot.

"Stop!" yelled Monahan when we hit the exact spot where he had seen the baby.

As the rotor wash pushed back the upper layers of water, I also saw the flash of white in the rushing stream. So did Fullove, who yelled to Cearley, "Hold right here!" Cearley kept a steady hover as Fullove prepared to lower me out on the short-haul rope.

Wearing a PFD, drysuit, goggles, and swiftwater rescue helmet, I was already connected to the rope by my rescue harness. After confirming that we were directly over the spot, Fullove lowered me off the skid.

The way to depart the helicopter in this evolution is to lean back while the air crewman lowers you with the brake bar rack. You keep your legs straight, slightly bent at the knees, with your feet planted on the skid, until you go perfectly upside down, looking at the ground (or the water). At this point, you remove your feet from the skid and gently swing back upright. This prevents you from *shock-loading* the single-rope system and possibly damaging the only thing that's keeping you from free-falling to earth. It's a bit unnerving at first, particularly if you're high in the air, but it's really no more dangerous than being lowered off any other object into clear air.

When I reached a predesignated point about 30 ft below the helicopter, Fullove locked off the brake bar and directed Cearley to descend until I was in the water. As I was lowered into the water, I was able to get my feet under me in the swift current, partially in the eddy formed by the tree. Once again I could see the white jumper flashing in the current.

I put my face into the water and groped around until I felt the baby's body. Sure enough, he was wedged between a rock and the tree. With just a little effort, I was able to free the baby and immediately I surfaced with the child in my arms. The baby's head showed signs of trauma from being swept nearly a mile downstream, and he was blue from hypoxia and cold. We had agreed before I left the helicopter that this was going to be treated as a cold-water near-drowning and worked as a full code, even though the baby had been submerged more than an hour.

I tucked the injured baby in one arm and with the other arm signaled that I was ready to be raised from the water. I quickly placed the Cearley strap around the baby's chest[12] and cinched it, clipping the strap to the figure-8 on a bight at the end of the rope on which I had been lowered into the stream. (My rescue harness was carabinered into the same bight, which is standard for the tethered rescuer evolution.)

Under different circumstances, I might simply have held the baby in my arms without the strap; he was that small. But I was about to be lifted into the air, and being suspended at the end of a short-haul line, meant that Cearley would have to find a suitable place to set me down. This meant I was going for a ride because we were at the bottom of a deep canyon without any clear spot in sight. The pilot needed to take me up-canyon to find a place to put me on the ground before I could climb into the cabin with the baby. I didn't want to chance losing my grip on the baby if something unexpected happened while I was performing CPR beneath the helicopter.

I came out of the water and was suspended in midair as Cearley lifted the helicopter higher above the creek and slowly started moving up-canyon. He was cognizant of the need to prevent the load (me and the baby) from beginning to pendulum below the helicopter, and his expert touch and control were evident as we gradually accelerated.

Fullove stood above me on the skid, harnessed and connected to the anchor points in the cabin with a pickoff strap. He alternately looked forward for obstacles and then down to ensure I was clearing the treetops and rocky outcroppings with sufficient space between me and the objects. Meanwhile, I was performing CPR on the baby, trying not to look at the terrain passing far below. By now, we were spinning very slowly below the helicopter. It was a slow, comfortable turning, typical of the wind and flight conditions.

In the cockpit, Cearley and the other air crewman scanned the terrain for a suitable set-down spot and for wires. Sure enough, they both spotted a set of high-tension wires draped from rim to rim across the canyon, no less than 400 ft above the creek. Cearley ascended to about 500 ft for his cargo (the baby and me) to clear the wires with an adequate safety margin.

The upward movement of the helicopter put a certain amount of tension in the rope, and the upward movement caused me to look up from the baby to see why we were suddenly ascending. I saw the high-tension towers off to one side of the canyon, stretching into the clouds on the mountainside, and knew instinctively that Cearley was taking us over a set of power lines stretched across the canyon.

There was no cause to look around any further, so I went back to the CPR as the terrain rushed by in my peripheral vision. I was dangling like a tetherball below the helicopter, unable to control my trajectory, completely vulnerable to whatever might happen next.

If Fullove's depth perception proved faulty, or if Cearley flew too low, the power lines would cut through the single ½-in. short-haul rope, and I would drop without warning into the bottom of the canyon, still holding the baby. The thought ran through my head (as it sometimes does under similar circumstances) that if we fall, at least it will be over fast. The terrain below was sufficiently rugged to pretty much ensure instant death, unless something soft (like a tree top) happened to break the fall. I continued concentrating on doing good CPR, keeping the baby's airway open, hoping that somehow we'd found him in time to make a difference. The power lines passed far below me and then, after a few more seconds, Cearley started a slow descent.

Apparently Cearley and the air crewman in the cockpit had spotted a place where they could set me down. The helicopter slowed gradually, and I could feel myself pendulum forward ever so slightly as I kept up the CPR. Eventually Cearley brought the helicopter to a hover and began to descend. Now I started to spin a bit faster, the rotor wash coming straight down and turning me a bit faster every second. It's something to be expected, no cause for alarm.

After another 15 or 20 seconds, I was spinning pretty quickly, and the ground was coming up. I knew that Fullove must be telling Cearley that I was spinning fast, that it would be important to get us on the ground pretty soon to prevent me from going around even faster. He did so, and we made the transition from a free-hanging spin to a stationary position on the ground. Then I stepped off to the side and pulled the slack from the rope to allow Cearley to set the helicopter down. Fullove helped me into the helicopter as I continued CPR, and Cearley immediately lifted off, heading for the trauma center. Fullove closed the cabin door and disconnected the rope system from me once I was seat-belted in.

In moments, we were scudding over the lake, the firefighter/paramedic in the cockpit radioing the hospital to announce our imminent arrival so they would have a team of personnel waiting in the emergency room. Some six or eight minutes later, Cearley was setting the helicopter down, and Fullove was sliding open the cabin door. He helped me out of the helicopter, and we handed off the baby to the emergency room staff, who continued to work furiously on him.

Tragically, the baby did not survive the ordeal. His death was attributed to a combination of head trauma and drowning.

In terms of search and extrication, the units that responded to this tragedy performed their jobs well, conducting the operations according to their training. The new HSWR evolution proved itself in this first-time use (it worked equally well in the Malibu swift-water rescue, which occurred at the same time). As often happens, the crews were innovative, using helicopters to part the water to increase the effectiveness of the search efforts. It's questionable that we would have found that baby in a timely manner using other methods.

In the years since this rescue, the new HSWR methods and equipment developed by LACoFD firefighters and pilots, as well as other agencies, have improved the way helicopters are used for water rescue, saving victims who might otherwise have perished. As time marches on and new innovations are developed, researched, tested, and shared, more victims will benefit from these improved methods. Firefighter/rescuer safety will also be improved, allowing them to carry on the tradition of helping others.

Static tethered rescuer evolution. The rescuer is lowered from the helicopter on the short-haul rope to a victim who is stranded on a stationary object or who is in a pond or lake (no current). The rescuer places a Cearley strap on the victim, attaches the victim to the rope, and signals to be lifted to safety.

Cinch harness rescue evolution. As mentioned earlier, the Cearley cinch harness is a victim-capturing device designed for static and dynamic helicopter rescues, including swift-water rescue. It is lowered from the helicopter at the end of the short-haul line to a victim. The victim places the cinch harness over his head and slips it below his armpits. The crewman standing on the helicopter skid jerks the rope to dislodge the Velcro on the harness, which allows the harness to cinch up on the victim. The crewman, talking on a hot mic (voice-activated microphone), instructs the pilot to raise the victim from the water. The victim is short-hauled to shore (or, in some cases, to a boat). If the victim is injured, unconscious, or is a small child, a rescuer may be lowered on the same rope to place the victim in the harness.

In addition to specialized harnesses, HSWR requires a variety of specialized equipment, including the following:

- Wetsuit or drysuit with reflective tape
- Swift-water rescue helmet with waterproof light and reflective tape *H* marking on top
- PFT with rescue knife, strobe light, and whistle
- Low-profile pelvic and chest rescue harness beneath the PFD
- Rescue strap from the harness up through the PFD to a carabiner for connection to the short-haul line
- Proper swift-water footwear
- Two Cearley straps with carabiners (one for backup in case the first strap is dropped during the rescue)
- Swift-water/rescue gloves and goggles

Of course, the most important piece of equipment the rescuer wears is beneath the helmet: a prepared, well-trained mind.

Case Study 3: Rubio Creek Rescue Ignites Efforts to Develop HSWR Methods

Among the most dramatic examples of the need for advanced HSWR capabilities came during the rescue of two children in the Rubio Creek flood-control channel in L.A. County. This 1992 El Niño incident occurred within days of other unsuccessful HSWR attempts by several southern California fire departments and several high-profile swift-water incidents resulting in fatalities. It served as the catalyst for the LACoFD to develop and implement the pioneering HSWR evolutions previously described.

On the day of the Rubio Creek incident, the sun was shining in the lower elevations of the Los Angeles Basin, but it was raining hard in the 11,000-ft San Gabriel Mountains, creating flash-flood conditions. Several 911 calls reported that two children, a boy and a girl playing in the water, had just been swept away by a flood surge. The LACoFD dispatched a first alarm water rescue assignment, and nearly a dozen units (including two helicopters) responded downstream to preassigned rescue points specified in the WRP for Rubio Creek.

The first victim to be spotted, the girl, was rescued three miles downstream by LACoFD engines 4 and 44, assisted by truck 4 and other units. The firefighters from these units used shore-based first responder methods (including throw bags and a tensioned diagonal line) to pluck the teenage girl from the current near a freeway overpass. She suffered a fractured femur and hypothermia, but survived after being airlifted by an LACoFD fire/rescue helicopter to a trauma center.

The second victim, an 11-year-old boy, was tracked by numerous ground-based fire units that attempted shore-based rescues to no avail. LACoFD helicopter 17 (a Bell 412 fire/rescue helicopter with a pilot and two USAR and swift-water-trained firefighter/paramedics) arrived on the scene, spotted the victim, and flew far enough downstream to set up for a rescue attempt.

Since no formal HSWR procedures had been established at the time, and the firefighters lacked the special equipment to snag the victim quickly and securely, the crew was basically doing the best it could under trying conditions, in a last-ditch effort to save the boy.

The boy was rapidly approaching a steep 20-ft high dropoff where the narrow Rubio Creek converges with the 400-ft wide Rio Hondo River, a place where the helicopter crew figured they would have to make a last-ditch attempt to rescue him. Firefighter/paramedic Rudy Mariscol was lowered on the hoist cable and dangled under helicopter 17, his feet just touching the water. The helicopter remained in place, meaning that Mariscol was stationary (not moving downstream with the current).

Unfortunately, Mariscol's stationary position ensured a collision between victim and rescuer. But at this point, no proven procedure had been established anywhere in the United States for plucking victims out of 40-mph water with helicopters, and the helicopter crew was willing to attempt a high-risk maneuver to save a life. As fire/rescue personnel and public safety pilots have done for centuries, they were improvising to the best of their abilities under highly unusual circumstances for which there was no formal protocol.

Pilot Karl Cotton was a veteran of hundreds of helicopter rescues from L.A. County to Alaska, recognized as having conducted one of the highest-altitude hoist rescue operations during the rescue of two mountain climbers on Denali. He lowered the helicopter to dip Mariscol into the water just before the victim was swept past. He positioned Mariscol immediately downstream of the boy. Mariscol was able to grasp the boy for an instant, but the current had slammed them together, and Mariscol was unable to get a good hold on the victim. At this point, Cotton had to lift high into the air to clear a freeway overpass and reposition the helicopter for yet another attempt further downstream on the roiling Rio Hondo. The boy was swept over the falls, and continued to be swept further downstream.

Helicopter 17 set up for another possible rescue attempt just downstream of the next rescue point, where several engines and truck companies established a tension-diagonal rope system and were waiting with throw bags and other shore-based rescue equipment. The plan that Cotton had quickly devised with the IC was to make another attempt with the helicopter *if* the victim went past the rescue point. As the boy emerged from beneath the freeway under which the Rio Hondo flowed, he was swept into the tension-diagonal line tensioned right above the water's surface, according to the plan. The diagonal rope system combined with the water's downstream force propelled the boy toward shore, where he was grabbed by firefighters and police officers who entered the water from shore to complete the rescue.

Although both victims were successfully rescued, Cotton and other members of the air operations, USAR unit, and the department's water rescue committee were convinced that it was time to devise a regimen of HSWR evolutions that would work in water moving 15 to 40 mph in narrow, concrete-lined urban streams and flood-control channels.

Soon thereafter, a committee of LACoFD pilots, air operations crewmembers, USAR companies, and water rescue committee members was convened by Fire Chief Michael Freeman to develop, test, and implement HSWR techniques. After six months of intense research, development, and testing, four innovative HSWR evolutions were unveiled by the group. These techniques, although high-risk in nature, proved extremely effective in extracting victims from fast-moving water, and soon, all pilots and firefighters assigned to the air operations, USAR, and airborne swift-water teams were being trained in the new methods.

Helo/High-rise (HHR) Operations

Intuition indicates that teams of specially trained and equipped firefighters prepared to be deployed onto burning or damaged high-rise buildings would offer a distinct advantage in high-rise emergencies. Critical tasks like rooftop and upper-floor size up, ventilation, SAR, firefighting, and rapid intervention are simplified if we don't have to climb many flights of stairs in heavy PPE and carry extra equipment just to reach the worksite. It's less fatiguing for firefighters to work their way *down* stairs than it is to walk up them. Helicopters provide us with options for size up, access, equipment and personnel transportation, rescue, and firefighting that simply wouldn't exist otherwise. In Figure 5–28, LACoFD Rescue Task Force members assigned as a helo high-rise team, practice a rappelling deployment onto the roof of a multi-story training tower.

Fig. 5–28 Rappelling Deployment

In the aftermath of the 9-11 attacks, we all recognize the potential for terrorists to attack additional high-rise buildings with airplanes, explosives, tanker trucks, and other means to cause worst-case scenarios. Therefore, the concept of helo/high-rise teams (HHRTs) begs to be addressed. This need is supported by the taped radio conversations of FDNY firefighters operating in the north tower of the WTC before it collapsed. The tapes,[13] released by the Port Authority of New York and New Jersey more than a year after the attacks, makes it clear that they were attacking fires and organizing rescues as high as the 78th floor. Additionally, they were advancing to even more floors above before the structure failed. Radio traffic indicated that ladder 15 and other companies were pushing their way into the fire floors and knocking down flame with success. They were organizing medical triage/treatment areas on the 40th floor, to which victims on the upper floors could be shuttled using elevators and stairwells.

Who is to say how much more success the firefighters might have had in saving lives if the FDNY had employed HHRTs on the roof during the early stages of the incident? Who is to say whether the topside approach might have allowed the FDNY to knock down enough fire to prevent the towers from collapsing, thereby saving thousands of lives? Who is to say whether it might have been possible to rescue dozens—or even hundreds—more people if firefighters had been operating from the roof? They might have been able to force open locked doors at the top of the stairwells and to facilitate helicopter-based extractions from the roof. Still other firefighters equipped with closed-circuit SCBAs might have been able to work their way down to attack the fires and protect egress through the stairwells.

We may never know if HHRTs would have helped. However, emerging information indicates that the readiness of local fire departments to place well-trained, properly equipped HHRTs (and, in some cases, other supporting companies) on the roof may be a key factor in future high-rise emergencies.

This sentiment is echoed by Vincent Dunn, 42-year veteran and retired deputy fire chief of the FDNY, who wrote the following after the 9-11 attacks:

The FDNY needs firefighting helicopters. There is a helicopter that can rescue people trapped on the roof of a high-rise building. Also, there is a helicopter that can spray 1,000 gal of water into the upper floors of a burning high-rise building (and return repeatedly with new loads of water after refilling by engine companies on the ground or from nearby helipad standpipe systems). This helicopter would allow firefighters to fight a fire caused by a terrorist bomb in the upper floors of a high-rise building from the outside, something we cannot do today. The people of New York need a fire department that has helicopters.[14]

Use of helicopters in past high-rise emergencies

Even without formally organized HHRTs, helicopters have repeatedly been called upon in high-rise disasters around the world to perform tasks that simply could not be performed in the same timely manner by firefighters climbing the stairs. Some of these helicopter operations have been ad hoc, using helicopter crews who were basically thrown into the incident in the midst of disaster, with little or no training and preparation for HHRT operations. Their success was achieved largely through guts and skill. Clearly, it is far safer and more effective to have a plan that includes fully-trained/equipped teams—preferably teams that are mentally and physically prepared to do the job, and have done it before in training, exercises, and perhaps in previous high-rise emergencies.

Indeed, helicopters have already been used to place firefighters on the roof to conduct ventilation, firefighting, SAR, and other critical tasks during high-rise fire emergencies. This includes (but is not limited to) the following examples.

The MGM Grand fire in Las Vegas. In this fire, helicopters from the U.S. Coast Guard, the Air National Guard, and other agencies rescued victims from the roof of the burning high-rise fire that killed more than 80 people.

The One Meridian fire in Philadelphia. In this fire, three Philadelphia firefighters became lost while conducting fireground operations in upper floors. In a desperate attempt to locate and rescue the firefighters, an ad-hoc HHRT was assembled using engine company personnel inserted on the roof. While these firefighters were unable to locate the original three missing firefighters, they did manage to direct other lost firefighters to safety.

The First Interstate fire in Los Angeles. Fire department personnel were confronted with four floors (levels 12 through 16) of a 63-story building fully involved. A pre-established LAFD HHRT was deployed to the roof in the early morning darkness to open the stairwells for improved ventilation, to search the upper floors for missing victims, and perform other tasks.

Baltimore high-rise fire. Using a plan established for years, Baltimore Fire Department's Rescue Company 1 was deployed by a Maryland State trooper helicopter to the roof of a burning high-rise hotel at night in 1999. The company managed to conduct topside size up, open the stairwells, and conduct SAR operations on the upper floors until the fire was knocked down by companies working their way up from the ground floor.

1993 WTC bombing. In the aftermath of the 1993 bombing of the WTC, the NYPD deployed a team of emergency service unit members to the roof of the north tower. In a prelude to the problems that complicated the response to the 9-11 attacks, the NYPD was roundly criticized for a number of cardinal sins at the 1993 bombing. These included the following:

- Failure to participate in a unified command structure at a major disaster
- Failure to coordinate the helicopter deployment with the FDNY fire command (after all, it was primarily a high-rise *fire* and rescue emergency)
- Deploying personnel without proper PPE
- Deploying personnel who were not experienced fighting fires (especially high-rise fires)

- Failure to establish and maintain communication with the IC and the many fire department units operating in the building

Puerto Rico terrorist attacks. When terrorists set fire to a high-rise hotel in Puerto Rico in the 1980s, helicopters were used to deploy firefighters to the roof to extract victims from atop the smoke-filled building.

Brazilian high-rise fires. In the capital city of Sao Paulo, Brazil, helicopter rescue operations played a significant role in two deadly high-rise fires that struck two years apart. The first, known as the Andraus fire, started on the fourth and fifth floors of a reinforced concrete high-rise with mixed residential and commercial use[15] in 1972. The fire made astonishingly rapid progress upward from floor to floor, cutting off escape via stairwells and trapping hundreds of people on the upper floors and roof (which was built with a fixed heliport).

According to witnesses, total involvement of floors four through seven occurred within minutes. The flames then took hold of the exterior of the Andraus building and raced to the top, with nearly 333 vertical ft of structure well involved. The first helicopter landed on the roof 1½ hours after the fire began. This was an obvious indication that there was no plan to place helicopter-based firefighting and rescue teams on the roof of high-rise buildings in the city; the effort was ad hoc, unplanned, and last-ditch. Nevertheless, it was the start of a round-robin helicopter rescue operation involving both civilian and military aircraft, which resulted in the aerial extraction of 350 people from the roof.

On February 1, 1974, yet another high-rise disaster occurred in Sao Paulo, this time at the Joelma building, a 25-story reinforced concrete office building with the first 10 floors in a parking structure configuration. A fire began at 9:00 A.M. on the 12th floor and quickly spread to the top, trapping 170 people on the roof and many other people on ledges and in windows. Eighty people were rescued off the roof, while 90 others lost their lives from exposure to heat and the products of combustion.

HHR: a controversial subject

The use of helicopters for high-rise emergencies is naturally a topic for which there is no easy answer, and it's a topic of debate in some quarters. Deploying firefighters on the roof via helicopter during high-rise emergencies is a tactic that's rarely employed for the following reasons:

- Relatively few fire/rescue agencies have their own helicopter fleet with which to develop, practice, and perfect HHR fire/rescue operations.
- Sometimes the coordination between fire departments and agencies that do have helicopters is sadly lacking or altogether absent.
- There are certain risks involved with any helicopter hovering and landing operation, and these risks may be exacerbated when operating above a burning high-rise building.
- Many of the world's fire services have a relative lack of experience dealing with HHR emergencies. Conversely, there is a corresponding shortage of proven tactics, strategy, guidance, and policy.
- Since helicopter firefighting and helicopter-based technical rescue operations are beyond the range of the personal experience of many fire department leaders, they may not be intimately aware of the capabilities and limitations of helicopters used by well-trained crews. Other fire/rescue authorities who might normally be open to the concept of HHR operations, find themselves dismissing such options as beyond the realm of feasibility because it's outside the range of their personal experience.
- Finally, the world simply lacks sufficient experience using helicopters on actual high-rise fires to definitively evaluate the effectiveness and ultimate parameters of HHR operations. This highlights the need for thorough and effective testing and studies that can provide better data and experience by which to develop rational conclusions about the HHRT concept.

Some firefighters and officers have argued that helicopters have very limited application in high-rise fires. They maintain that the time-tested method of entering at the ground floor and working their way up the stairwells (sometimes using elevators to transport equipment and/or personnel to levels below the fire) remains the most effective approach. They question the reliability of helicopters and their ability to perform the role of high-rise rescue. In his 1977 book, *High-rise/Fire and Life Safety*, John T. O'Hagan, former fire commissioner of the FDNY, made the following observations about the use of helicopters at Sao Paulo's deadly Joelma building:

> *The primary rescue effort at the roof level consisted of repeated attempts by helicopters to reach the victims. But the conditions in the vicinity of the roof were not favorable to a helicopter landing because the updraft created by the fire adversely affected the ability of the pilots to maneuver. Further, the products of combustion diluted the oxygen content in the air to the point where the internal combustion engines were unable to fire properly and the pilots ran the risk of stalling out.*
>
> *(Some of) the victims waited in vain*[16] *and this experience raises a point that places the helicopter in the proper perspective as a rescue device during a serious fire. When a fire develops in a high-rise building to the point where it cuts off the escape of a large number of people and seriously threatens their lives, the conditions around the building will make it difficult, if not impossible, to approach the building for landing.*

O'Hagan also comments on the questionable reliability of helicopters when describing the 1972 Andraus fire in Sao Paulo, where 350 people were rescued from the roof by helicopters:

> *The successful removal of such a large number of people by helicopter from the roof prompted many people to hail this means of escape as the solution to the life and safety problem in high-rise buildings. . . .*
>
> *In circumstances where there is immediate need for rescue for the occupants on a roof of a building seriously threatened by fire, the reflex time (notification to arrival) for the rescue team has to be minimal if they are to be successful. Response within the time constraints that exist under fire conditions would require the services of trained helicopter personnel on an almost standby basis. Our ability to meet this commitment is questionable. This does not mean that the potential of the helicopter should be ignored, but rather that it should be recognized that it is a means of last resort.*

Commissioner O'Hagan makes some good points. He was among the most authoritative voices in the fire service and a world expert on high-rise firefighting, but nearly 30 years have passed, and perhaps it's time to challenge certain time-honored views.

There have been many dramatic innovations in helicopter technology and flight capabilities since the 1970s. Helicopters fly higher, longer, with more reliability, and with bigger payloads than they did 30 years ago. Innovations such as automatic hovering, improved night-vision systems, FLIR, ground-link systems, and GPS allow helicopters to complete difficult missions with a higher degree of reliability and safety. Hoist rescue systems and related equipment have also vastly improved.

In addition to improved technology, fire and rescue services have taken better advantage of the benefits of helicopter flight to improve technical rescue, EMS, command and control, and firefighting capabilities. New helicopter-based rescue methods have been developed to pluck people from predicaments that once might have led to death. We have seen the expansion of concepts like rescue companies and USAR companies whose members are trained and equipped for helicopter-based technical SAR and firefighting operations. And there is a recent trend in some areas of the world to establish helicopter-based rescue and firefighting teams for both urban and wildland situations.

Commissioner O'Hagan's position can be used to make a case for both sides of the argument about using helicopters for high-rise emergencies. One side might argue that the traditional view of helicopters as a last resort during high-rise emergencies is correct, and that most high-rise emergencies are controlled effectively by firefighters working their way up stairways and attacking from the core, as practiced almost universally.

In response, the other side might counter that the insistence that helicopters are a last resort has contributed to a certain *malaise* that has actually stifled research, development, and experimentation with helicopters for high-rise disasters. This indifference may have dissuaded those who might otherwise have developed pioneering methods of using helicopters during high-rise emergencies. They might also argue that the assumption about helicopters as a last resort has become so ingrained in fire service culture that there has been little impetus to challenge it—at least until the 9-11 attacks. These showed the public that traditional means of escape from high-rise buildings may be impossible when a catastrophic event like a plane crash, earthquake, or explosion makes stairways and elevators unusable.

One thing is certain: with evermore deadly threats of high-rise disaster from fires, earthquakes, terrorist attacks, and other events—and the possibility that these threats could expand—traditional views about high-rise emergency operations are being challenged in some quarters. There is a growing concern about the need to revise some time-held concepts to meet the challenges of the world we face today and in the coming decades. Here are a few examples.

Rooftop landings. In reference to problems of landing on the roof, it's important to remember that it's not always necessary for helicopters to actually land on the roof of a burning high-rise to conduct rescue. Progressive agencies using HHRTs have trained their personnel to employ alternative means of reaching and extracting victims from roofs. These methods, which may allow the helicopter to avoid or reduce adverse effects from thermal columns, smoke, and other aerial hazards, include hot-skid touchdowns, rappel insertions, hoist insertions and extractions, and short-haul extractions. All of these methods can be performed without landing the helicopter on the roof.

Flex-time concerns. In some areas, the concern of extended flex time (the time between requesting resources and their arrival on the scene) has discouraged the use of helicopter rescue teams. It can now be said however that fire departments across the United States and some other nations have demonstrated the ability to get teams of firefighters in the air in helicopters within minutes.

In places like Florida, Maryland, and California, some fire departments have helicopter-based firefighter teams on the first alarm for many types of incidents including brush fires, cliff rescues, ocean rescues, plane crashes, vehicles over the side, swift-water rescues, etc. Some of these teams are flying to various emergencies in helicopters literally every day.

Some departments (including the LAFD and LACoFD) have HHRTs consisting of rescue/USAR companies and other designated units automatically responding on multiple-alarm fires in high-rise buildings. This approach greatly reduces flex time, allowing these teams to get to the roof within minutes of the first alarm. In many cases, they arrive before conditions worsen to the point that helicopters may have trouble landing on the roof or hovering above it (Fig. 5–29).

Fig. 5–29 PPE Worn by HHRT Members

In Figure 5–30, this USAR company captain donned an NFPA-certified seat harness beneath his turnout pants (to protect it from the products of combustion during high-rise fire operations), with the carabiner connected to the front hard-point and poking through the front opening of the turnout pants. The blue pickoff strap is used to secure the HHRT member inside the cabin of the copter whenever he is not seat-belted into one of the flight seats (as per LACoFD safety SOG's for flight operations). In his right hand is a steel 8-plate friction device, with which he will rappel from the copter. LACoFD has found that by stacking 8-plates on the rappel rope, HHRT members can expedite their deployment from the copter, with each member clipping in to his pre-rigged 8-plate as he moves to the door for his rappel. This member is wearing a 1-hr SCBA for a training session. However, LACoFD HHRT members wear closed-circuit SCBA (also used for mine and tunnel rescue operations and long-distance confined space operations) during actual helo high-rise operations. This is because it gives them hours of breathing air without the need to carry extra air cylinders and without the need to change them out.

Fig. 5–30 A Seat Harness Can Protect from the Products of Combustion During High-rise Operations

The rigging (Fig. 5–30) can also be used to connect the HHRT member to a hoist cable if conditions dictate that hoisting is more appropriate than rappelling and/or short-hauling personnel onto (and from) the roof. The seat harness was donned beneath the turnout pants, with the carabiner poking through the front opening of the pants and coat. The blue "pickoff strap" provides an adjustable connection to secure the HHRT member when he is unbelted in the cabin of the copter, in order to prevent an accidental fall while the doors are open.

Some personal equipment carried by individual HHRT members help in conduct operations like forcible entry, searching for victims in heavy smoke conditions, etc. The HHRT member in Figure 5–31 carries a thermal imaging camera attached to his harness via a prussic and a sawz-all to be belted around his waist prior to rappelling from the copter. It's important to carefully select these items and practice with them to ensure they do not impair the member's ability to safely rappel. Additional (heavier) HHRT equipment (e.g., a rescue saw and irons, a rabbit tool, etc) will be deployed in a large bag (lowered to the roof by the air crewperson using the same rappelling rope) after the team is deployed to the roof.

Fig. 5–31 Thermal Imaging Camera and Sawz-all

In some cases, these teams can be on the roof so quickly they can help conduct better size-up, open up stairwells, and conduct *protect in place* operations until other firefighting teams climb from the ground floor. They may also be able to control crowds on the roof to prevent unnecessary jumpers and other helpful tasks during the initial stages of a high-rise fire. Thus, the dynamics of the discussion about flex time with regard to high-rise fires have changed in some places.

Flight hazards. One factor affecting the success of HHR operations is the experience the pilot has flying in close vicinity to large fires, with their inherent dangers like thermal columns, heavy smoke, erratic winds, and hostile terrain.

Intuitively, we can assume that pilots who fly in wildland fire conditions on a regular (or daily) basis have an advantage over those who mainly fly under non-fire conditions with good lighting and air-traffic control. Pilots with extensive wildland fire experience are routinely subjected to hostile flight conditions as they drop water on fires and deploy flight crews near the head of the fire. They also have experience flying EMS and technical rescue missions in hostile mountainous terrain, including snowstorms, rain, and high winds. If that same pilot regularly practices landing or hovering on high-rise buildings or receives flight simulator training intended to approximate the challenges of a high-rise environment, the pilot's readiness for an HHR mission is naturally reinforced.

The point is this: the ability to get helicopters safely on top of a high-rise building (and safety off of it) is largely dependent on the ability of the individual pilot and the helicopter the pilot's flying. Simply put, some are better prepared than others to get crews onto the roof of a burning building and extract trapped victims from the roof. But if helicopters are only used after everything else fails, there is little impetus to practice these events. In other words, the prevailing philosophy about helicopters being a last resort may actually be a sort of self-fulfilling prophecy. It is certainly one in need of breaking if we are to use these specialized tools to the best advantage of the IC during a high-rise emergency.

The establishment of HHRTs has provided fire departments that have them with a constantly on the ready to conduct HHR operations. This concept works better if the HHRTs are based in rescue companies, USAR units, truck companies, and other units that have additional specialized missions that lend themselves to the skills and experience that make high-rise operations safer and more effective.

Increased dangers resulting from helicopter use. Some fire authorities maintain that the rotor wash from helicopters may fan the fire. Others argue that the thermal columns and smoke may prevent helicopters from reaching the roof or cause them to crash. Some have even suggested that there is no guarantee the helicopter can get back to the roof to retrieve the firefighters or that conditions may be untenable for firefighters to operate above the fire.

However, skeptics are hard-pressed to cite specific studies, experience, or examples that support their distrust in the effectiveness of HHRT operations. One must ask how many of them have personally responded to a working high-rise fire, explosion, earthquake, mid-rise plane crash, or other disaster where the conditions were so bad that people were leaping from upper-story windows and the roof? In a disaster where stairwells and elevators have been destroyed, there may be little hope of reaching trapped victims except by helicopter. The 9-11 attacks and other recent disasters have demonstrated that future worst-case scenarios like these are not only possible, but perhaps even probable.

Proponents of HHRT operations argue that helicopters are an under-utilized tool that could—if properly used—expedite placing properly trained and equipped teams of firefighters on the roof. They could then assess the conditions and report them to the IC, who armed with more data, could make better decisions. They contend that transporting specially trained companies of firefighters to the roof could help accomplish the following goals:

- Control crowds escaping onto the roof from smoke-choked stairwells and fire-involved floors (and potentially, from damage related to terrorism, earthquakes, and other causes). This could provide a hedge against people leaping from the roof or windows.

- Organize and conduct rooftop helicopter evacuations.
- Open stairwell doors and penthouse doors to help facilitate vertical ventilation to make the upper floors more tenable for firefighters and trapped victims.
- Use thermal-imaging systems, closed-circuit SCBA, and other equipment to search for people missing and trapped in heavy smoke conditions.
- Conduct interior fire attacks from above the fire to reduce vertical extension, protect routes of escape, and perhaps even cool structural steel to delay or prevent collapse until other firefighters climbing the stairs could attack the fire.
- Protect victims who are being *sheltered in place.*
- Conduct RIOs and other essential tasks.

All of these activities could be performed in a timelier manner than may be possible by firefighters hiking up from lower floors.

The argument for HHRTs. Anyone who has climbed the stairs of a high-rise building wearing full turnouts and carrying a spare SCBA bottle in one hand with a hose pack and other equipment over a shoulder, can attest to the demands and time required just to reach the fire floor(s). We have all heard of firefighters who were so physically spent just reaching the fire, that fire attack was hampered by exhaustion, heat stress, and even cardiac problems related to the stair climb. This is one reason some agencies establish a staging area and rehab area two floors below the fire floor. Some otherwise healthy firefighters, struggling under the combined effect of the weight of their gear, the heat-retaining properties of their equipment, the climb, and adrenaline, have been known to suffer chest pains and cardiac problems. They may basically become incapacitated or even in need of medical assistance themselves before they reach the fire floor. Many of these complications could be alleviated by inserting firefighters on the roof and allowing them to walk *down* the stairs (Fig. 5–32 through 5–34).

Fig. 5–32 Practicing Rappelling Maneuvers

Fig. 5–33 Reducing the Shock-loading

Fig. 5–34 Rappelling to the Roof

In Figure 5–32, USAR company members, work with Air Ops crewpersons and practice rappelling with full PPE from the helicopter mounted atop the LACoFD's Air Ops training tower. Even though LACoFD USAR companies (which staff the HHRT's) are accustomed to rappelling in every condition from the mountains to the urban environment, they are required to practice these basic skills from the tower at least twice annually, in addition to "live flight" deployment from hovering helicopters during HHRT exercises.

The HHRT member inverts. This means that as he rappels over the helicopter skid, he leaves his feet on the skid until his body is below the skid. This reduces shock-loading on the rope and anchor systems.

The HHRT member continues his rappel to a point about 10 ft off the roof, where he stops momentarily to quickly survey his landing point. This is important because he must ensure that he doesn't rappel onto a previously unseen object that could cause injury before continuing to the roof deck and then disconnecting himself quickly from the rope system to make room for the next rappeller.

Intuition suggests that helicopter deployment of well-trained companies of firefighters and officers has the potential for getting a faster assessment of conditions above the fire. It should also allow evaluating structural integrity and getting fresh personnel in position to establish effective vertical ventilation faster. It may help safeguard or remove trapped occupants more readily and perhaps get water on the fire faster (thereby preventing fire spread and reducing the potential for structural collapse). Generally, helicopter deployment could improve the situation for those unlucky to be trapped above the fire and for those attempting to command the incident and control the fire.

Unfortunately, few (if any) fire departments in the world have sufficient experience using helicopter-deployed firefighting teams at working high-rise fires to definitively prove or disprove the concept. No one can say in exactly which situations it's appropriate (or not) to deploy firefighters to the roof with helicopters. One would be hard-pressed to state categorically that the concept of HHRTs is viable in every place that has high-rise buildings. Few can state authoritatively when helicopter-deployed teams will work and exactly when they are contraindicated. The fact is, more research, development, exercise, and experience in actual working high-rise fires is needed to make positive or negative definitive statements about the efficacy of helicopter-deployed firefighting and rescue teams.

Until these questions can be answered with certainty, fire departments with high-rise buildings in their jurisdictions are advised to be prepared to try it when a worst-case scenario makes it necessary. This could mean practicing with local outside agencies, including law enforcement and the military, who can provide helicopters for this purpose. Fire departments with their own helicopters and high-rise hazards should certainly develop the capability to deploy specially trained firefighters to the roof when conditions clearly call for this tactic.

The move toward adoption of HHRTs

Some fire departments have already determined that there is sufficient experience and data to justify deploying specially trained companies of firefighters to the roof during high-rise emergencies. Some departments have identified rescue or USAR companies, truck companies, or other designated units to be inserted by helicopter onto burning and/or damaged high-rise buildings. Thus far, the experience has been favorable.

Baltimore Fire Department's Rescue Company 1, under the supervision of Captain Joe Brocato, was inserted as an HHRT during a fatal fire in 1999. Working under nighttime conditions with heavy fire and smoke in a high-rise hotel, Rescue 1's crew was inserted by a

Maryland state trooper rescue/EMS helicopter, something they had practiced for years. The operation was successful in terms of placing highly trained firefighters above the fire floor to assess the conditions, opening the stairwell doors to assist with vertical ventilation, and conducting SAR.

The LAFD inserted specially trained firefighters to the roof of the burning 63-story First Interstate building in the late 1980s. HHRT members encountered unexpectedly severe heat and smoke while opening the stairwells, but they managed to get the job done. Fortunately, the fire occurred late at night when only the maintenance and security forces (and some late-working employees) were present in the building. Even so, there were multiple fatalities. If the same fire had occurred during business hours, there may have been thousands of people trapped above the fire, which fully involved floors 12 to 16 before its upward progress was halted by extraordinary firefighting operations that nearly killed several firefighters.

During the MGM Grand Hotel fire in Las Vegas, helicopters from the U.S. Coast Guard and other agencies rescued people trapped by deadly smoke in the upper floors. Helicopters have also rescued people in a number of high-rise fires around the world in recent years. These examples indicate that organized teams of firefighters, properly trained and equipped to be deployed to the roof by landing, one skid, rappelling, hoisting, or other means, can be an advantage during major high-rise fires.

The 9-11 attack naturally raised questions about helicopter high-rise operations. But the conditions that confronted the FDNY were so extraordinary that it's difficult to draw definitive conclusions about the ultimate efficacy of using HHR operations. To some, the 9-11 attack raised more questions than conclusions:

- What were the actual limitations caused by thermal columns and smoke?
- Were the NYPD helicopter teams, staffed by rescue-trained police officers, properly equipped and trained to be working above a raging fire in a high-rise building?
- What kind of high-rise firefighting and rescue experience did they have?
- If the NYPD pilots had been trained and experienced to fly in proximity to large fires, could they have found a reasonably safe way to insert teams on the upwind edge of the roof of either tower?
- Were police officers prepared to force open stairwell doors apparently locked from the inside (potentially allowing people trapped at the top of the stairwells to escape) and to deal with the heat that might have emanated from within?
- Were NYPD personnel prepared to conduct SAR operations on the upper floors for long periods in smoky, oxygen-deficient conditions (e.g., using closed-circuit SCBA and wearing full turnouts)?
- Were NYPD helicopter crews prepared to insert FDNY personnel who were properly equipped, trained, and experienced to fight a high-rise fire?
- Were police officers prepared to hook up their own fire hoses to operational standpipes to fight fire and protect means of egress?
- Was there a plan to begin rooftop helicopter evacuations in the early stages of the fire when conditions might have been more favorable for extracting unknown numbers of people?
- If and when did the NYPD helicopter crews notice signs of structural instability, and were appropriate efforts made to transmit that information to the fire department IC? And if not, why not?

One thing we *can* definitely say about deploying personnel to the rooftops of burning high-rises is this: personnel without full turnouts and SCBA, without extensive high-rise firefighting experience, without fire department communications, and operating outside the fireground command system, are ill-prepared to operate effectively on the floors above a high-rise fire, and they are more likely to be a liability.

We can also definitively say that police and other non-fire department agencies with helicopters on the scene of a burning high-rise, have a moral and societal responsibility to coordinate all their efforts and operations with the jurisdictional fire department IC. They need to establish open radio and other communications with the firefighting forces operating inside the building (and their commanders), to immediately share critical intelligence with the fire department, and to operate under the aegis of the fire department. To do anything less is to endanger the lives of citizens, firefighters, and other people affected by the event. Law enforcement agencies that refuse to operate by these rules, simply confirm the argument that more fire departments should operate their own helicopter fleets to execute the fire department mission effectively without interference of uncooperative law enforcement agencies.

HHRT requirements

The firefighters assigned to an HHRT should be experienced in fighting difficult fires (Fig. 5–40). High-rise firefighting experience is obviously an advantage. All HHRT members should wear full turnouts and SCBA, preferably with a re-breather, closed-circuit SCBA typically used for mine and tunnel rescue operations. This SCBA allows at least four hours of rated working time, and up to eight hours if the wearer becomes trapped and conserves air/oxygen. At the minimum, members should have a one-hour SCBA bottle and a class III rescue harness (either internally worn beneath or built into the turnouts, or donned over the turnouts.

They should also have the following:

- One or more thermal-imaging systems
- Night-vision devices where available
- Forcible-entry tools (including tools to take down antennae and other flight hazards on the roof)
- An officer's high-rise hose pack and a rope pack
- A water extinguisher
- Victim harnesses
- The availability of additional hose lines and other tools for firefighting should they become necessary.

HHRT operational procedures

The sample procedures and protocols discussed here are intended for HHRT operations associated with Bell 412, Sikorsky Firehawk helicopters, or other Type II (medium) helicopters where the team is transported in the cabin and deploys from the cabin or the skids.

En route. En route to the high-rise emergency, the HHRT and its leader should be monitoring the radio to ascertain the overall situation and to begin preplanning potential needs. The leader should establish contact with the IC or the air operations officer to provide the ETA, determine the mission, and discuss specifics.

The team leader should brief the team on conditions and the assignment and coordinate with the pilot and flight crew to ensure they are clear on the nature of the mission requested. While en route, team members should double-check each other's PPE (including rappelling gear) using the buddy system and pre-established checklists. Team members should also double-check the crewman's equipment before the cabin door is slid open on arrival to ensure the safety of the crewman (who should be tethered to the cabin of the helicopter via harness and appropriate connections).

Upon arrival. The HHRT should perform the following common tasks after arriving at a high-rise emergency:

1. Perform an aerial survey, looking for fire location(s), fire and smoke behavior, victims in windows and on the roof, and any flight hazards on the roof. In darkness or heavy smoke, use night-vision goggles and thermal imaging.
2. Observe for signs of structural instability that might lead to local or catastrophic collapse.
3. Determine the feasibility of landing on the roof or conducting rappel or hoist operations.
4. Report these and other conditions to the IC.

If conditions indicate the need for HHRT operations, the HHRT should be directed by the IC to deploy onto the roof.

Getting the HHRT onto the roof. The HHRT will generally deploy to the roof of a high-rise building by the following methods, in order of preference.

- ***Helicopter landing insertion.*** The helicopter lands on the roof (if it has a designated helipad or a roof that's certified to accommodate the helicopter landing) and firefighters step out of the helicopter with their equipment. If there is no helipad, or if the ability of the roof to accommodate the weight of a helicopter landing is uncertain, one of the other following methods should be used.
- ***One-skid insertion.*** The helicopter conducts a one-skid, placing a skid on the edge of the roof while the pilot maintains power to hold the helicopter steady at roof level and HHRT members step off the helicopter, unload their equipment, and move out of the way.
- ***Rappel insertion.*** HHRT members rappel from the helicopter onto the roof. Their equipment is lowered down to them by a helicopter crewperson using the rappel rope; then the rope is dropped onto the roof for firefighters to use for potential rescues or searches.
- ***Hoist insertion.*** The HHRT members are lowered onto the roof using the hoist cable as the helicopter hovers over the roof. Their equipment is lowered to them in the same manner. As the HHRT member rappels past the step, he begins to invert to reduce shock-loading onto the rope system. For those unaccustomed to inverting in this manner, it may be disconcerting to do it 60 ft above the roof of a high-rise building that may be another 500 ft (or more) above the ground. This is one reason why live-flight practice is important to acclimatize HHRT members with rappelling from dizzying heights.

On the roof. Once on the roof (Fig. 5–35), HHRT members can perform the following tasks:

- Belay other members to the roof.
- Remove rooftop obstructions to clear for helicopter access. This will facilitate additional firefighters and equipment being delivered to the roof and ensure rapid extraction of trapped victims and firefighters as necessary.
- In coordination with command, open stairwells to help facilitate vertical ventilation to improve the atmosphere above the fire. This improves the victims' chances of survival and improves visibility for firefighters.
- Perform other tasks as assigned to improve ventilation.
- Search upper floors for victims (using thermal imagers, night vision, etc.) and remove them to a safe atmosphere or shelter in place as necessary.
- Conduct fire attacks on upper floors to reduce *lapping* and other vertical extensions, as well as, horizontal extensions. This will also protect stairwells and victims.
- If necessary, conduct crowd control on the roof and upper floors; conduct *shelter in place* operations as directed. With proper coordination with command, rescue and evacuate victims either to the roof or down the stairwells. Stairwells should only be used if they are secure and moving down is the best option at the time.
- Conduct helicopter evacuation of victims trapped on the roof if necessary. Unless secondary terrorist attacks may occur, it may be advantageous to move victims from the rooftop of one high-rise to the rooftop of another. In this case, transfer at least one member of the HHRT to the roof of the second building *with* or *before* the first group of victims.
- Conduct RIC operations from above and facilitate (as necessary) the rooftop evacuation of downed firefighters.
- Conduct forcible entry to allow victims trapped in stairwells and offices that may be damaged by fire, collapse, explosion, or other forces.
- Effect rescue of victims trapped in elevators.
- Conduct high-angle rescue of victims trapped in situations requiring those tactics.
- Coordinate the rooftop deployment of additional companies and equipment as deemed necessary by command.
- Conduct other tasks as directed by the IC or other officers.

Once the equipment bag is lowered to the roof, the air crewperson will disconnect the rappel rope from the anchor in the copter, and drop the rope to the roof, where it may be used by the HHRT for various other tasks (including an "operational retreat").

Fig. 5–35 Bottom Belay Conducted by Member after He Reaches the Roof

Rooftop evacuation of victims by helicopter. In order of preference, the following rooftop evacuation tactics can be employed by the HHRT:

Roof landing extraction. As the HHRT controls the crowd and the team leader communicates with the pilot of the incoming helicopter, it lands on the roof. HHRT members assist people into the cabin of the helicopter (the number of people on each flight to be determined by the pilot). An air crewperson or member of the HHRT secures the door before takeoff.

One-skid extraction. As the crowd is controlled and kept at a safe distance, the HHRT leader signals the helicopter to approach the building and perform a one-skid. Next, HHRT members assist a prescribed number of people into the cabin, with the air crewman securing the door before the helicopter departs.

Short-haul extraction. As the crowd is controlled, one or more citizens is placed in an evacuation harness (or, if injured, an appropriate rescue litter). The helicopter is signaled in and the pilot hovers overhead while a short-haul line is lowered. Members of the HHRT attach the victim(s) to the loop(s) in the rope and signal the air crewman (standing on the skid above) to lift.

In turn, the air crewman, on hot mic inside the cabin, directs the pilot to ascend. As the victim(s) are short-hauled to the other building or to the ground, the air crewman maintains a close watch to communicate the victim's position to the pilot until the victim is set down at the designated point. At that time, the remaining member of the HHRT assists the victim's touchdown, disconnects the rope from the victim(s), and signals the helicopter to depart back to the burning high-rise to evacuate more victims.

Back on the roof of the burning high-rise, the HHRT should already have harnessed up the next citizen(s) to be extracted, using additional harnesses they carry to set up a round-robin that will expedite the process of removing victims.

At the set-down point, the remaining member of the HHRT removes the harness from the victim(s) just extracted. This harness is clipped to the end of the short haul when the next person is dropped off, so the air crew can transfer the harness back to the roof of the burning high-rise, where it will be placed on another citizen.

Hoist Extraction. During this process the HHRT performs the following steps:

1. As members of the HHRT control the crowd, one or more citizens is placed in an evacuation harness (or, if injured, an appropriate rescue litter).

2. The helicopter is signaled in, and the pilot hovers overhead while the hoist cable is lowered.
3. Members of the HHRT attach the victim(s) to the cable and signal the air crewman (standing on the skid above) to begin hoisting.

The air crewperson, on hot mic inside the cabin directs the pilot to very slowly ascend until the power of the helicopter itself is lifting the victim(s) off the rooftop. This is standard protocol for hoist operations.

Normally, once the victim(s) are actually suspended in the air under power of the helicopter, the pilot tells the air crewman, "Okay, you've got the load," at which time, the air crewman activates the hoist motor to begin lifting the victim(s) toward the helicopter. Eventually they are brought down all the way to the level of the cabin and assisted into the cabin where they are secured.

However, in the case of a high-rise fire with multiple victims waiting to be evacuated, the hoist cable might be used to short-haul victims at the end of the extended cable directly to the roof of an adjacent high-rise or to the ground. This would reduce the time between each rescue cycle by eliminating the time it takes to retract the hoist cable for each victim.

As the victim(s) is set down, the designated member of the HHRT disconnects the victim's harness from the cable and signals the helicopter to return to the burning high-rise to evacuate more victims.

Back on the roof of the burning high-rise, the HHRT should already have harnessed up the next citizen(s) to be extracted, using additional harnesses. Finally, back at the set-down point, the designated member of the HHRT removes the harness from the victim(s). The harness is connected to the cable after the next person is dropped off, and the cycle continues until all victims—and possibly the remaining firefighters from the HHRT—are extracted from the roof of the burning high-rise.

Operational retreat. In all these cases, the HHRT and the team leader must be cognizant of the potential for an operational retreat to be ordered by the IC. This means it's time to bring the helicopter back and evacuate all the firefighters off the roof *if* the smoke and other conditions allow it to return.

As some HHRT members have been reminded by pilots, a rooftop evacuation under those circumstances might resemble a battlefield evacuation under fire in the sense that there is urgency to get in and out without delay. The helicopter swoops in, touches down *hot* (or perhaps with one skid just off the deck or the edge of the building) for a few moments as the firefighters bail into the cabin. Immediately, the helicopter takes off, getting away from the building as quickly as possible.

Every member of an HHRT recognizes that it's possible for conditions to deteriorate to the point where it becomes impossible for the helicopter to return to the rooftop at all, effectively stranding them above a raging high-rise fire. There is also the potential for structural failure—a grave situation indeed. This emphasizes the need for fireground commanders to be prepared to deploy HHRTs when conditions dictate. In addition, members of the HHRTs must be fully aware of these risks and consider how to reduce them to the most reasonable level. It also emphasizes the need for timely operations and constant evaluation of fire, smoke, and structural conditions.

Case Study 4: Why Helicopter Rescuers Should Carry Dramamine

Sometimes passengers in airplanes and helicopters wonder why airsickness bags are stored in full view. It's a rather unpleasant reminder that rough flight conditions can do strange things to your stomach. It's something that can also happen to rescuers assigned to helicopters, which illustrates one potential problem: we can hardly expect to help others under emergency conditions if we become incapacitated by airsickness. This was clearly demonstrated by a true and sometimes comedic story of two firefighters assigned to a HSW rescue team one day during the deadly floods in southern California in January 1994.

At the height of a major storm that blew down from the Pacific Northwest, several flood-control channels in southern L.A. County overflowed their banks. Floodwaters raged through city streets, rapidly inundating several dozen blocks. Intersections quickly flooded and some motorists barely had time to climb to the roofs of their vehicles. In a matter of minutes, some homes began to go underwater and people were forced to run for high ground. Dozens of people became trapped, and two dozen fire/rescue units were dispatched from three departments, including two fire/rescue helicopters staffed with two USAR-trained firefighter/paramedics and a two-person swift-water rescue team consisting of USAR company personnel on special assignment.

Among the swift-water rescue team members assigned to the helicopter that day was a certain firefighter specialist (whom we'll call Dave) and a firefighter/paramedic (Tom). Dave, normally assigned to a USAR company, was on special detail to the HSW team during the storms. Tom had been permanently assigned to the helicopter unit for a decade and was no stranger to the effects of air turbulence.

Rain was pouring from an ominous black sky as the helicopter lifted off. Immediately, it was buffeted aloft by severe winds. Along the route to the rescue operations, the pilot (whom we'll call Ben) was faced with poor visibility and low-level storm clouds blanketing the foothills along the Pacific Coast. To reach the scene, he was forced to skirt the ceiling of the visible storm cells and maneuver through the deepest passes, then race along the coast before turning inland once again to find the incident. This meant even more turbulence.

Sitting in the rear cabin of the helicopter and looking out the window at the thunderous black clouds in the flight path, Dave recognized the conditions as an omen: He knew he was prone to bouts of motion sickness, and already he could feel his stomach beginning to protest. He began to wish he had taken some Dramamine that day (although it was too late now because the drug would be of little help in time for this mission).

As Ben flew the helicopter in a scud-run between the clouds and the hilltops, they were buffeted by downdrafts and turbulence, and Dave's stomach was churning. Fifteen minutes later and green with nausea, he knew he was in trouble.

In preparation for the inevitable, Dave reached for an airsick bag, a set of which are normally stored in a particular nook in the wall of the cabin. His hand fumbled in the nook, but found only emptiness. A wave of panic came over him as he glanced around the cabin, looking for the hidden bags. Tom noted the look on Dave's face and immediately recognized the problem. There was an airsick bag container on Tom's side of the cabin, and he reached in to pull out a bag. Dave, now becoming desperate, watched Tom put his hand into the container and pull out . . . nothing! Tom could not believe it. There were *always* spare sick bags stored in the helicopter. Somehow, they must have been misplaced.

Dave was now looking for *somewhere* to spill his guts, and the options were becoming increasingly limited. Tom looked again for the missing bags, but found nothing. Finally, after coming up empty-handed, Tom signaled to Dave: Hold on a few more seconds! They were nearing the rescue site. Dave's cheeks were puffed out, trying to hold it back.

Tom knew it was time for drastic action. They were out of options, and the cabin of a helicopter is no place to be after such options run out. Reluctantly, with a resigned look, Tom took off his insulated flight jacket, which had survived countless missions under the most severe conditions. The pilot glanced back to see what all the commotion was about. Tom took the sleeves of the jacket and tied them into a knot, zipped up the front of the jacket, and twisted the neck into another knot. He had just created the world's most expensive airsick bag. At this point, Dave would have been happy to use Tom's boots.

When Dave was finished, he immediately felt better. And a good thing too, because they were now on the scene of their first rescue—a pregnant woman stranded on top of a car in the middle of the flood. The crew was equipped and trained to conduct a variety of rescue operations, including a hoist evolution with the helicopter's 230-ft cable, a one-skid hover, a tethered swimmer evolution, or a swimmer free short-haul rescue.

After evaluating the conditions, it was decided that a one-skid hover would be the safest and most effective method. Ben held the helicopter steady with the right skid just off the roof of the car. Tom (attached to the cabin of the helicopter via his full body harness and safety strap) stood on the skid and pulled the woman into the cabin. As Ben ascended and turned toward the hospital to drop her off, the crew buckled her into the litter and began taking vitals. Fortunately, the hospital was only two air minutes away.

At the hospital, the characteristic *whop-whop* of rotor blades beating the wet air alerted the crowd of the incoming helicopter. Two dozen people braved the rain to gather around the helispot to watch the large yellow helicopter land.

Unfortunately, inside the helicopter, Dave was green again. As his stomach began to fail him, Tom looked frantically for another container. His flight jacket was already gone. About the only thing that even remotely resembled a fluid container was his flight helmet, and he was not about to give that up. The woman, who was stable and uninjured, watched in puzzlement as Dave and Tom motioned to each other over the roar of the helicopter.

It became a race against time in which seconds counted. Dave wasn't sure he could hold on long enough. As the helicopter touched down, Dave unlatched the cabin door and slid it open. The onlookers, now facing a violent rotor wash, squinted to watch the firefighters unload their victim.

The scene played out in slow motion. First, Dave hopped out of the helicopter and looked up helplessly at the crowd. Their faces wore a puzzled look. What was that nice fireman doing? Tom, watching the scene from the helicopter, noticed a look of alarm sweep across the faces in the crowd as Dave bent over to vomit on the ground just outside the right skid. In another instant, alarm turned to panic as the onlookers realized that the drops being driven at them by the rotor wash might not be pure rain. Some of them turned to run, trying to duck the spray.

Tom escorted the woman into the emergency room and immediately returned to prepare for the next rescue. By now, Dave felt just fine. They proceeded to rescue several other citizens from various perches in the flood, until everyone was removed from danger.

Dave swore that he would never fly again under such conditions without taking his Dramamine; the experience was etched in his mind. Apparently, however, it wasn't etched deep enough just yet. Not more than a month later, he was back at his normal assignment as engineer of a USAR company participating in air/sea rescue training off the coast of Venice Beach. This training was part of the air/sea disaster plan for an international airport in southern California. As part of a training exercise, Dave once again found himself a member of a rescue swimmer team being deployed into the ocean from the same helicopter.

This time the weather was calm, and Dave knew the helicopter ride would be fairly smooth. Unfortunately, he overlooked the choppy seas until it was too late. He was about to have another bad day.

As the helicopter approached the jump zone (where 100 lifeguards were now floating in the open sea to simulate victims of a plane crash after takeoff over the water from the airport), the cabin door slid open and Dave stepped out onto the skid with his jump partner. As the helicopter slowed to 10 knots forward speed and 15 ft over the water, the jumpmaster gave the signal and Dave and his partner went into the water with a 10-person inflatable rescue platform. They inflated the platform and began pulling mock victims from the water. When they had filled the rescue platform with lifeguards, it was time to wait for a fireboat to retrieve them.

As they clung to the side of the platform, Dave gave his anti-motion sickness advice to the others: "Whatever you do, don't look at the shoreline. It will just make the motion of the water more pronounced. See how those buildings over there go up and down as the choppy ocean lifts and drops us?" He went on, "See how the horizon seems to shift if you watch the beach as the sea rolls under us? Look at how the boat masts in the marina seem to sway in the wind," he said. "Well, it's kind of an optical illusion. See, the masts aren't really leaning; it's just us bobbing in the water here that makes it *look* like they're leaning."

The next thing he knew, Dave was chumming. Unfortunately, Dave also managed to swallow a good amount of seawater in the process, which made his chumming efforts that much more furious—and painful. After two hours of chumming, Dave was a wet, limp noodle of a firefighter. We had to send him home to recuperate on dry land.

Today, it can safely be said that Dave heeds the lessons learned from these harsh experiences. He checks the weather before flying on duty, and he always keeps a few anti-sickness patches in his duty bag.

Helicopter Transportation of USAR Units and Rescue Companies

Some fire departments and rescue agencies find it advantageous to use helicopters to expedite the response of rescue companies, USAR units, and SAR teams. This is particularly true for departments whose jurisdiction includes large geographical areas, hostile terrain, special hazards requiring special deployment considerations, or impediments to travel, such as heavy traffic or topographic features (e.g., mountains, lakes, deep canyons, rivers, or the ocean).

Helicopter transportation of rescue companies and USAR units gives the IC more options for getting specially trained and equipped personnel on the scene in a timelier manner. It allows them to be placed in strategic locations that might otherwise take hours if ground transportation (including long hike-ins) were used. Helicopter transportation also allows units to combine their rescue efforts. Helicopters may provide better access to remote rescue sites, allow rescuers to perform RIOs, support aerial SAR operations, and provide much-needed support for other special operations assignments at the scene.

For many years, the Chicago fire department has used helicopters to fly its squad companies to dive rescue operations and other special assignments. The Baltimore Fire Department has teamed with Maryland state troopers to deploy firefighters to the roof of burning buildings. The Miami-Dade Fire and Rescue Department has a fleet of helicopters flown by fire department pilots and crews consisting of firefighter/paramedics trained in USAR and dive operations. These units conduct aerial rescue operations and MedEvacs across southern Florida and off the coast.

The Phoenix Fire Department has a well-oiled system to use police helicopters to deploy its USAR-trained firefighters for cliff rescues and swift-water rescue operations. The Orange County fire authority has a fleet of helicopters to drop water on wildland fires, transport USAR personnel, and conduct technical rescue operations. Likewise, the city of Los Angeles has its own fleet of helicopters to cover the city's 500-plus square miles. In addition, to provide helicopter fire and rescue service to 3,500 or so square miles of L.A. County, the LACoFD operates a force of seven helicopters, including two Firehawks, four twin-engine Bell 412s, and one Bell Long Ranger for command-and-control operations In Figure 5–50, USAR company members load into an LACoFD copter as part of their assignment to be rescue swimmers during marine disaster operations in the Pacific Ocean or large lakes in southern California.

Since the 1980s, the LACoFD has used these helicopters to transport its USAR and hazmat units over mountains, across the desert, and over lakes and rivers. They have been used to reach islands dozens of miles off the Pacific Coast and to bypass heavy urban traffic during long-distance responses. They are also used to reach damaged areas after landslides or collapsed structures (bridges/overpasses) that have interrupted normal routes of travel. Not only do helicopters help ensure timely technical support to first responders and ICs on a daily basis, but they often serve to reduce the time required to locate and extricate trapped or injured victims and personnel, thereby relieving pain and suffering.

The ever-present potential for disasters resulting from earthquakes, floods, landslides, mud and debris flows, dam failures, and tsunamis is another reason LACoFD commanders find it advantageous to retain the option of transporting USAR and hazmat task forces via helicopter. In this way, they can provide timely technical SAR and hazmat service to areas otherwise isolated by downed bridges, washed-out roads, blocked or damaged highways, massive flooding, and other transportation complications.

Figure 5–36 shows an aerial survey from a helicopter only moments after the Northridge earthquake struck L.A. County. Dozens of out-of-control fires and many collapsed structures and bridges were observed, assessed, and reported by airborne USAR teams, expediting response to the worst problems. Smoke from dozens of out-of-control structure fires erupted within minutes of the Northridge quake.

Fig. 5–36 USAR Company Aerial Survey After the Northridge Quake

For rescue operations, the obvious advantages of helicopter transportation include faster response to distant incidents, better access to hard-to-reach sites, and the ability to perform aerial observation and size-up of disaster sites and technical rescue situations. In addition, helicopters provide rapid over-water access to the various islands and the ability to bypass gridlocked traffic on the ground. The LACoFD's two decades of experience with USAR helicopter transportation has demonstrated the efficacy of this approach. In innumerable cases, USAR response times have been slashed, patients have been extricated much sooner than otherwise possible, disasters have been managed in a more timely manner, and rapid intervention capabilities have been improved (Fig. 5–37).

Fig. 5–37 Over the Side Rescue, With USAR Firefighter Lowered onto an Ice Chute at the 10,000 Foot Level to Pick Off the Victim

In Figure 5–37, USAR company personnel and first responders are carrying the victim of a vehicle over the side rescue to a waiting copter for extraction from the bottom of a canyon and to a trauma center. The vehicle went 400 ft over the side and landed near this reservoir in the San Gabriel Mountains. The helicopter expedited USAR operations by carrying extrication equipment to the site and rapidly evacuating the patient after he was freed from his car. It saved time that would normally be used securing the victim to a rope rescue system and hauling him and a litter team up the canyon wall to the road.

Another advantage is the multidisciplinary training and equipment of the LACoFD's USAR units that serve to augment those of the helicopter crews during helicopter-based hoist rescues, short-haul operations, HSWRs, high-rise fireground operations, ice and snow rescues, and other special operations. USAR task force members are also trained to conduct RIOs under each of these conditions, working from helicopters to provide timely assistance to firefighters who become lost, trapped, or injured during the course of rescue operations.

There are some conditions under which helicopter transportation of LACoFD USAR task force members is considered *automatic* (weather and other conditions permitting) because of their responsibility to conduct certain helicopter-based tasks. These include HHRT, HSWR (in the absence of HSWR teams), and SAR operations at the site of marine emergencies and disasters. Naturally, inherent limitations to the use of helicopters for transporting USAR and hazmat task forces may at times include lack of helicopter availability, inclement weather, poor visibility, hostile topography, weight and space limitations, and other considerations.

Many of the factors used to determine whether to transport USAR or rescue firefighters are *situational*, defying absolute rules that attempt to predict every possible condition. Therefore, the decision whether to fly remains one largely of judgment and experience, based on common-sense parameters, past history, and recognition of mission objectives and incident needs. Naturally, it's difficult—if not impossible—to quantify all these factors into a single policy or procedure without leeway for good judgment and experience. In an effort to provide an example of how a policy for USAR/rescue firefighter helicopter transportation might work, a sample is included in appendix VII.

LACoFD rescues approximately 10 to 12 horses every year from the bottoms of canyons, from rivers and streams, from wells and other entrapment situations, and other predicaments. The integration of helicopters and a special team from the L.A. County Department of Animal Care and Control has greatly improved the effectiveness and safety of large animal rescue operations in the past decade.

Case Study 5: Rescues in Malibu Illustrate Challenges and Improvements

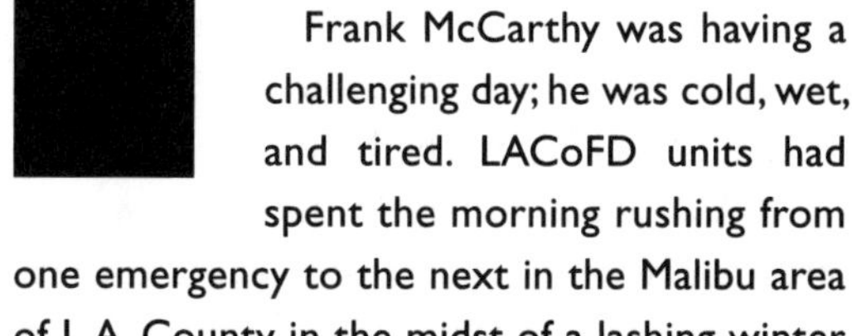

Frank McCarthy was having a challenging day; he was cold, wet, and tired. LACoFD units had spent the morning rushing from one emergency to the next in the Malibu area of L.A. County in the midst of a lashing winter storm, and the end was nowhere in sight. It was the winter of 1993, the middle of a dramatic El Niño cycle that was wreaking havoc with the west coasts of North and South America.

Firefighters and swift-water rescue teams were rescuing people from atop homes, cars and islands amid the swirling and bone-chilling floodwaters that rampaged down from the mountains. For McCarthy and the crew of Air Squad 8, an HSW rescue team assigned to a fire/rescue helicopter, the day was to prove even more challenging before it was over.

Just 10 minutes earlier, McCarthy and the crew of Air Squad 8 had finished rescuing stranded motorists from the Big Rock area of the Pacific Coast Highway, using the hoist rescue system. As the helicopter lifted off, the Malibu IC radioed a request for Air Squad 8 to respond to a report of an elderly couple trapped in their home in the middle of the flood.

Back in the cabin of the rescue helicopter, while preparing to be lowered into the flood, McCarthy listened to radio reports from ground units whose paths were blocked by flash flooding, rocks, mud, and power lines. The Pacific Coast Highway was now impassable in several places after having been ravaged through the night by torrents of debris, mud, and rock. More than 24 fire engine companies, organized into groups of five (known in ICS parlance as *strike teams*), had been dispatched from county fire stations as far away as Pomona and Lakewood to assist the Malibu-based firefighters. The spreading flood zones had isolated some of the strike team units. Those that were capable of getting into the impacted areas were extremely busy with emergencies along the 27-mile stretch of coast between Santa Monica and the Ventura County line.

Some of the engine companies and rescue units were attempting to maneuver through mud-choked canyons, stopping to rescue people stranded on cars and homes in the middle of floodwaters. As the storm worsened, others were moving people to high ground before it was too late. Multiple reports were coming in of people trapped in areas that had become inaccessible to ground units. Some homes were being buried in mud, while others were being eaten away by debris-choked floodwaters. The day was becoming a fully formed disaster, complete with communication failures, power outages, structure fires, and people in severely threatened neighborhoods refusing to leave their homes under evacuation orders. In some places, firefighters resorted to climbing into the buckets of large front-end loader tractors to be driven upstream through floods of mud and rock in otherwise impassable canyons to reach people trapped in their homes and cars.

The ground units were augmented by three rescue helicopters, two fire department USAR units with swift-water rescue capabilities and special extrication equipment, and several ground-based swift-water rescue teams, all staffed by specially-trained firefighters with expertise in flood rescue operations. Several swift-water-trained lifeguards from the fire department's lifeguard division were assigned to Malibu as yet another swift-water rescue team.

As the storm began to reach a crescendo, all of the teams suddenly became inundated with calls for help. Reports of people trapped on rooftops, cars, and islands amid the streams were coming in faster than rescuers could be dispatched. Many fire units simply responded from one emergency to the next, without respite, for hours on end.

As they circled the mouth of Malibu Canyon, the crew spotted a home awash in flood waters near Sera Retreat. The house itself was breaking the current of the flood, much like a boulder in the middle of a stream, with roiling waters rushing by on both sides and leaving an eddy on the downstream side of the structure where the backyard was located. The fenced yard was covered by a sheet of waist-deep, mud-colored water. On the raised rear porch of the house stood two people, frantically waving their arms at the yellow and white helicopter. Standing with them in the driving rain was a large, black Labrador retriever.

Finnerty discussed a rescue plan with the rest of the crew over the intercom. They decided to use the rescue hoist mounted on the side of the helicopter to lower McCarthy as a rescue swimmer into the backyard. McCarthy would then make his way to the porch, wading through waist-deep water, and secure the victims in harnesses that would be attached to the cable. Then they would be lifted to the safety of the helicopter.

After ensuring that everyone was clear on the plan, Finnerty brought the helicopter to a hover over the house. Monahan, wearing rain gear and a rescue harness attached to the wall of the cabin of the helicopter to prevent him from falling, ran through the safety checklist used each time a hoist operation is conducted. His partner double-checked the rigging on his harness and all other components of the system that would prevent him from falling. Both firefighters were connected to the helicopter by special straps attached to their full-body rescue harnesses that allowed them to lean out to look for victims beneath the dense canopy of trees below.

Then Monahan slid open the cabin door and stepped out onto the wet helicopter skid, holding the hoist controller in his hand. He leaned out into open air so that he could look straight down at the target.

Obviously, helicopter rescue is not a good job for people uncomfortable working at great heights. The slightest mishaps can be catastrophic in this unforgiving environment. LACoFD Firefighter Jeff Langley had been killed just one year earlier when he fell 90 ft from a helicopter while rescuing a young man in a remote part of Malibu. Langley had been performing the same job that Monahan was now performing—standing on the skid of a hovering helicopter and controlling the hoist while another rescuer dangles below on the end of the cable. Last year's tragedy was still fresh in the minds of these rescuers.

McCarthy, dressed in an orange drysuit, helmet, rescue harness, goggles, and PFD, followed Monahan onto the skid, the hoist cable attached to his harness. He faced the helicopter, his back to the rescue scene 70 ft below, and prepared to be lowered out over the skid. Rain fell within the circular swath of air covered by the spinning rotor blades and was driven downward by the hurricane-force rotor wash. Hard drops of water, thus propelled, stung his skin and made it difficult to see. Both firefighters squinted through goggles that provide some measure of protection from the powerful blast of water and air.

McCarthy signaled his readiness, and Monahan turned the knurled wheel of the hoist controller with his right thumb, slowly unspooling the cable that descended from the external hoist arm. His descent now controlled by Monahan, McCarthy was lowered beneath the helicopter, gently swinging beneath the skids to reduce any shock-loading to the cable and the helicopter. Monahan verbally guided the pilot into position by giving him a constant stream of directions on the hot microphone. This allowed the two to talk freely without interference and without having to activate the microphone manually.

The helicopter was perfectly centered over the backyard now, a position that would allow Monahan to lower McCarthy to the ground without becoming tangled in the canopy of trees and power poles below. Monahan noted that the entire backyard of the house appeared to be 3-ft deep in muddy water. McCarthy, also looking down as he descended, braced himself to hit the solid ground that he assumed was beneath the layer of water.

But an instant before McCarthy reached the water, he saw something that made him instinctively signal Monahan to stop lowering him. Mounted on the block wall of the yard was a long aluminum pole with a net attached to one end. McCarthy recognized it as the kind of net used to skim leaves from a swimming pool. McCarthy signaled to be raised by moving his outstretched arm in an upward arc, his thumb pointed outward to enhance the visibility of the signal for Monahan, who was now looking straight down at him. It was too late, however, and he was suddenly completely submerged in churning brown water.

Monahan, standing on the skid of the helicopter and looking down as he controlled McCarthy's descent, had been having difficulty judging McCarthy's exact position in relation to the water. This is a common problem when darkness and bodies of water throw off depth perception. Monahan had done well in positioning McCarthy over the backyard and just above the surface of the water. Then he gave McCarthy enough slack to stand up in what he anticipated would be a sheet of waist-deep water over solid ground.

So Monahan, also, was quite surprised when McCarthy suddenly disappeared below the surface. He squinted to look closer and instructed Finnerty to hold the helicopter right there. Finnerty held the helicopter steady. An instant later, McCarthy's white-helmeted head popped up from water. He gave the OK hand signal and began swimming toward the victims on the porch of the house. It still hadn't occurred to Monahan that there might actually be a swimming pool beneath the layer of running water, because from his position, there was no visible indication of its presence. At this point, however, it didn't matter. McCarthy had unhooked himself from the hoist cable and was swimming toward the porch. Monahan turned the knob and up came the cable.

As Monahan continued to monitor the situation on the ground, Finnerty moved the helicopter away from the house to reduce the rotor wash hitting McCarthy and the victims, and to give McCarthy a chance to explain the operation to his new charges. As McCarthy fought his way to the house through churning, chest-deep water, he squinted against the stinging rain, whipped to a frenzy by the rotor wash of the big yellow helicopter that hovered overhead. He dragged himself from the water onto the raised porch.

After checking the condition of the elderly couple and finding them to be in good shape, McCarthy surveyed the overall scene to determine his options. At this point, he explained to them that the helicopter was the safest method of rescue. The man insisted that his wife be evacuated first; McCarthy concurred, and began placing a rescue harness on her. She was quite scared as McCarthy rigged her up in what appeared to be a parachute harness. He explained the safety features of the harness to her and told her what to expect when the helicopter returned overhead. When he assured her that he would accompany her on the cable during the ascent, the woman seemed to calm somewhat.

After double-checking the harnesses, McCarthy signaled for Air Squad 8 to begin the next phase of the operation. Finnerty guided the ship back into position over the porch and Monahan lowered the cable, which swung crazily in the rotor wash. McCarthy reached up and grabbed the stainless steel clip at the end of the cable, known by the air squad crew as the *hook*. Working quickly, he attached the woman's harness to the hook with a steel carabiner, a mountaineering tool adapted for rescue operations such as this. After double-checking the connection and instructing the woman to grasp the appropriate part of the hook device, McCarthy signaled Monahan to begin the lifting operation.

With Monahan guiding Finnerty, the pilot slowly lifted the helicopter higher, until the feet of McCarthy and the woman came off the ground and their full weight was suspended on the cable. It is standard procedure to gently lift the rescuer and victims off the ground using the skill of the helicopter pilot, which helps ensure that they are not abruptly snapped into the air by the hoist cable motor. This motor moves at a predetermined speed and cannot be feathered like a clutch. Once they are fully suspended on the cable and dangling in midair, the motor is used to reel them into the helicopter.

These procedures reduce shock-loading in the cable and the hoist motor, lessen jarring that can disrupt a helicopter's stability, and make for a reasonably comfortable ride up the cable.

"You've got the load!" said Monahan as McCarthy and the woman swung gently into the air. McCarthy put his feet out to kick them away from the roof of the house.

"Okay, go ahead and raise the load," answered Finnerty, working to keep the helicopter steady. In response, Monahan turned the knob. The woman ascended toward the ship under power of the hoist.

"Seventy feet....Sixty feet," called out Monahan to let Finnerty know exactly where the load was in relation to the helicopter.

"Okay, I've got them at the skid. . . helping them over the skid. . . and I'm moving the woman into the cabin," said Monahan as he adjusted the cable to bring the woman into the helicopter's interior, with McCarthy helping. "Frank is securing her in the seat," said Monahan as the woman was buckled into the helicopter seat and the hook was disconnected from her harness. Through the entire process, she had been secured either by the cable or the restraint. McCarthy remained attached to the hook throughout the operation. This, also, was standard procedure, intended to prevent anyone from falling from the helicopter.

"Okay, Frank's back on the skid, and I'm lowering down for victim number two," said Monahan, guiding McCarthy over the skid and toward the porch. Finnerty acknowledged and kept the helicopter hovering steady.

By this time several television news helicopters were circling in the sky above; the rescue was being aired live.

"We've got a lot of metal in the sky," observed Finnerty over the intercom. Finnerty clicked over to the aviation frequency and radioed a message to the news pilots, requesting them to stay high to avoid unnecessary noise that might interfere with communications on the ground. As is usually the case, the news media pilots cooperated. In the skies of L.A., this has become a daily dance over the scenes of emergencies. Even new media pilots understand the delicate nature of such operations and generally keep a respectful distance as they circle. Today was no exception.

After harnessing and hoisting the woman's husband to safety, McCarthy watched as the helicopter banked and flew off to the west. The plan, as previously agreed upon by the rescue crew, was for the helicopter to transport the elderly couple to the helispot at the Malibu Incident Command Post (ICP), located near Pepperdine University. There, they could be examined for medical problems or trauma, given warm, dry clothing, and moved to the evacuation center (or, if necessary, to a hospital). Then Finnerty would return to pluck McCarthy from the porch.

McCarthy stood looking out at the flood as the sound of the helicopters faded. The news helicopters followed Air Squad 8 to Pepperdine University to film the two flood victims being transferred to ground-based fire department units for medical evaluation. He turned to see the dog sitting on his haunches, looking up intently at McCarthy. There was no canine rescue harness immediately available, so McCarthy considered leaving the dog right where he was. However, as he viewed the flood that raged around both sides of the home, it became apparent that the house might be under water shortly. The black Lab sat and stared at McCarthy as he pondered the situation. Resignedly, McCarthy turned and went into the house, followed closely by the dog, who signaled his intent to remain at the side of his new friend throughout the ordeal.

McCarthy knew he would have trouble hoisting the dog into the helicopter, which hovered over the treetops approximately 70 ft above. If the dog struggled—or if it bit him—McCarthy knew he might accidentally drop the animal. What he needed was something to hold the Lab securely, something that could be connected to the end of the hoist cable for the 70-ft ascent to the helicopter without allowing the dog to wriggle free or bite members of the rescue crew.

McCarthy rummaged through the house with the Lab at his heels. He went thought the garage, several bedrooms, and finally a portion of the attic. He emerged with a large black suitcase, which he set down on the porch. The dog sat, as McCarthy unzipped the case and visually compared the dog's mass to the interior space of the case. As long as his new friend was near, the dog was interested but unconcerned. Finally, as the *whop-whop* sound of the returning helicopters grew, McCarthy decided to give it a try.

At first, stuffing the big Lab into a suitcase did not seem to be an insurmountable objective. McCarthy had little trouble getting the animal to sit in the opened case. It was only when the dog realized that McCarthy intended to zip the case up *with him inside* that the dog decided he would no longer be a willing participant in the plan. With one hand, McCarthy tried to hold the dog inside the suitcase, while the other hand zipped the bag. Paws kept appearing and the dog suddenly panicked when he realized that the zipper was coming closed. McCarthy struggled to push the Lab's head back into the case as he closed the final inches of zipper. The suitcase now bulged in different places, as if a catfight were occurring inside.

As Finnerty brought Air Squad 8 to a hover, McCarthy secured the suitcase with nylon rescue webbing and a carabiner. Monahan lowered the cable, and McCarthy attached the suitcase and his own harness to the hook, securing both carabiners and signaling his readiness to be raised. On the way up the cable, McCarthy held the suitcase securely and spoke soothingly to the dog over the sound of the rotor wash. Upon reaching the level of the skid, McCarthy was assisted into the cabin. The dog was kept in the suitcase until they landed; they decided that it would not be a good idea to have an 80-lb, panicked dog running loose in the helicopter. Five minutes later, the Lab was reunited with his masters, and the crew of Air Squad 8 was sent to the next rescue.

As the day progressed, Air Squad 8 performed a number of other rescues, extracting about 20 victims from various predicaments under increasing dangers from the weather and other hazards. The weather was extremely dangerous for any kind of flight and grew even worse as the helicopter slipped between the grey-black clouds that blanketed the rugged Santa Monica Mountains. Low clouds hugged the ridges all the way to the sea. Hidden within them were flight hazards that included mist-veiled peaks, periodic whiteout conditions, and power lines that stretched like spider webs, virtually unseen, across the canyons.

Power lines are treated as the mortal enemy of helicopters because of their demonstrated ability to bring helicopters down in an instant. Helicopters have the agility and power to fly low to the ground and deep within rugged canyons, which has proven to be a great advantage when fighting wildland fires and performing technical rescues in mountainous terrain. Unfortunately, many canyons are crossed by high-tension electrical lines, and helicopters run into them with disturbing frequency, often with fatal results.

One is tempted to use the metaphor of a spider's web to describe the relationship of power lines to helicopters—a characterization that would not be entirely inaccurate. Every year, helicopters strike transmission wires and become entangled in them or slam to the ground, killing crews and passengers. Because power lines are often invisible when viewed from the air against the backdrop of the land, *wire strikes* occur in all types of weather, in darkness, and even in bright sunlight. High-tension wires are such a commonplace hazard that many helicopters are equipped with blade-like structures capable of literally slicing through heavy lines, hopefully allowing them to remain airborne until they can safely land. On this day, the wires were of special concern because the driving rain colluded with darkness and thick clouds to make them practically invisible.

The situation that day wasn't much better for firefighters and rescue teams on the ground. They were struggling to keep up with calls from people trapped in worsening predicaments across a swath of mountains that runs 30 miles along the coast and 15 miles inland. Some roads were being washed out, while others were covered in moving brown masses of mud that swept cars away and blocked access for fire engines and rescue units. Homes were being buried or washed away in several locations.

In another part of Malibu, power lines were causing trouble for ground units, including USAR-2, one of the LACoFD's swift-water rescue-trained USAR companies assigned to Malibu for the storm rescue operations. While responding to a report of people trapped by floodwaters in the rugged Fernwood area of Topanga Canyon, engineer Dave Norman suddenly found the route blocked by high-voltage wires dangling low over Topanga Canyon Road.

Just uphill, a utility pole had snapped and left its top half suspended by high-voltage wires that refused to snap. The whole mess dangled precariously at a level just above the windshield of the big rescue truck, whose emergency lights and high body would certainly become snagged if they attempted to drive through without regard for the wires. Cutting or raising the lines was out of the question because their voltage was not known and special equipment was needed to do it safely. Also, a curtain of falling rain and streams of water flowing down the road greatly increased the potential for electrocution.

The captain (this author) decided that safe handling of the wires would normally require the response of the local power utility, in this case, an Edison field response unit. I knew that Edison crews were already swamped with other emergencies, and it might take hours for them to reach our location. Already wet in our swift-water rescue gear from previous rescue operations, we sat in the cab watching the wires between rapid-fire swipes of the windshield wiper arms as a torrent of rain fell outside. On the radio, we heard reports indicating that conditions for the stranded victims were worsening. It was imperative to get through to assist the local engine company with what might prove to be a harrowing rescue operation.

We observed the following phenomenon: each time the wind gusted, the tensioned power lines bounced up and down with the weight of the broken top half of the pole, leaving brief openings that *might* be just high enough for the big rescue truck to squeeze through. The rain was beating down heavily on the cab of our apparatus. We looked at each other in one of those moments that says, "How much do we want to risk here, and if we screw up, how bad might be the consequences?" We both knew there was no other way to reach the rescue site from the coastal side; to turn around to attempt another route might delay response by another 40 minutes if we were lucky and the roads held. Even with a mechanical axe on board, we lacked the proper tools to cut and maneuver the high-voltage lines in a driving rainstorm, standing on a steep muddy slope.

"Well," I asked Norman, "Do you think you can make it?"

"If I gun it, I might be able to squeeze through. *If* I time it right," he replied.

"Give it a shot if you think you can. Otherwise, we'll have to turn back and find another route," I responded.

Norman revved the engine of the apparatus, feathered the clutch, and practiced lurching forward, mentally gauging the timing required to squeeze through the gap when a wind gust lifted the wires sufficiently high. The time of the maneuver would be akin to entering a school yard jump rope game; a false move would result in a wire strike and possible entanglement with deadly hot wires. It would be a gamble, but of course, many fire department emergency operations include a certain level of calculated risk. The trick is to ensure that the risk is commensurate with the need. In this case, the reported predicament of the victims led us to believe it was worth risking potential entanglement to get through. And I was confident in Norman's ability to time it right and get the job done.

Norman inched the rescue apparatus forward for a final push beneath the wires. The wind was erratic and gusting from different directions all the while. The timing had to be precise, and Norman would have to make an instant decision to punch it through when the moment of opportunity presented itself. There was no guarantee that any particular wind gust would lift the wires sufficiently high for the apparatus to pass beneath without snagging them.

Finally, after several false starts, it was apparent that the wires were not lifting high enough. The clincher was the knowledge that if our apparatus became entangled, it would only complicate things at the original rescue by further delaying our arrival. It would require another USAR unit or swift-water rescue team to assist the engine company now committed to the original rescue. Finally, resigned to the futility of trying to get though this snag without worsening our situation, I instructed Norman to back down the road and turn it around. We would find another way to get to the rescue.

I radioed engine 69 to inform them of our situation. We responded back toward the coast to find an alternate route up an adjoining canyon that would allow us to cut back into Topanga Canyon to reach the rescue site. Engine 69 replied that they might be able to stabilize the situation and possibly rescue the victims. They really had no other option at that point, because they were essentially isolated. Eventually, engine 69's crew found a spot at which they could line the creek and extract the victims one by one, with a PFD-equipped firefighter escorting them back to safety. Engine 69's successful rescue allowed our unit to then respond to yet another rescue situation in Topanga Canyon.

It went on like that for three days. Approximately 80 victims were rescued from flash floods, mudslides, and mud and debris flows in the Malibu area alone. A dozen other people were successfully rescued in other areas of L.A. County during that period. No victims or rescuers were lost, an indication of the effectiveness of the multi-prong, ground-based, and helicopter-based approach used to manage flood, mud and debris flow, and swift-water rescue problems.

Endnotes

1 The Pavehawk is a military rescue version of the renowned Blackhawk helicopter.

2 Meanwhile, at precisely the same time, another group of climbers was dead or dying in the middle of a raging storm on the slopes of Mt. Rainier, and another dramatic helicopter-based rescue operation was in progress there.

3 Air squads are helicopters equipped with belly-mounted 360-gal water/foam tanks, external hoists with 220 ft of cable, and crews consisting of one pilot and two firefighter/paramedics trained in USAR and swift-water rescue. These crews are augmented with personnel from the LACoFD's USAR task force fire stations for technical rescues.

4 One member of this group, firefighter/paramedic Jeff Langley, a pioneer in the field of helicopter and swift-water rescue, died during the course of a helicopter hoist rescue operation in Malibu in 1993.

5 A number of other agencies in southern California and other parts of the United States have also developed innovative HSWR methods and equipment. Countless fire/rescue agencies are now using some variation of these new methods and equipment.

6 Ironically, another helicopter-based LACoFD swift-water rescue team, pre-deployed to the Malibu area for the storms that day, was forced to use the tethered rescuer HSW evolution to pluck two of four teenagers from an island in the middle of a raging creek. The helicopter's hoist motor overheated after rescuing the first two victims. This occurred at virtually the same time we were flying to the hospital with the baby we rescued at the Creek Incident. That same year, an L.A. County sheriff's department rescue helicopter, using a similar tethered rescuer method, plucked an unconscious child from beneath the roaring waters of the rugged San Gabriel River. That victim had been wedged between rocks after being swept downstream, and Deputy Terry Ascherin was lowered via hoist cable in a rescue harness to snatch the child from certain death.

7 The closest receiving hospital to this rescue also happens to be a trauma center capable of handling pediatric trauma.

8 Since this rescue, the LACoFD has provided hand-held thermal-imaging systems to its USAR task force fire stations and has installed FLIR to its command helicopter, as well as, providing night-vision systems to these units. These systems might have been helpful in reduced light conditions during a storm.

9 This is one reason a helicopter should be dispatched on every submerged-victim rescue whenever feasible, because the helicopter crew may spot the victim even when other rescuers standing or floating nearby can't.

10 Much like a cliff rescue or other high-angle rescue operation, in this HSW evolution, the rescuer is lowered over the side by an air crewman using a brake bar rack, which is then locked off when the rescuer reaches the desired distance beneath the helicopter. This version of a short-haul rescue allows the helicopter to fly along and immediately deploy a rescuer at the desired moment, rather than having a rescuer dangling beneath the belly of the helicopter for the duration of the flight. Once the rescuer is in position and the brake bar rack locked off, the pilot (on the air crewman's commands) dips the rescuer into the water and—if necessary—moves downstream in the current, thereby eliminating the situation where victim and rescuer slam together because the rescuer is stationary. The main difference between this operation and a standard rope rescue operation is this: there is no belay line for the rescuer; he is attached only to a single rope. Multiple ropes in this situation would simply foul as soon as the rescuer began his inevitable spin below the helicopter, so a single rope is the only viable alternative.

11 The incident was divided geographically into a river-right division and river-left division; functionally, it consisted of the existing search group, with the addition of a rescue group, a medical group, an upstream safety, and a downstream safety.

12 All victims normally are secured with a Cearley rescue strap.

13 See "9/11 Tape Raises Added Questions On Radio Failures," by Jim Dwyer and Kevin Flynn, *New York Times*, page A1.

14 Vincent Dunn, "An Open Letter to Tom Ridge, Homeland Security Director," *Firehouse Magazine* (April 2002).

15 John T. O'Hagan, *High-rise/Fire & Life Safety*, Fire Engineering Books (PennWell Publishing, 1977).

16 Of those trapped on the roof, 80 were rescued and 90 perished.

Appendix I: FDNY Deputy Chief Ray Downey (Deceased) Congressional Testimony

Federal response to domestic terrorism involving WMD training for first responders witness statement.

Chief Raymond M. Downey
Special Operations Command
Fire Department
City Of New York

March 21, 1998

Good morning Mr. Chairman and committee members. My name is Ray Downey and I'm the chief of rescue operations for the fire department in the City of New York. First I would like to thank you for holding this hearing today. The fire service greatly appreciates the fact that your committee has concerns regarding WMD. The fire service is greatly concerned, being first responders that will eventually have to deal with the issues of WMD.

My intent today is to speak not only as a member of the FDNY, but for the entire fire service. I have had the unique experience to respond to incidents of WMD both as a first responder and as a federal asset that arrived on the scene some twelve hours after the incident. This is one of many concerns that the fire service has about the training and expectations of both the fire service and federal support that is being promised in the event of an incident involving WMD.

During the WTC bombing I was on the scene immediately after the bombing had taken place. Unless you have been there, you cannot fully appreciate what firefighters face during an incident of WMD terrorism. The fire service has always been respected by the public for their immediate response capability to the calls of those in danger. With that response comes the dedication and fearless courage displayed by these firefighters. It wasn't any different at the WTC. As a result of this terrorism incident, firefighters that operated at this incident still question what would have happened had that bomb been a dirty bomb.

Would we take the same actions today if a major bombing were to occur in our response district? Five years later, we are better prepared, have more knowledge about WMD, but still see many shortfalls in the area of first responder capabilities for dealing with and mitigating incidents of WMD. The fear of chemical or biological terrorism is foremost in the minds of every firefighter. What we read and hear about regarding the nations preparations and training for these incidents does not go far enough.

During the WTC bombing, firefighters not only faced a difficult fire operation, but had to evacuate almost 50,000 occupants from the Trade Center complex. More than 500 victims were treated for various injuries, while another 600 responded to hospitals on their own. What would have happened if they had been contaminated by a chemical agent?

In 1993, not a fire department in the United States could have handled an incident involving a chemical agent affecting this many victims. Can the fire service handle the same potential incident in 1998 after five years of additional preparation and training? The answer in most cases is a No. Why? Lack of sufficient funding and training for WMD.

Two years after the Oklahoma City bombing, Chief Gary Marrs of the Oklahoma [City] Fire Department providing witness testimony stated that we, the fire service, are no better prepared than we were back in 1995. Why? The training that has been given with federal funding is not being directed to the first responder, and the lack of providing funding for the necessary equipment for these responders is directly related to the lack of our preparedness. My experience of working for sixteen (16) days as the operations chief for the USAR teams in Oklahoma City only reinforces my feelings about the needs of the first responders. The Oklahoma City Fire Department has not received the real credit they deserve for the heroic actions they took during those first few hours before any help or support arrived from outside their jurisdiction. As was the case in the rescue effort during the WTC bombing, the same results occurred during the Oklahoma City bombing. Not one victim in either incident died as a result of awaiting rescue by the firefighters after the bombing.

The first responders (AKA the firefighters in both cases) performed these heroic actions only because they were able to be on the scene within minutes and were properly trained and equipped. But, what if that bomb had an additional chemical agent dispersed with the explosion? Would the success rate [have] been the same? Not likely. What would happen if it occurred today? Would we be prepared? Some fire departments have increased their capabilities, but the majority of the country is still not prepared for these type incidents.

In Atlanta, the experience gained by the fire service after the bombing of the family planning center undoubtedly saved lives of firefighters and other emergency personnel that responded to the bombing of the gay and lesbian nightclub. They had learned their lessons from the previous incidents. We have not had this opportunity when dealing with chemical, biological, or nuclear terrorist incidents. What is it that the fire service needs to be prepared for these type incidents? The preparation, training, and equipment requirements should be approached from a bottom-up planning process. Permit the first responders to get involved with the many various agencies at the federal level that are preparing terrorism training programs that will ultimately affect them. This can be accomplished by reaching out to the first responder and finding out exactly what the needs of the fire service are. The federal government needs to provide assistance and funding for training, detection equipment, personal protective equipment, and mass decontamination capabilities. The realization by the federal government that the resources that they will supply to local jurisdictions during a WMD incident will be of support role and work under the direction of the local IC. If these goals can be reached, the fire service will be much more capable of dealing with WMD incidents.

I want to thank you again for this opportunity to appear as a witness before you today and express to you on behalf of the entire fire service our sincere gratitude for all the work and accomplishments that have benefited the fire service through your efforts. It is the first responder that will be facing the challenges that WMD presents, they are the ones that need the funding and assistance that the Federal Government can provide.

Appendix II: Levels of Personnel Protective Garments

The advent of NBC agents used in terrorist attacks makes it necessary for firefighters and rescuers to seriously consider what level of protection is offered by their PPE. This has become a prominent issue in the aftermath of recent terrorist attacks that have placed firefighters, EMS personnel, and other rescuers in situations where they are vulnerable to the effects of WMD agents. The 9-11 WTC attacks crystallized the risk to firefighters, police officers, EMS personnel, and other rescuers. In Figure A–1, members of FEMA Virginia USAR Task Force 2 conduct a solemn memorial ceremony before departing the site of the Pentagon collapse at the end of their mission there.

Fig. A–1 Memorial at the Pentagon

The U.S. Fire Administration (USFA) defines *critical infrastructures* of the emergency services as "those physical and cyber assets that are essential for the accomplishment of missions affecting life and property. They are the people, things, or systems that will seriously degrade or prevent survivability and mission success if not intact and operational."[25] Included in the category Critical Infrastructures are firefighters/rescuers/paramedics; fire and EMS stations, apparatus, and communications; and public safety points, or 911 communications centers and their personnel, etc. So it's apparent that protection of the health and safety of firefighters, rescuers, EMS personnel, hazmat personnel, and others who respond to rescue-related terrorist attacks has been designated a priority.

What does this mean to you, the firefighter or rescuer? To begin, it means that you should be wearing the appropriate attire when responding to an actual or threatened terrorism event. The trouble is that this is (relatively speaking) a newly recognized issue as it relates to WMD and NBC agents. There is debate about what constitutes appropriate PPE for first responders dealing with the consequences of modern terrorism.

There is even more debate about who will fund the research, development, and procurement of the personal attire required to operate effectively in WMD/NBC-contaminated environments. Consequently, as of this writing, the typical first responder simply does not have access to the appropriate level of PPE required to ensure a reasonable degree of safety and health when dealing with a WMD-related attack. This section is not intended as a definitive answer to the question: "What PPE should be worn by rescuers?" But it may help illuminate some of the critical issues.

Protective garments for firefighters and other rescuers

Number 29 Code of Federal Regulations (CFR), Part 1910.120, Appendix B, Part A, lists four levels of protective garments, defined as Levels A, B, C, and D. Level A offers the most protection, and Level D, offers the least protection. Before listing the levels, it's important to recognize the following basic facts:[26]

- No fabric thus far developed is impermeable forever.
- Fabric used for PPE should not tear or rip easily.
- Fitting is an obvious factor that cannot be ignored.
- All potential leakage areas or areas of weakness (wrists, ankles, and mask if not encapsulated) should be secured with duct tape or chemical-resistant tape.

Level A protective garments

Level A garments are fully encapsulated suits worn with a SCBA (closed circuit, open circuit, or supplied air). These garments are gas-tight, vapor-proof, and chemical resistant. Level A protection is required when firefighters and rescuers may be exposed to the worst contaminants. They protect the skin, respiratory tract, eyes, and mucous membranes. Level A protection *must* be worn when entering a zone containing one or more unknown or questionable contaminant, especially airborne agents. It is important to wear Level A PPE when there is an indication that an aerosol-generating device may have been used. Key components of Level A PPE include (but are not limited to) the following:

- Reusable or disposable fully encapsulated chemical-resistant suit (tested and certified against chemical and biological agents)
- SCBA and spare (escape) bottle
- Chemical-resistant gloves
- Inner gloves
- Personal cooling system, vest, or full suit with support equipment
- Helmet
- Inner chemical-resistant suit or garment
- Chemical-resistant tape
- Chemical-resistant boots and outer booties
- Secure two-way communications (radio, hard wire, cell phone, etc.)

Level B protective garments

Level B PPE is used when the highest level of respiratory protection is required, but a lower level of skin and eye protection will suffice. The U.S. Center for Disease Control (CDC) mandates Level B protection when there is a suspected biological aerosol that is no longer being dispersed, or if there is a splash hazard. Key components of Level B protection include (but are not limited to) the following:

- Splash-resistant chemical clothing and hood
- SCBA and escape bottle
- Chemical-resistant gloves
- Inner gloves
- Personal cooling system
- Helmet
- Chemical-resistant tape
- Chemical-resistant boots and outer booties
- Secured two-way communication

Level C protective garments

Level C garments consist of a splash-resistant chemical suit and the same level of skin protection as Level B, with an air-purifying respirator (APR). Level C is appropriate when the concentration and types of airborne substances are known and the criteria for using APRs is met. Key components of Level C PPE include (but are not limited to) the following:

- Splash-resistant chemical clothing and hood
- Air-permeable or semi-permeable chemical-resistant clothing
- Full-face APR with appropriate cartridges (SCBA can be used)
- Chemical-resistant gloves, boots, tape, and outer booties
- Helmet
- Personal cooling system
- Inner gloves
- Secure two-way communications

Level D protective garments

Level D is a basic work uniform that can be used when no respiratory protection is required and there is minimal need for skin protection. There can be no known hazard, and the job cannot involve potential for splashes, immersions, or unexpected inhalation of hazmats or chemicals. Firefighter turnouts with SCBA on the wearer's person meet or exceed Level D protection. Key components of Level D PPE include the following:

- Coveralls
- Safety boots or shoes
- Safety glasses
- Helmet
- Gloves
- Emergency escape breathing apparatus
- Face shield

Firefighter garments and SCBA in chemical or biologically contaminated atmospheres

As of this writing, there is debate in some circles about the level of protection that should characterize full firefighting turnouts or bunker gear with SCBA. Some authorities contend that the standard NFPA-compliant firefighting ensemble with SCBA is sufficient for brief entry into contaminated zones with certain requirements. One requirement is the application of chemical-resistant tape or duct tape to seal all openings in the garments. The firefighter must also wear chemical-resistant gloves and outer booties, etc. The final requirement is performing gross decontamination and secondary decontamination upon exiting the hot zone. Others believe that even with these precautions, the risk is not worth the gain. Suffice to say that the issue of the firefighter ensemble in contaminated places has not been fully resolved.

We know that full turnout/bunker gear with SCBA provides respiratory protection and limited splash protection. However, it will not protect firefighters against skin absorption of certain chemical substances and their vapors. Therefore, this ensemble clearly does not meet Level A or Level B protection.

The CDC has developed interim recommendations for firefighters and rescuers operating in potential biological agent-contaminated environments. The recommendations require (at a minimum) the use of half-mask or full-face piece APRs with particulate filter efficiencies ranging from N95 (for hazards such as pulmonary tuberculosis) to P100 (for hazards such as hantavirus).

According to some early research on fireground PPE as protection for WMD events, the hierarchy of firefighting PPE, listed in order of increased protection, includes the following combinations:

- Least protection—standard firefighter turnout ensemble with SCBA and no taping of cuffs and openings with duct tape
- Next-lowest level of protection—standard firefighting turnout ensemble with SCBA and self-taped cuffs and openings (using duct tape)
- Second-highest level of protection—standard firefighting ensemble with SCBA and buddy-taped cuffs and openings
- Most protection—standard firefighter ensemble over Tyvek undergarment

Standard turnout gear with SCBA provides a first responder with sufficient protection from nerve agent vapor hazards inside interior or downwind areas of the hot zone to allow 30 min rescue time for known live victims. Self-taped turnout gear with SCBA provides sufficient protection in an unknown-nerve-agent environment for a 3-min reconnaissance to search for living victims.

During USAR operations at the Pentagon after 9-11, it took several days for all the worker safety agencies and the EPA to agree on the level of required protection. The result was a ratcheting-up of the level of protection. Protection now includes APRs and Tyvek suits for some personnel in addition to other normal PPE such as helmets, Nomex3 jump suits, safety boots, gloves, goggles, etc. These PPE requirements apply to firefighters and rescuers actually making contact with the debris and bodies.

As a member of the IST coordinating the multi-division operations of the USAR task forces at the Pentagon, this author was involved in enforcing the PPE requirements within the collapse zone. The mandated level of protection was adjusted up and down during the 12-day SAR operation.

The fluctuating requirements created some operational problems that were overcome through good command and coordination. The consternation expressed by many of the firefighter/rescuers for the changing requirements was understandable. In the end, we were left with one overall impression about the issue of PPE: the level of protection we had agreed upon and required was not consistent and grew more stringent as time went on.

The course of events begs two questions: If the level of protection had to be increased as the incident became more controlled and contamination hazards seemed to lessen, what was happening in the early stages of the incident? Did we require an appropriate level of PPE from the beginning?

Responding to the first question, rescuers probably were exposed to more contaminants with less protection during the early phases of the operation. That exposure occurred after the fires were being controlled, the smoke was dissipating, and personnel had begun doffing their SCBA. The answer to the second question seems to be no, probably not.

As a result of this situation, there has been informal agreement in some circles that the following basic protocols should be adopted nationwide in the case of explosions, structure collapses, obvious terrorist attacks, and suspicious incidents:

- If there is collapse, fire, smoke, dust, or mist, or if there is potential for contamination by chemical, alpha particle, or biological agents, firefighters should approach the scene and operate at the scene wearing their full firefighter ensemble with SCBA, hood, gloves, etc.
- Non-firefighter rescuers should wear their full ensemble with SCBA or APR. In fact, in suspicious cases and after explosions of unknown origin, it's prudent also to use radiological devices to monitor for radiological hazards while approaching the scene. If readings are excessively high (e.g., 200 roentgens per hour or higher), responders should stop and reverse course until the radiation readings decrease to a safe operational level.

As an incident becomes more controlled, responders can assess the atmosphere and eliminate potential contaminants to determine whether to maintain strict PPE controls or reduce them. Those decisions can be made in a way that ensures a reasonable level of protection for personnel engaged in SAR operations. In other words, first responders initially gear up with the highest level of protection, then gradually back down to lower levels as serious hazards are eliminated or determined to be nonexistent. The opposite approach potentially exposes first responders to more severe hazards in the early stages when there is less chance to determine exactly what contaminants may be present. In short, it is the reverse of the approach used at the Pentagon, in Oklahoma City, at the WTC, and other rescue incidents where potential or actual hazardous contamination existed.

Appendix III: ICS Position Considerations for Collapse Operations

The following positions are generic to some collapse SAR incidents. Although these positions are consistent with the ICS and SEMS, the following recommendations are open to interpretation and adjustment based on local protocols, local resources, and the conditions at the scene of the collapse. It is strongly recommended that all fire/rescue agencies employ ICS/SEMS in their daily and disaster operations. Properly used, it makes life easier for ICs, officers, and line workers.

When in doubt about specific positions required for collapse search and rescue operations, consult with the rescue/USAR company officers to determine which positions are needed for your incident. Here is a sampling of typical positions for these operations:

- IC
- Operations—to supervise the implementation of strategic goals developed by the IC
- Division supervisors—to supervise specific geographical areas of the incident, each of which is given a letter designation (e.g., division A, division B, etc.). Division A is generally in the front of the building or area (when this is possible), and the other letters follow in order in a clockwise direction around the incident site. For very large incidents or those that threaten to broaden, the IC can designate the left side (front) as division A, the right side (front) as division Z, and then fill in the corresponding letters as necessary. This is a typical approach to wildland fires, which have a tendency to multiply exponentially as the incident progresses. It will work for large-area disasters.
- Group supervisors—to supervise and coordinate functional groups (e.g., search group, rescue group, etc.) that have no specific geographical boundaries, which in fact may be assigned to conduct their functional tasks in one or more divisions
- Safety officer—trained in collapse search and rescue operations, prepared to recognize and mitigate safety problems as they occur
- Rescue group leader (USAR/rescue company officer or other USAR-trained member—to recommend, evaluate, and carry out the strategic goals of the IC/OIC through the development and use of appropriate tactics
- Four- to five-member rescue squads—made up of first responders or rescue specialists from rescue companies or USAR units. Each squad should be supervised by a USAR-trained rescue squad officer, for a total of six members (similar to the FEMA USAR task force configuration, and depending on personnel availability and local protocol)

  or

  Six-member void space search/rescue teams—depending on manpower availability and local protocol

  or

  Two-member search/rescue teams—depending on manpower availability and local protocol
- Search group leader—USAR/rescue company officer or other qualified member
- Two-member technical search teams
- Add two-member canine search teams—one handler, one canine each to the search group as they arrive, and integrate technical search capabilities. Consider a canine search coordinator if multiple canine teams are to be used.
- Medical group leader

- Two-member medical teams—two paramedics (PM), or one PM and one EMT, to treat trapped victims until they are extricated
- One medical director—a physician to oversee the treatment of trapped victims, including the capability to perform field amputations and other advanced treatment options
- Equipment pool manager
- Rehab group—to establish and operate a rehab system to keep firefighters and rescuers operating at peak capacity for long-term and/or highly personnel-intensive operations
- Technical group leader
- Heavy-equipment/rigging officer—a USAR-trained heavy-equipment operator to advise on cranes and other heavy equipment, and to act as liaison
- Shoring officer—to supervise shoring operations
- Cutting team—to set up a cutting station and cut specified shoring
- Structures specialist—generally a qualified structural engineer or other member having construction/engineering training, advised about structural integrity and options for stabilizing structures
- Hazmat specialist
- Plans officer—to work with USAR-trained personnel to develop an IAP specific to collapse operations
- Logistics officer
- Equipment pool manager—to supervise pool of rescue equipment and materials
- Staging manager
- Technical information specialist—to deal with mapping and GIS issues as needed
- Personnel accountability officer
- RIC—with RIC leader

In the case of a major collapse operation, the need for resources will rapidly outstrip the first alarm response. The IC should not hesitate to reduce the *reflex time*. Reflex time as defined by Ray Downey is the time between recognizing the need for a resource and the time that it arrives on the scene. Reflex time can be reduced by requesting additional resources early in the incident if people are trapped and missing. Consult with USAR/rescue company officers to determine what resources are required.

Appendix IV: Debris Pile Tunneling and Trenching

In collapses involving piles of masonry, soil, grain, and other materials, some rescue teams have at times advocated driving tunnels or trenches laterally and sometimes upward or downward into the collapse area. This has been used in Europe, especially the UK. Rescuers seek places where live victims may be trapped deep in the building within hidden void areas like basements, bathrooms, closets, and other places. These areas may provide protection because of room size or sturdy furnishings. Is trenching and tunneling effective for collapse rescue operations? As of this writing, trenching/tunneling is controversial and there seems to be no consensus on its advisability.

One school of thought says that—depending on the collapse situation and the expertise of the rescue teams—trenches and tunnels may be efficient ways to move in a direct line. Rescuers can move toward victims detected by hail searches and other means. This concept could allow rescuers to remove debris in a systematic manner to expedite access to trapped victims and create a path of egress. Some experienced rescuers swear by this approach in certain types of collapse such as un-reinforced masonry structures that leave huge mounds of debris segmented by floors and even cellars, with people trapped within.

Another school maintains that tunneling and trenching are inherently too hazardous, not just for the rescuers but also for some victims trapped inside the collapse. These individuals maintain their view for primarily two reasons: (1) they believe that some shoring options are unreliable or too unwieldy, and (2) they feel that traditional lateral/vertical void-space penetration and search or delayering from the top down by selective debris removal offer higher degrees of success and more protection for rescuers.

Despite the reservations that some readers might express, a discussion of trenching and tunneling is included here for consideration, discussion, and hopefully debate. Although this author has been assigned to a number of structure collapse-related disasters over the years, debris pile tunneling and trenching are not within the author's personal experience. Much of the information contained herein was gleaned from personal discussions with firefighters and rescuers from the United Kingdom, which developed and honed some of these methods during World War II. Some of these personnel have used tunneling and trenching operations during disasters in Turkey and other disaster-prone nations. Further testing and experimentation may provide more definitive answers about the efficacy of these methods.

According to firefighters and rescuers who have done it, tunneling and trenching differ in subtle ways from standard void-space searches (part of stage 3 of collapse rescue operations). Void-space search generally involves rescuers following existing voids, cracks, and passageways through a collapse, removing debris that's in the way and breaching the occasional obstacle. Tunneling and trenching, on the other hand, are intended to create more direct routes to places where victims have been detected or where they are suspected to be waiting for assistance. In tunneling and trenching, rescuers essentially burrow through whatever debris might be in their path. They snake around hardened areas when necessary, taking care to follow the path of least resistance. They take advantage of soft spots for forward movement and hardened areas to help keep intact the openings they create.

Tunneling and trenching are hazardous undertakings that are not universally approved and used but which are strongly advocated by some teams. These operations should only be attempted by highly trained rescuers under the supervision of highly experienced officers who have done it before. The dangers associated with these measures should not be taken lightly. The potential for secondary collapse is substantial, especially in post-earthquake situations where aftershocks are sure to rock the area for days or weeks.

Naturally, the size of the debris to be searched is a factor in determining whether to trench or tunnel. According to those who have performed trenching and tunneling, if the height of the debris pile is less than 10 ft, a trench is a more viable alternative than a tunnel. However, if the pile exceeds 10 ft, it may take more time, effort, and danger to trench into the places where victims are awaiting help. The higher the debris pile, the more a tunnel makes sense to follow a direct line to reach trapped victims.

Although both operations carry significant dangers, trenching is naturally less hazardous than tunneling for the obvious reason that in tunneling there is debris directly overhead through which rescuers are attempting to bore a hole. Therefore, relatively speaking, trenching is in many cases a more palatable option if the situation allows. Once again, the two basic rescue questions should be asked before embarking on either journey: *Is it (reasonably) safe?* and *Will it work?*

Debris trenching

According to rescuers and firefighters who have performed debris pile tunneling, the following are basic parameters that should be considered or observed:

- Determine the path to be taken. Base it on obvious and probable structural conditions and the likelihood of trapped victims. Consider the floor plan, occupancy, time of day when the event occurred, reports from survivors, team reconnaissance, hail searches, canine search operations, technical search results, and other indicators. Deviations may be necessary when it's determined to be faster and safer to detour around load-bearing walls, beams, structural steelwork, etc.
- Start at the edge of the debris pile and work your way to the ground, then drive the trench horizontally through the debris.
- Using hands, shovels, and other hand tools, remove the largest pieces from atop the area to be trenched.
- After the larger pieces are removed, begin burrowing downward by removing other pieces.
- Remember that trapped victims may be underfoot in some cases, so step lightly and avoid disturbing material whenever possible. All bystanders and nonessential personnel should be removed from the debris piles and damaged structures to avoid secondary collapse and injury or death to people trapped beneath. To prevent inward collapse, the top of the trench should be wider than the base. In this sense, the walls of the trench should slope outward.
- Using available materials, shore the side walls as you would a regular trench, with obvious differences due to the nature and configuration of the materials through which you are trenching.
- Through the use of bucket brigades or other methods, transport debris from the trench to another part of the site where it will be out of the way of SAR operations.
- Do not pile the debris on roads or other places that impede the access of emergency vehicles, supplies, ambulances, fire and rescue units, etc.
- Do not pile the debris atop fire hydrants or cisterns, for you may need firefighting water at any time!
- Do not pile debris in gullies or waterways because it can dam water and cause flooding.
- Do not pile debris atop unsearched piles of debris or collapsed buildings where victims may still be trapped inside.
- If the event is potentially related to terrorism or there are other crime scene considerations, place the debris in a secured and well-marked area that will allow investigators and law enforcement officials to conduct post-collapse investigations.

- Mark piles of debris so that other SAR teams will not attempt to re-search the debris piles as if they were original collapse debris piles.
- Rotate personnel as necessary to maintain efficiency and to prevent injury and exhaustion among rescuers.
- As live victims are heard or detected (determining as best you can), do not lose contact with them. Determine whether your actions are making the situation worse for them (e.g., are they being crushed, debris is pushing in on them, etc.).
- If deceased victims or body parts are encountered, they should be removed to a secure area, preferably by coroner or morgue detail personnel (if available). In disasters where these resources are not yet available, rescuers should remove the victims or parts if they are in the path of the trench and secure them for later identification, investigation, and eventual repatriation with families.
- Monitor the atmosphere and provide adequate ventilation.
- Maintain fire suppression and rapid intervention capabilities throughout the process.
- Maintain a personal accountability system

Debris tunneling

According to those who have done it, the process of debris tunneling differs markedly from tunneling through rock or earth. The loose, unconsolidated, unstable nature of the debris pile makes it extremely vulnerable to shifting and secondary collapse. The situation is made all the worse by firefighters and rescue team members tunneling straight through the pile. As forward progress is made, bracing and shoring is required. Again, because of the nature of the material being tunneled, this is a complicated process. Aftershocks or other secondary effects do not make matters any better for rescuers. Use the following guidelines:

- Avoid disturbing beams, columns, walls, and other items determined to be load bearing. If it is difficult to determine whether some timbers, beams, walls, and columns are carrying heavy loads, it's often safer and more efficient to brace them and avoid cutting or moving them.
- Cutting/breaching of walls and slabs determined to be non-load bearing should be conducted by firefighters/rescuers familiar with these operations. They should use appropriate equipment (hopefully equipment that reduces vibration). They should be cognizant of the complications that can arise from excess vibration, breaching load-bearing walls, and aftershocks.
- Shore as you move forward into the building. Use available materials, including those that can be obtained from the equipment cache, local fire/rescue agencies, and local suppliers. Street signs, posts, and other materials from the actual collapsed building can also be used.
- Similar to the tactics of void-space search squads, lead rescuers must push their way into the collapse, picking their path, choosing which material can be moved or breached, shoring it as necessary, and passing debris back to the others. Rescuers following behind can be assigned to enlarge the opening, pass debris back down the line to the outside, etc.
- Monitor the atmosphere and provide adequate ventilation.
- Maintain fire suppression and rapid intervention capabilities throughout the process.
- Maintain a personal accountability system

Appendix V:
Structure and Victim Marking System[1]

Table A–1 Structure/Hazards Markings

Description	Marking	Information
Make a large (2′ x 2′) square box with orange spray paint on the outside of the main entrance to the structure. Put the date, time, hazardous material conditions and team or company identifier outside the box on the right hand side. This information can be made with a lumber marking device.	☐	9/12/93 1310 hrs. HM - nat. gas SMA - E-1
Structure is accessible and safe for search and rescue operations. Damage is minor with little danger of further collapse.	☐	9/12/93 1310 hrs. HM - none SMA - E-1
Structure is significantly damaged. Some areas are relatively safe, but other areas may need shoring, bracing, or removal of falling and collapse hazards.	⧄	9/12/93 1310 hrs. HM - nat. gas SMA - E-1
Structure is not safe for search or rescue operations. May be subject to sudden additional collapse. Remote search ops may proceed at significant risk. If rescue ops are undertaken, safe haven areas and rapid evacuation routes should be created.	☒	9/12/93 1310 hrs. HM - nat. gas SMA - E-1
Arrow located next to a marking box indicates the direction to a safe entrance into the structure, should the marking box need to be made remote from the indicated entrance.	←	

Search markings

Search markings must be easy to make, easy to read, and easy to understand (Fig. A–2 to A–4). To be easily seen, the search mark must be large and of a contrasting color to the background surface. Orange spray paint seems to be the most easily seen color on most backgrounds. Line marking or downward spray cans apply the best paint marks. Lumber chalk or lumber crayons should be used to mark additional information inside the search mark itself because they are easier to write with that spray paint.

A large distinct marking will be made outside the main entrance of each building or structure searched. This *main entrance* search marking will be completed in two steps. (1) A large (approximately 2-ft) single slash is made near the main entrance at the start of the search. (2) When a search of the entire structure is completed, a second large slash is drawn in the opposite direction forming an X. Specific information is placed in all four quadrants of the main entrance X, summarizing the entire search of the structure. The left quadrant is for the rescue team identifier. The top quadrant is for the date and time the search was completed. The right quadrant is for any significant hazards located in the structure. The bottom quadrant is for the number of *live* or *dead* victims still inside the structure. Use a small x in the bottom quadrant if no victims are inside the structure.

During the search inside the structure, a large single slash is made upon entry of each room or area. After a search of the room or area is completed, a second large slash is drawn in the opposite direction forming an X. The only information placed in any of the X quadrants while inside the structure pertain to significant hazards or the number of live or dead victims.

Main entrance search marking

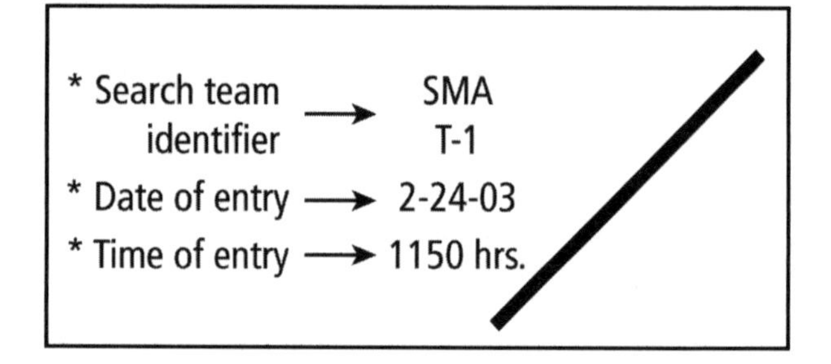

Fig. A–2 Main Entrance Search Marking: When You Enter

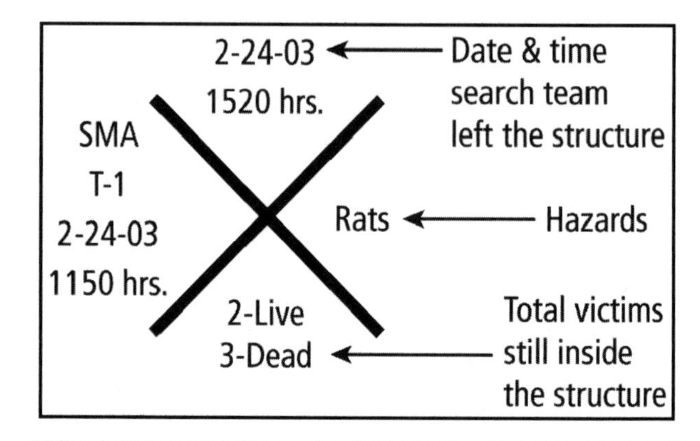

Fig. A–3 Main Entrance Search Marking: When You Exit

Interior search markings

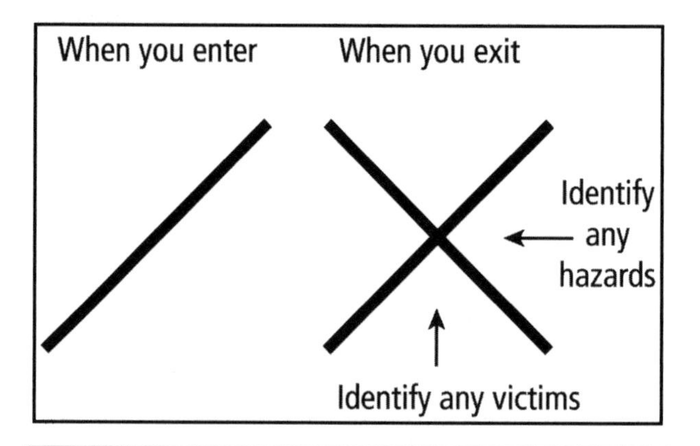

Fig. A–4 Interior Search Markings: Each Room or Area

Table A–2 Interior Search Markings

Make a large (2′ x 2′) "V" with orange spray paint near the location of a *potential* victim. Mark the name of the search team or crew identifier in the top part of the "V" with paint or a lumber marker type device.

Paint a circle around the "V" when a potential victim is *confirmed* to be *alive* either visually, vocally, or hearing specific sounds that would indicate a high probability of a live victim. If more than one confirmed live victim, mark the total number of victims under the "V".

Paint a horizontal line through the middle of the "V" when a *confirmed* victim is determined to be deceased. If more than one confirmed deceased victim, mark the total number of victims under the "V". Use both the live and deceased victim marking symbols when a combination of live and deceased victims are determined to be in the same location.

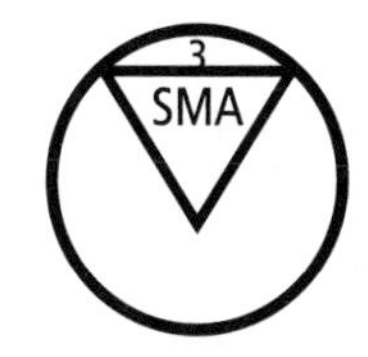

Paint an "X" through the confirmed victim symbol after the *all* victim(s) have been removed from the specific location identified by the marking.

An arrow may need to be painted next to the "V" pointing towards the victim when the victim's location is not immediately near where the "V" is painted.

Appendix VI: More About Search Operations

Search is not considered a separate stage of collapse operations for a very important reason. We want to begin searching for signs and indications of hidden victims from the first report of a collapse, and we want to continue searching until the incident has been cleared of all victims. Search operations (in various forms) should occur from stage 1 through stage 5 of most collapse emergencies.

The information gathered during stage 1 helps set the stage for the physical search operations that follow. This includes interviews with witnesses, family members, and friends. It includes a review of floor plans, seating plans, apartment occupancy lists, hotel guest lists, and other information about occupancy. It includes a rapid size up of the overall situation, the building occupancy, and a consideration of the time of day and the day of the week (including holidays). It includes observation of the collapse area, with an eye for evidenced of trapped victims and other relevant factors. The intelligence gleaned from this process about where victims are likely to be found (and where *not* likely to be found) helps guide the search-related actions of firefighters and other rescuers through the next four stages.

Stage 2 is generally where we begin conducting physical searches for hidden victims. Some will argue that this is also a stage 1 function, and this author has no dispute with that view. The argument of whether physical search begins in stage 1 or 2 is rather academic. What is truly important is that search operations begin as early in the emergency as possible, that they are effective and timely, and that they continue throughout the operation until there is no hope of locating live victims.

Here, then, are some of the search methods that have proven effective for search operations in collapsed structures:

Hail search[2]

The hail search is a visual and verbal evaluation of whether victims may be trapped at or near the surface of a collapse or in a deeper space where victims' voices may carry to the outside. It is conducted by arranging firefighters and rescuers around the collapse and conducting a series of callouts in "round robin" fashion, with all eyes and ears trained on the collapse. We are listening for voices or tapping emanating from the debris, and we are looking for any visual clues that may indicate victims buried in the collapse.

1. Assign firefighters and other rescuers to positions around the perimeter of the collapse in calling and listening locations. If the collapse area is very large, surround selected locations with personnel.
2. The search group leader, IC (or OIC) calls for total silence. All machinery is shut down, radios turned down, bystanders and news media are quieted, and personnel listen for voices emanating from the collapse zone.
3. Moving in a clockwise direction around the perimeter, each firefighter or rescuer calls out instructions to anyone trapped within the collapse who might hear them (e.g., "Hello! Knock three times with your hand or a hard object if you can here us!"). This informs trapped victims that people are listening for them *now* and also instructs them to make a discernable signal that might be picked up by electronic search devices. Knocking three times is consistent with SOS and some other emergency signals; therefore, it's easy for people to remember. Obviously it's not important that victims knock a particular number of times but rather that they make some type of signal that can be detected, verified, differentiated from background noises, and distinguished as coming from trapped victims.

4. Other rescuers listen carefully for any sounds and their vector. If technical search devices are being used, their operator likewise remains alert for any signals emanating from within the collapse.
5. If a sound is heard coming from within the collapse, ascertain that it is from a trapped victim as opposed to a generator, stereo, or other source within the collapse not previously heard. Attempt to trace its source. It may be necessary to triangulate by repeating the hails while moving rescuers on the pile to more closely pinpoint the location of the trapped victim(s).
6. If one or more victims are located, maintain verbal contact with them. Determine their situation, injuries, location, location of other victims, and other factors as they are able to describe them to you. For example, in some multi-story buildings, there may be different colored carpet or paint on different floors. If these buildings pancake, it may be possible to locate victims more closely by asking them to describe the color of the carpet or walls, and then match that information with what is visible from outside the collapse. In that manner, rescuers have more closely pinpointed the location of trapped victims.

Acoustic and seismic sensing devices

Many modern fire departments and USAR teams/task forces are equipped with extremely sensitive acoustic and seismic sensing devices that can detect trapped victims calling for help or tapping. These devices may be used alone, but they are far more effective when used in combination with hail search, canine search, remote visuals devices, and other search options.

Remote visual search devices

Victims hidden from sight may be spotted by remote devices such as search cameras, snake-eye systems, or other devices. These devices may be used alone, or they may be used to confirm indications from other forms of search. They are also helpful in determining the exact position of a victim's body in relation to the rescue efforts.

Thermal imaging systems

Thermal imaging systems have evolved to a state where they are extremely sensitive, user friendly, with clear resolution, and rugged. They may be used to identify victims trapped in rubble or other situations, especially in the stages where the victim's body is warmer or colder than the surrounding material. However, once the victim's body takes on the same temperature as the surrounding material, thermal imaging loses much of its advantage.

Night vision

Night vision has certain applications for locating victims under poor light conditions. They are durable and effective in allowing rescuers to move about in low-light situations. However, the typical disaster uses spotlights, flashlights, and other forms of bright light, which makes the use of night vision devices somewhat problematic.

Search canines

USAR-trained canine search teams are without a doubt the best way to locate people that you cannot see or hear in collapsed buildings and other USAR-related disasters and emergencies. Today, most states have an inventory of USAR-trained canine search teams, and the United States has dozens of such teams, many of them attached to the nation's FEMA USAR task forces. For major collapse disasters anywhere in the United States, there should be USAR-certified canine search teams available just minutes or hours away.

Wilderness search canines may also be used, but they are not accustomed to the conditions found at a collapsed building disaster. Other dogs (i.e., police canines) may also be used if USAR-trained canine teams are not available.

In general, canine search operations at a disaster should include two dogs and two handlers. The second dog and handler are used to confirm the *alert* of the first canine team.

The future of search operations

The future of search operations will see rapid advances in the following technologies:

- GPS
- Ground-penetrating radar
- Robotics
- Alternate animal search
- Smart buildings
- Personal alert devices

Appendix VII: Sample Structure Triage Forms

STRUCTURE TRIAGE FORM DATE/TIME: ________ BY: ________ PAGE ___ OF ___

TEAM: ______ STRUCTURES SPECIALIST: ______	DATE/TIME OF CATASTROPHE: ________

AREA MAP

BLDG. ID: ______	1. ZERO VICTIMS POSSIBLE	(WRITE ZERO)	GO TO NEXT BUILDING	
FLOOR AREA: ______	2. POTENTIAL No. OF TRAPPED / 5	(MIN=1 MAX=50)		
No. STORIES: ______	3. CONDITION OF VOIDS	1 >VERY COMPACT>>SEPARATE LAYERS>>PARTIAL COLLAPSE> 20		
OCCUPANCY: ______	4. TIME TO GET TO VICTIM	1 > ONE DAY >>>>>>>>>>>>>>>>>>>> 2 HOURS > 20		
MATERIAL: (CIRCLE ALL THAT APPLY)	5. CHANCE OF FURTHER COLLAPSE	-1 > LOW CHANCE >>>>>>>>>>>>>>> HIGH CHANCE > -20		-
WOOD CONCRETE STEEL UNREINFORCED MASONRY PRECAST CONCRETE	6. SPECIAL INF.:	SCHOOL / HOSPITAL = +25 KNOWN LIVE VICTIM = +5 EACH		
CALCULATE AREA NUMBER TRAPPED:	NO GO (CIRCLE, WRITE NO-GO & WHEN / IF TO REVISIT)			
	FIRE HAZARDOUS MATERIALS	OTHER: ________		BLDG. TOTAL

BLDG. ID: ______	1. ZERO VICTIMS POSSIBLE	(WRITE ZERO)	GO TO NEXT BUILDING	
FLOOR AREA: ______	2. POTENTIAL No. OF TRAPPED / 5	(MIN=1 MAX=50)		
No. STORIES: ______	3. CONDITION OF VOIDS	1 >VERY COMPACT>>SEPARATE LAYERS>>PARTIAL COLLAPSE> 20		
OCCUPANCY: ______	4. TIME TO GET TO VICTIM	1 > ONE DAY >>>>>>>>>>>>>>>>>>>> 2 HOURS > 20		
MATERIAL: (CIRCLE ALL THAT APPLY)	5. CHANCE OF FURTHER COLLAPSE	-1 > LOW CHANCE >>>>>>>>>>>>>>> HIGH CHANCE > -20		-
WOOD CONCRETE STEEL UNREINFORCED MASONRY PRECAST CONCRETE	6. SPECIAL INF.:	SCHOOL / HOSPITAL = +25 KNOWN LIVE VICTIM = +5 EACH		
CALCULATE AREA NUMBER TRAPPED:	NO GO (CIRCLE, WRITE NO-GO & WHEN / IF TO REVISIT)			
	FIRE HAZARDOUS MATERIALS	OTHER: ________		BLDG. TOTAL

BLDG. ID: ______	1. ZERO VICTIMS POSSIBLE	(WRITE ZERO)	GO TO NEXT BUILDING	
FLOOR AREA: ______	2. POTENTIAL No. OF TRAPPED / 5	(MIN=1 MAX=50)		
No. STORIES: ______	3. CONDITION OF VOIDS	1 >VERY COMPACT>>SEPARATE LAYERS>>PARTIAL COLLAPSE> 20		
OCCUPANCY: ______	4. TIME TO GET TO VICTIM	1 > ONE DAY >>>>>>>>>>>>>>>>>>>> 2 HOURS > 20		
MATERIAL: (CIRCLE ALL THAT APPLY)	5. CHANCE OF FURTHER COLLAPSE	-1 > LOW CHANCE >>>>>>>>>>>>>>> HIGH CHANCE > -20		-
WOOD CONCRETE STEEL UNREINFORCED MASONRY PRECAST CONCRETE	6. SPECIAL INF.:	SCHOOL / HOSPITAL = +25 KNOWN LIVE VICTIM = +5 EACH		
CALCULATE AREA NUMBER TRAPPED:	NO GO (CIRCLE, WRITE NO-GO & WHEN / IF TO REVISIT)			
	FIRE HAZARDOUS MATERIALS	OTHER: ________		BLDG. TOTAL

Fig. A–5 Sample 1 Structure Triage Forms

STRUCTURE TRIAGE FORM DATE/TIME: ________ BY: ____________ PAGE ____ OF ____

BLDG. ID: ________	1. ZERO VICTIMS POSSIBLE	(WRITE ZERO)	GO TO NEXT BUILDING	
FLOOR AREA: ________	2. POTENTIAL No. OF TRAPPED / 5	(MIN=1 MAX=50)		
No. STORIES: ________	3. CONDITION OF VOIDS	1 >VERY COMPACT>>SEPARATE LAYERS>>PARTIAL COLLAPSE>	20	
OCCUPANCY: ________	4. TIME TO GET TO VICTIM	1 > ONE DAY >>>>>>>>>>>>>>>>>>>> 2 HOURS >	20	
MATERIAL: (CIRCLE ALL THAT APPLY) WOOD CONCRETE STEEL UNREINFORCED MASONRY PRECAST CONCRETE	5. CHANCE OF FURTHER COLLAPSE	-1 > LOW CHANCE >>>>>>>>>>>>>>> HIGH CHANCE >	-20	-
	6. SPECIAL INF.:	SCHOOL / HOSPITAL = +25 KNOWN LIVE VICTIM = +5 EACH		
CALCULATE AREA NUMBER TRAPPED:	NO GO (CIRCLE, WRITE NO-GO & WHEN / IF TO REVISIT) FIRE HAZARDOUS MATERIALS OTHER: ________			BLDG. TOTAL

BLDG. ID: ________	1. ZERO VICTIMS POSSIBLE	(WRITE ZERO)	GO TO NEXT BUILDING	
FLOOR AREA: ________	2. POTENTIAL No. OF TRAPPED / 5	(MIN=1 MAX=50)		
No. STORIES: ________	3. CONDITION OF VOIDS	1 >VERY COMPACT>>SEPARATE LAYERS>>PARTIAL COLLAPSE>	20	
OCCUPANCY: ________	4. TIME TO GET TO VICTIM	1 > ONE DAY >>>>>>>>>>>>>>>>>>>> 2 HOURS >	20	
MATERIAL: (CIRCLE ALL THAT APPLY) WOOD CONCRETE STEEL UNREINFORCED MASONRY PRECAST CONCRETE	5. CHANCE OF FURTHER COLLAPSE	-1 > LOW CHANCE >>>>>>>>>>>>>>> HIGH CHANCE >	-20	-
	6. SPECIAL INF.:	SCHOOL / HOSPITAL = +25 KNOWN LIVE VICTIM = +5 EACH		
CALCULATE AREA NUMBER TRAPPED:	NO GO (CIRCLE, WRITE NO-GO & WHEN / IF TO REVISIT) FIRE HAZARDOUS MATERIALS OTHER: ________			BLDG. TOTAL

BLDG. ID: ________	1. ZERO VICTIMS POSSIBLE	(WRITE ZERO)	GO TO NEXT BUILDING	
FLOOR AREA: ________	2. POTENTIAL No. OF TRAPPED / 5	(MIN=1 MAX=50)		
No. STORIES: ________	3. CONDITION OF VOIDS	1 >VERY COMPACT>>SEPARATE LAYERS>>PARTIAL COLLAPSE>	20	
OCCUPANCY: ________	4. TIME TO GET TO VICTIM	1 > ONE DAY >>>>>>>>>>>>>>>>>>>> 2 HOURS >	20	
MATERIAL: (CIRCLE ALL THAT APPLY) WOOD CONCRETE STEEL UNREINFORCED MASONRY PRECAST CONCRETE	5. CHANCE OF FURTHER COLLAPSE	-1 > LOW CHANCE >>>>>>>>>>>>>>> HIGH CHANCE >	-20	-
	6. SPECIAL INF.:	SCHOOL / HOSPITAL = +25 KNOWN LIVE VICTIM = +5 EACH		
CALCULATE AREA NUMBER TRAPPED:	NO GO (CIRCLE, WRITE NO-GO & WHEN / IF TO REVISIT) FIRE HAZARDOUS MATERIALS OTHER: ________			BLDG. TOTAL

BLDG. ID: ________	1. ZERO VICTIMS POSSIBLE	(WRITE ZERO)	GO TO NEXT BUILDING	
FLOOR AREA: ________	2. POTENTIAL No. OF TRAPPED / 5	(MIN=1 MAX=50)		
No. STORIES: ________	3. CONDITION OF VOIDS	1 >VERY COMPACT>>SEPARATE LAYERS>>PARTIAL COLLAPSE>	20	
OCCUPANCY: ________	4. TIME TO GET TO VICTIM	1 > ONE DAY >>>>>>>>>>>>>>>>>>>> 2 HOURS >	20	
MATERIAL: (CIRCLE ALL THAT APPLY) WOOD CONCRETE STEEL UNREINFORCED MASONRY PRECAST CONCRETE	5. CHANCE OF FURTHER COLLAPSE	-1 > LOW CHANCE >>>>>>>>>>>>>>> HIGH CHANCE >	-20	-
	6. SPECIAL INF.:	SCHOOL / HOSPITAL = +25 KNOWN LIVE VICTIM = +5 EACH		
CALCULATE AREA NUMBER TRAPPED:	NO GO (CIRCLE, WRITE NO-GO & WHEN / IF TO REVISIT) FIRE HAZARDOUS MATERIALS OTHER:			BLDG. TOTAL

BLDG. ID: ________	1. ZERO VICTIMS POSSIBLE	(WRITE ZERO)	GO TO NEXT BUILDING	
FLOOR AREA: ________	2. POTENTIAL No. OF TRAPPED / 5	(MIN=1 MAX=50)		
No. STORIES: ________	3. CONDITION OF VOIDS	1 >VERY COMPACT>>SEPARATE LAYERS>>PARTIAL COLLAPSE>	20	
OCCUPANCY: ________	4. TIME TO GET TO VICTIM	1 > ONE DAY >>>>>>>>>>>>>>>>>>>> 2 HOURS >	20	
MATERIAL: (CIRCLE ALL THAT APPLY) WOOD CONCRETE STEEL UNREINFORCED MASONRY PRECAST CONCRETE	5. CHANCE OF FURTHER COLLAPSE	-1 > LOW CHANCE >>>>>>>>>>>>>>> HIGH CHANCE >	-20	-
	6. SPECIAL INF.:	SCHOOL / HOSPITAL = +25 KNOWN LIVE VICTIM = +5 EACH		
CALCULATE AREA NUMBER TRAPPED:	NO GO (CIRCLE, WRITE NO-GO & WHEN / IF TO REVISIT) FIRE HAZARDOUS MATERIALS OTHER:			BLDG. TOTAL

Fig. A–6 Sample 2 Structure Triage Forms

U.S.&R. STRUCTURE / HAZARDS EVALUATION FORM

BY: ____________________

Where required, circle all the information or items that apply.

NOTE: AFTERSHOCKS MAY CAUSE ADDITIONAL DAMAGE OTHER THAN NOTED.

STRUCTURE DESCRIPTION:	BUILDING MARKING:
Building ID: ____________________	Date/Time of Evaluation: ____________
No. Stories: No. Basements:	Date/Time of Catastrophe: ____________
MATERIALS:	**TYPE OF COLLAPSE:**
Wood Concrete Steel Unreinforced Masonry Precast Concrete	Pancake Soft 1st Floor Wall Failure
Other: ____________________	Torsion Middle Story Overturn
	Other:
FRAMING SYSTEM:	**LOCATION OF VOIDS:**
Shearwall Moment Frame Braced Frame	Between Floors Basement Shafts
Other:	Other:
OCCUPANCY:	**DESCRIPTION OF UNSAFE AREAS:** ____________
Hospital Police Station Fire Station	
Emergency Operations Center Office Building School	
Public Assembly Industrial Hotel	
Apartment Retail Store Other:	
VICTIM & OTHER INFORMATION: ____________	
LOCATION OF BEST ACCESS & SAR STRATEGY:	**DESCRIPTION OF HAZARDS:** ____________

SKETCH:

Fig. A–7 Sample 3 Structure Triage Forms

U.S.&R. STRUCTURE / HAZARDS EVALUATION FORM BY: ______________________

Where required, circle all the information or items that apply. NOTE: AFTERSHOCKS MAY CAUSE ADDITIONAL DAMAGE OTHER THAN NOTED.

Sketch, Page 2

Fig. A–8 Sample 4 Structure Triage Forms

Appendix VIII: Sample HSW Rescue Operations (Swimmer-free and Tethered-swimmer Short-haul Evolutions)

This is a sample checklist for helicopter crews conducting helicopter-based swift-water rescue operations, using two different evolutions: The first involves the rescuer entering the water from the shoreline and securing the victim. The copter lowers a rope to which the rescuer clips himself and the victim for a short haul pickout. The second requires a rescuer to be lowered from the copter by means of a rope anchored in the cabin of the copter with the rescuer descending under control of a fractioning device. Then the rescuer is dipped in the water to make contact and secure the victim for a pickout.

1. Cabin prep and lowering procedures same as belay insertion.
2. When rescuer in positioned at or above water, crew chief ties off brake bar rack and informs pilot.
3. As victim comes down river, crew chief calls distance.
4. Crew chief anticipates contact rescue and tells pilot to *turn and go* just prior to victim contacting rescuer.
5. Pilot turns and heads downstream.
6. Crew chief calls height, giving rescuer enough slack in rope to reach victim, capture, and secure.
7. Crew chief describes rescuer contact and hookup to pilot.
8. When victim is secure, rescuer signals crew chief *OK*.
9. Crew chief tells pilot "I have the OK sign, up helicopter with ___ ft of slack.
10. Pilot acknowledges.
11. As rescuer and victim clear water, crew chief calls height, stopping at 5 to 10 ft above the water.
12. Crew chief tells pilot "They are 5 ft above the water, load is free and clear, you are clear for forward flight."
13. Pilot acknowledges and flies to the nearest safe drop-off point.
14. Crew chief calls distance and adjustments for any obstacles.
15. When over drop-off point, crew chief tells pilot "Down helicopter ___ ft."
16. Pilot acknowledges and lowers helicopter.
17. Crew chief calls off distance and tells pilot when "Load is on the ground."
18. Rescuer disconnects victim first and then himself (in case a second rescue is necessary).
19. If no other rescue necessary, rescuer disconnects and holds rope away from body and gives OK sign.
20. Crew chief tells pilot, "I have the OK sign, up helicopter with ___ ft of slack."
21. Pilot acknowledges and raises helicopter.
22. Crew chief tells pilot, "Load is off the ground, load is free and clear, you are clear for forward flight."
23. Pilot acknowledges.
24. Crew chief coils rope, stows in cabin and advises pilot that the rope is stowed.
25. Pilot acknowledges and lands in the appropriate area.
26. Victim may be in need of air transport to hospital.

Appendix III

Appendix IX:[3] Sample Helicopter Rappel Guidelines[4]

Bell 205 helicopter

This is a sample checklist for helicopter crews and rescuers conducting rappel operations from hovering helicopters.

1. Place aircraft in rappel configuration.
2. Crewmembers don harnesses and complete safety check (same as hoist operations procedures).
3. Crew chief secures to alternate anchor point in right well via pick-off strap on harness and sits in right well area.
4. Crew chief attaches 8-plate safety aitrie, pick-off strap, or webbing to most forward single ring.
5. Crewmember sits in right aft facing seat, secures to hard point via pick-off strap or fastens seat belt.
6. Crew chief and crewmember double-check each other's connections; crew chief closes door and notifies pilot they are ready.
7. Rope is secured to anchor point with a figure-8 knot on a bite with a carabiner gate out, down and locked.
8. Crewmember places 8-plate onto rope then connects it to harness.
9. Crew chief attaches aitrie, webbing or pick-off strap to large opening of 8-plate with carabiner as safety for crewmember's transition to skid.
10. Crewmember then disconnects from hard point or releases seat belt.
11. Crewmember moves to right well area passing in front of the crew chief.
12. Crew chief begins swift-water checklist with pilot.
13. Pilot notifies crew chief when below 40 knots.
14. Crew chief opens right cabin door.
15. Crewmember steps onto skid facing crew chief. The crewmember can hold onto the aitrie, webbing or pick-off strap attached to the large opening of the 8-plate.
16. The pilot and crew chief coordinate on target area.
17. Once in position, the crew chief gives the ready sign a *thumbs up* to crewmember.
18. The crewmember gives ready sign back to crew chief.
19. The crew chief drops the rope bag.
20. The crew chief removes the safety strap from the 8-plate when the crewmember has his rappel hand locked to the rear and is ready for rappel.
21. The crewmember begins to rappel.
22. Crew chief keeps the pilot informed of crewmember's distance from the helicopter to the ground.
23. The crewmember gives the OK sign upon reaching the ground.
24. The crewmember disconnects from the rope and indicates to the crew chief that they are off rappel by holding the 8-plate over their helmet and holding the rope out away from their body.
25. Crewmember ties figure-8 on a bight 10 ft from the point of landing towards rope bag.
26. The crew chief will then deploy necessary equipment according to procedures. Once equipment is deployed or if no equipment is necessary, crew chief will bring rope back up into cabin after the crewmember has attached the figure-8 on a bight to rope bag to prevent excess rope from being deployed.

Bell 412 helicopter

Rappel process is the same as for the Bell 205 helicopter with the following changes:

1. It is not necessary to move equipment from the tail boom.
2. The crew chief secures to the primary anchor point rather than the alternate anchor point.
3. The attachment point for the rappel rope is on the hoist arm. A doubled piece of 1-in. tubular webbing tied with a water knot is wrapped around the attachment point and is secured with a carabiner gate down and locked.

Appendix X: Sample Short-haul Operational Guidelines[5]

Short-haul setup is the same as for the belay insertion:

1. Crew chief lowers adequate rope and locks off on break rack or holds bight below break rack with hand.
2. Crew chief guides in the pilot by same method used in hoist operations.
3. When the carabiner or steel ring is in the crewmember's hand, the crew chief provides some slack.
4. The crewmember attaches to the rope and signals by giving the OK and signaling to raise.
5. The crew chief advises the pilot when the load is free and clear.
6. The crew chief directs the pilot to the designated short-haul location.
7. Once on the ground at the location, the crewmember follows the same guidelines as the original insertion.

Lowered short-haul procedures:

Crew chief prepares cabin as follows:

1. Checks anchor rings, webbing, and screw link.
2. Attaches carabiner with brake bar rack to screw link.
3. Ties figure-8 follow through around a collection ring and finishes with a safety tie-off.
4. Weaves rope in brake rack (minimum 4 bars) and ties off.
5. Attaches three wrap Prusik as stopper/adjuster.
6. Crew chief attaches to the alternate anchor.
7. Crew chief signals for rescuer to approach.
8. Crew chief attaches rescuer to collection ring.
9. Crewmember steps on skid.
10. Crew chief double-checks rescuer.
11. Crewmember double-checks crew chief.
12. Crew chief signals rescuer with *thumbs-up*.
13. Rescuer returns thumbs-up if ready to go.
14. After crewmember signals ready, crew chief tells pilot "Ready in the back."
15. Pilot acknowledges and takes off for target area.
16. Crew chief calls distance and guide's pilot to target.
17. When over target, crew chief tells pilot "Over target, hold."
18. Crew chief unties brake bar rack and gives crewmember thumbs-up.
19. Crewmember returns thumbs-up when ready.
20. When crewmember returns thumbs-up, crew chief begins to lower crewmember.
21. Crew chief lowers slowly until crewmember is clear of skid.
22. Once below skid, crew chief may increase lowering speed.
23. Crew chief tells pilot "Crewmember going down" and calls distances.
24. Crew chief tells pilot when rescuer is almost on ground and when he touches down.
25. Crewmember disconnects from rope, signals OK, and holds rope away from body.
26. Crew chief tells pilot when crewmember is off rope and then begins to coil rope.
27. Crew chief tells pilot when rope is coiled and stowed.
28. Crew chief gets back in cabin and closes door.
29. Crew chief then tells pilot, "Cabin secured, door closed, you are clear for forward flight."
30. Pilot acknowledges.

Appendix XI: Sample Helo/High-rise Team Ops Guidelines–Rappel Insertion

Loading:

1. Helicopter high-rise team approaches in reverse order: first in/last out.
2. Crew chief stows equipment; packs under seat in well as team loads.
3. Team is seated in reverse order with team leader in last.
4. Team leader carries pre-set rope bag containing:
   - 300 ft of rope
   - Figure-8 on a bight
   - Carabiner through the bight
   - A green webbing loop attached to the carabiner
5. All team members seatbelt in or tie-in to floor safety line.
6. Team leader goes on intercom.

Size-up:

1. As helicopter approaches building, the crew chief and pilot go through hoist checklist.
2. Below 40 knots, the door is opened.
3. As helicopter circles building, the pilot, crew chief, and team leader evaluate and discuss options.

Rappel setup:

1. Team leader hands crew chief green webbing from rope bag with carabiner and rope still attached.
2. Crew chief receives green webbing loop and wraps it once around the hoist arm and then connects the carabiner into the webbing bight gate down and locked completing the rappel anchor point on the Bell 412. On the Bell 205, carabiner attaches to the main anchor point on wall of well.
3. Team leader visually double-checks the anchor point installation.

Rappel:

1. Team leader goes off intercom.
2. Team leader attaches 8-plate to rope in full rappel configuration and then attaches to harness, locks carabiner, and removes all slack.
3. Team leader holds up rope and hardware for pre-rappel visual inspection by the crew chief.
4. The crew chief hands the team leader a carabiner that is attached to a hard point with an aitrie, a loop of tubular webbing, or a pick-off strap. This safety is attached to the center of the large opening in the 8-plate and is used for the transition to the skid.
5. Team leader attaches rope bag to rescue harness for rappel or slides rope bag to crew chief to drop prior to rappel.
6. Crew chief directs pilot to target.
7. Crew chief gives the signal to team leader to unbuckle or disconnect.
8. Team leader unbuckles or disconnects and scoots across the cabin floor removing slack from the system as he moves to the ready position on the skid.
9. The crew chief centers the helicopter over target with direction to the pilot.
10. Crew chief deploys drop bag if not attached to rescue harness.
11. The crew chief does final visual check of rappeller.

12. Once the team leader is in position on the skid with hand in a lock-off position and ready for immediate deployment, the safety attached to the 8-plate will be removed.
13. Crew chief notifies pilot: "Rappeller is ready."
14. Pilot notifies crew chief: "Start rappel."
15. Crew chief gives thumbs-up to rappeller.
16. Rappeller returns thumbs-up and starts rappel.
17. Crew chief notifies pilot rappeller "on rappel."
18. Crew chief notifies pilot distance to ground, "Rappeller is at __ ft."
19. Rappeller stops 6 ft above the roof momentarily to avoid ankle injury due to fast rappel or sudden wind gust affecting helicopter, then finishes rappel.
20. Crew chief notifies pilot, "Rappeller on roof."
21. Once on the roof, the team leader disconnects the 8-plate and removes from rope. The rope bag is removed from harness if connected for rappel.
22. The team leader holds the 8-plate over helmet to signal that he is off rappel and off rope.
23. Crew chief notifies pilot, "Rappeller off rope."
24. Crew chief gathers a bight of the rope and gives it to the next rappeller.
25. Rappeller hooks in 8-plate and insertion is the exact same process as above.
26. Crew chief deploys equipment.
27. If no equipment deployed, crew chief unlocks carabiner from anchor point, removes from rope, and drops rope to roof.
28. Crew chief notifies pilot: "Rope is free and clear; you are clear for flight."
29. As helicopter leaves, crew chief clears webbing and carabiner from hoist arm on Bell 412 if no further operations are anticipated and closes door.
30. Webbing and carabiner are stowed on well wall for future use.

Equipment Deployment:

1. If team cannot take all equipment, crew chief will lower bags to team when last rappeller is on roof.
2. Crew chief unlocks carabiner attached to webbing on hoist arm on Bell 412 or from main anchor point on well wall of Bell 205 and removes rope with figure-8 on a bight knot in end.
3. Crew chief places rope in carabiner holding knot in hand.
4. Crew chief takes knot and attaches to carabiner on bag #1.
5. While crew chief is rigging topside, team member takes 15 ft of slack and tie an in-line figure-8 knot in rope.
6. Crew chief notifies pilot, "Lowering equipment." Informs pilot of progress.
7. Crew chief lowers bag #1 to roof.
8. Team member disconnects bag #1 from rope.
9. Crew chief attaches bag #2 to in-line figure-8 knot.
10. Crew chief notifies pilot, "Lowering equipment." Informs pilot of progress.
11. Crew chief notifies pilot: "Equipment on roof."
12. Crew chief lowers bag #2 to roof.
13. Team member disconnects bag #2 from rope.
14. Crew chief notifies pilot: "Equipment on roof."
15. Crew chief removes rope from carabiner and drops to roof.
16. Crew chief notifies pilot: "Rope is free and clear; you are clear for flight."
17. As helicopter leaves, crew chief clears webbing and carabiner from hoist arm on Bell 412 if no further operations are anticipated and closes door.
18. Webbing and carabiner are stowed on well wall for future use.

Short-haul, high-rise procedures:

1. Crew chief attaches rope to helicopter in swift-water or rappel configuration.
2. While on the ground, the end of the rope is stretched 100 ft to the front of the helicopter and tied-off at brake bar rack or hoist arm on bell 412 or main anchor point on Bell 205.
3. Two team members attach to end rope at a 10-ft interval.
4. Team attachment made with in-line figure-8 knot at the 10-ft mark and a figure-8 on a bight knot is tied at the end of the rope.
5. When the aircrew is ready to lift off, pilot asks crew: "Ready for short haul?"
6. Crew chief replies: "Ready."
7. Pilot gives team thumbs-up signal.
8. Both team members return the thumbs-up signal.
9. Pilot states: "Coming up." Then lifts the helicopter straight up.
10. As helicopter rises, team members walk toward the helicopter maintaining their 10-ft interval.
11. As team member at 10-ft mark centers under rope, the crew chief calls distance to pilot.
12. Crew chief notifies pilot when load #1 is on line and when load #1 is off ground.
13. As second team member centers under helicopter, crew chief notifies pilot when load #2 is on line and when load #2 is off ground.
14. When team clears all obstacles, crew chief notifies pilot: "Load is free and clear; you are clear for flight."
15. As helicopter approaches target rooftop, crew chief calls off distance and height to pilot.
16. Crew chief tells pilot to hold forward flight when over target, then tells pilot, "Down __ ft".
17. Crew chief calls distance and when team member #1 touches roof says: "Load #1 on roof."
18. First team member moves away from landing spot and disconnects from rope.
19. Crew chief calls distance and when team member #2 touches roof says: "Load #2 on roof."
20. Crew chief tells pilot to continue down a few feet to allow slack in rope.
21. Second team member disconnects from rope.
22. Both team members hold rope away from bodies to signal crew chief that they are clear.
23. Equipment may be delivered via a second rope at this time.
24. Crew chief informs pilot that, "Team is clear; up helicopter."
25. Crew chief coils rope and stows in cabin as helicopter rises.
26. When rope is stowed, the crew chief informs pilot that: "Rope is stowed; you are clear for forward flight."
27. Crew chief clears hoist arm and cabin door is closed and secured. Crew chief informs pilot: "Door is secured."

Emergency evacuation of civilians:

1. Setup is the same for short-hauls except that a carabiner, a blue one-in. tubular webbing, and a CMC victim rescue harness are attached at both the 10-ft in-line knot and the end knot on the rope.
2. A receiving team will be needed at the landing site.
3. The rope can be set up and coiled inside helicopter.
4. Crew chief directs pilot into target area.
5. Crew chief lowers rope to full extension.
6. Crew chief calls distance.
7. Roof team readies civilians by controlling their movement on roof.

8. As line touches down, team takes control at the 10-ft, in-line knot and at the end knot.
9. One evacuee is attached to CMC victim rescue harness at the 10-ft knot and one at the end knot.
10. Roof team tells them to hold onto rope above knot.
11. Team leader gives crew chief OK signal.
12. Crew chief tells pilot, "Ready for short-haul; up helicopter with __ ft of slack".
13. Pilot raises helicopter.
14. Crew chief informs pilot when: "Load is coming on line; load is off ground; load is free and clear; you are clear for forward flight."
15. Pilot acknowledges and flies to landing area, crew chief calls distance.
16. Crew chief calls distance and when victim #1 touches ground, says: "Load #1 on ground."
17. When first victim reaches landing spot, the receiving team member takes physical control and moves victim away from landing spot and disconnects victim from rope then harness is removed from victim and reattached to rope.
18. Crew chief calls distance and when victim #2 touches ground says: "Load #2 on ground."
19. Crew chief tells pilot to continue down a few feet to allow slack in rope.
20. When second victim reaches landing spot, the receiving team member takes physical control and moves victim away from landing spot and disconnects victim from rope then harness is removed from victim and reattached to rope.
21. Both team members hold rope with CMC victim rescue harness and blue webbing away from bodies to signal crew chief that they are clear.
22. Crew chief informs pilot, "Team is clear; up helicopter."
23. Crew chief coils rope and stows in cabin as helicopter rises.
24. When rope is stowed, the crew chief informs pilot that: "Rope is stowed; you are clear for forward flight."
25. Crew chief clears hoist arm and cabin door is closed and secured. Crew chief informs pilot: "Door is secured."

Appendix XII: Sample SOGs for Helicopter Transportation of Rescue/USAR Companies

Introduction

Purpose: To define the parameters and establish guidelines for the use of helicopters to transport (USAR or rescue company) firefighters and equipment to USAR emergencies, high-rise fires, marine emergencies and disasters, swift-water rescue incidents, technical rescues in mountainous areas and offshore islands; local disasters, transportation accidents, and other emergencies to which (USAR or rescue) companies are normally dispatched.

Background: For decades, the fire service has used helicopters to reduce the response time of fire and rescue resources, to conduct aerial reconnaissance, to enhance command and control, to provide better access to remote sites, to perform rapid intervention duties, and to support other special operations during the course of USAR and technical rescue emergencies. Some fire departments have combined rescue units with helicopter units to create multidisciplinary airborne rescue teams; others have used helicopters to reduce the response time of ground-based teams to difficult-to-reach incidents.

Considering the extreme nature of some incidents and the long response times of some specialized rescue units, it is sometimes prudent to expedite the response of USAR-trained or rescue-trained firefighters using helicopters. This serves to ensure timely technical support to first responders and ICs on a daily basis and often reduces the time required to locate and extricate trapped or injured victims and (thereby relieve pain and suffering). It expedites search and rescue operations for lost or trapped personnel and also helps to mitigate other emergency situations in ways that are not available without the use of helicopters.

The ever-present potential (in some locales) for disasters resulting from earthquakes, floods, landslides, dam failures, terrorist attacks, etc. is yet another reason to retain the option of transporting USAR and rescue companies by helicopter. Helicopters can provides timely technical SAR service to areas otherwise isolated by downed bridges, washed-out roads, blocked or damaged highways, massive flooding, and other transportation complications.

The obvious advantages of helicopter transportation include faster response to distant incidents, better access to hard-to-reach sites, the ability to perform aerial observation and size up, rapid over-water access, and the ability to bypass gridlocked traffic. In innumerable cases, response times have been slashed, patients have been extricated much sooner than otherwise possible, disasters have been managed in a more timely manner, and rapid intervention capabilities have been improved through the use of helicopters.

Inherent limitations to the use of helicopters for transporting rescue units may include lack of helicopter availability, inclement weather, poor visibility, hostile topography, weight and space limitations for passengers and cargo, and other considerations.

Many of the factors used to determine whether to transport USAR or rescue units via helicopter are *situational*, defying absolute rules that attempt to predict every possible condition. Therefore, the decision of whether to fly remains one largely of judgment and experience, based on common-sense parameters, past history, and recognition of departmental mission objectives and incident needs. Although it's difficult—if not impossible—to quantify all of these factors into a single policy or procedure, without leeway for good judgment and experience, this instruction provides basic guidelines from which to begin.

To ensure prudent use of helicopters for the transportation of USAR and rescue firefighters and equipment, it's imperative that all field personnel, ICs, and dispatchers understand and employ the time-tested parameters and guidelines established by this instruction.

Scope: This instruction applies to all members of the agency.

Author: The *(fill in the blank)* shall be responsible for content, revision, and annual review of this instruction.

Responsibility

ICs: Shall be responsible for evaluating and deciding on requests for helicopter transportation of USAR or rescue units (except those situations designated *automatic helicopter transportation*) based on the parameters established by this instruction.

USAR and rescue company officers: Shall be responsible for the following:

- Evaluating the need for helicopter transportation and making the request to the IC based on the parameters established by this instruction.
- Determining when conditions meet the *automatic helicopter transportation* parameters based on special deployment conditions that require USAR or rescue units to operate from helicopters to accomplish tactical missions.
- Selecting the USAR or rescue company members and equipment to be transported via helicopter, based on the guidelines herein, in consultation with the helicopter pilot regarding weight and space restrictions.
- Directing the actions of the USAR or rescue apparatus operators. The choices are (1) fly with the rest of the crew if the apparatus itself is not needed at the incident and if the apparatus operator is needed as part of the helicopter rescue crew (2) respond in the USAR or rescue apparatus, and report to the command post, to a staging location, to a helispot, or another designated location (3) keeping the IC informed about the mode of transportation and related issues, and (4) coordinating the reunification of the USAR or rescue company members, equipment, and apparatus in a timely manner at the conclusion of the incident.

Helicopter pilots: Shall be responsible for informing the USAR or rescue company officers of the number of personnel and the amount of equipment that may be carried on the mission at hand, based on current flight considerations (e.g., temperature, altitude, terrain, fuel, mission, etc.). They shall be responsible for coordinating with the USAR or rescue company officers to select the appropriate helispots for rendezvous, to select the appropriate tactics upon arrival, and other mission-related tasks.

Helicopter crewpersons: Shall be responsible for helping USAR or rescue company members with the loading and offloading of equipment, coordinating entry and departure of the aircraft, and other tasks to expedite the response and ensure standard safety parameters.

Dispatchers: Are responsible for supporting the requests for helicopter transportation of USAR units or rescue companies through their normal means of communications and coordination.

All personnel: Shall be responsible for familiarity with these operational guidelines.

Policy

Essential guidelines for transportation of USAR units or rescue companies: The decision to request helicopter transportation of USAR or rescue companies is a dispatch and transportation issue. The decision should generally be made within seconds of an alarm and be based on a variety of factors that must be quickly evaluated by the USAR or rescue company officers. Generally, helicopter transportation of USAR or rescue company personnel and equipment is justified under each of the following conditions:

- If conditions allow for helicopter operations, and the response time of a USAR or rescue company may be reduced by at least 30 minutes through the use of helicopters.
- If quicker response of USAR or rescue company personnel and equipment is likely to materially improve public or personnel safety (including rapid intervention capabilities).
- If USAR or rescue companies are responding from the same location as the first-in helicopter (e.g., responding from public displays, training exercises, and other situations where these units are commonly co-located).
- For incident types that generally require USAR or rescue company personnel to operate from helicopters to accomplish tactical operations like helo/swift-water rescue, rescue swimmer operations at marine emergencies and disasters, and helo/high-rise operations.
- When helicopter crews determine that the support of USAR or rescue company members is required to accomplish certain helicopter-based operations like hoist rescues, short-haul rescues, ice and snow rescues, etc.

Deciding whether to request helicopter transportation for USAR units and rescue companies: Upon receipt of the alarm, USAR or rescue company officers shall consider the following factors to determine whether it may be appropriate to request helicopter transportation of task force personnel and equipment:

- *The response type* and other information that describes the basic nature of the incident. Combined with other factors, the incident's nature is strongly indicative of the need to expedite the response of USAR units or rescue companies to manage the typical life-safety problems, logistical complications, and legal requirements such as respiratory protection, shoring, and other capabilities provided by USAR or rescue companies.
- *Radio traffic and reports* from dispatchers that indicate special hazards, needs, and transportation considerations.
- *The time of day*, which is indicative of potential traffic congestion, occupancy factors (for fires and rescues in certain structures), and the ambient light conditions, which affect the ability of helicopters to safely operate in the mountains, over the open ocean, and other hostile terrain.

  It should be noted that nighttime conditions are sometimes a contra-indication for helicopter transportation, especially deep in the mountains where it may not be feasible for helicopters to insert USAR or rescue company personnel and equipment due to flight risks such as the lack of ambient light. At night, ground transportation of USAR units and rescue companies into the mountains is generally a more reliable option than flying.
- *The location of the emergency*. High mountain slopes, deep canyons, open ocean, high-rise buildings, city streets, freeways, industrial sites, and other typical incident locations have inherent risks that affect the ability of helicopters to perform an assigned mission.
- *Weather conditions*, including fog, rain, cloud cover, wind, hail, thunderstorms, high heat, snow and sleet, and other factors that affect the ability of helicopters to fly and hover. This is especially applicable in mountains and canyons where weather conditions may change rapidly and where escape routes in case of mechanical difficulty or other flight problems are few.
- *The location of the USAR or rescue company in relation to the emergency*. For example, if the USAR unit or rescue company is on one side

of a mountain range and the incident is on the other side, ground response may entail driving completely around the mountains, which greatly delays the response. Even driving directly over the mountains is extremely time-consuming due to the steep and twisting nature of many mountain roads. The use of helicopters may cut the response time in half (or even more) by providing the most direct line to the incident. If the incident is in a remote area, helicopter transportation can provide a direct route to the scene, bypassing city streets and twisting rural roads, perhaps reducing the response time by half.

- *Whether it will be necessary to divert the primary helicopter or request a second copter to transport the USAR unit or rescue company.* Generally the first-in air squad should not be diverted to pick up a USAR or rescue company. The intent of this guideline is to ensure the timeliest response of helicopter-based ALS, hoist rescue, and trauma center transportation capabilities.
- If the nature of the incident places it in the *automatic helicopter transportation* category.
- *High-rise fires.* Helicopter crews are generally attired in Nomex® jumpsuits, but they generally aren't equipped with full turnouts, Nomex® hoods, structural helmets and gloves, or SCBA. In short, they may lack the PPE required to be inserted onto the rooftop of a burning building. To address this situation and provide for rooftop rescue and interior search, rescue, ventilation, and fire suppression operations during high-rise emergencies, some departments have designated USAR or rescue companies as helo/high-rise teams, trained to be inserted via helicopter.
- *Marine emergencies and disasters.* Some helicopter crews aren't equipped with wet suits, fins, snorkels, and other essential PPE for entry into the ocean or lakes to conduct surface SAR operations. Meanwhile, some USAR and rescue companies are trained and equipped as rescue swimmers, prepared to be deployed from helicopters during marine emergencies and disasters.
- *Swift-water rescues.* Since few helicopter crews are attired in dry suits, wet suits, swift-water helmets, and other standard PPE for entry into the water, making them vulnerable to the effects of moving water. USAR units and rescue companies may be the only helicopter-deployable swift-water rescue teams available.
- *Ice and snow rescues.* Some USAR and rescue companies are trained and equipped to conduct technical rescues in the ice and snow, including ice chutes, snow-covered slopes, etc.
- *Trench rescues, confined-space rescues, structural collapse, mud and debris flows, mine and tunnel rescues*, and other technical rescue for which helicopter crews typically are not equipped or trained for entry operations.
- When the location of the USAR or rescue company in relation to the helicopter is such that it makes sense to fly. This is particularly true when the USAR or rescue company is at the same location as the first-in helicopter.
- When the USAR or rescue company's starting location is within the general flight path of the first-due helicopter.

Procedures

Requesting helicopter transportation of USAR and rescue companies. After evaluating all the relevant factors upon receipt of the alarm, the USAR or rescue company officers should decide whether it's appropriate to arrange helicopter transportation. They shall determine whether the situation falls into the category of *request through the IC* or *automatic helicopter transportation* and make the appropriate requests and notifications based on this instruction:

Request through the IC. Requesting permission of the IC (usually the captain of the jurisdictional company or the jurisdictional battalion chief) to arrange helicopter transportation of USAR or rescue companies is appropriate whenever preemptive conditions for *automatic helicopter transportation* are not present. Generally speaking, the following sequence of events should follow:

1. Upon receipt of the alarm, the USAR or rescue company officer should evaluate the pertinent factors and determine that a request for permission to arrange helicopter transportation is appropriate.

2. The USAR or rescue company officer immediately contacts the IC face-to-face (if possible), or via radio, cellular phone, landline, MDT, or other means, and makes the request for permission to arrange helicopter transportation.
3. At this incipient stage of the response (within seconds or minutes of the alarm), the IC is usually the first-due company officer or the battalion chief.
4. The USAR or rescue company officer notifies the IC of his unit's location, gives an estimate of its ground response ETA based on current conditions, and provides other factors that indicate helicopter transportation is appropriate.
5. In some cases, the helicopter crew may determine that helicopter transportation of USAR or rescue company personnel is needed to provide specialized equipment (i.e., night-vision goggles, thermal-imaging systems, extrication equipment, etc.) or specially trained personnel to perform helicopter-based tactics, or to provide rapid intervention during high-risk helicopter rescue evolutions. In these instances, the helicopter crew should make the recommendation to the IC.
6. The IC approves or denies the request.
7. If the request is denied, the USAR or rescue company responds via its apparatus (on the ground).
8. If the request is approved, the USAR or rescue company officer determines the most appropriate helispot at which to rendezvous with a helicopter and notifies the dispatcher to respond a helicopter to that location. Ideally, the USAR or rescue company would have a pre-identified, approved helispot near their fire station to eliminate delays when they are dispatched from their home quarters.
9. The USAR or rescue company responds to the helispot and offloads appropriate equipment and personnel for transfer to the helicopter.
10. Using the appropriate air-to-ground frequency, the USAR or rescue company officer contacts the helicopter to confirm its assignment, its ETA, the number of USAR/rescue firefighters that can be transported, and the appropriateness of the selected helispot.
11. If necessary, the USAR or rescue company officer requests the response of a helicopter safety engine to respond to the helispot. This is generally necessary when there is a crowd-control problem, for crash/fire rescue, or some extra hazard.
12. While the designated USAR/rescue firefighters prepare for transfer to the helicopter, the remainder of the company responds to the incident via their apparatus, as directed by the officer.
13. Upon landing at the helispot, the helicopter crew assists USAR/rescue personnel with loading and securing of equipment and personnel prior to takeoff.
14. En route to the incident, the helicopter crew makes standard notifications regarding ETA, ensuring that the IC is aware of the presence of the USAR or rescue company aboard the copter.
15. At the conclusion of the incident, the USAR or rescue company officer coordinates the reuniting of all company members and the apparatus.

Automatic helicopter transportation. Upon receipt of the alarm, the USAR or rescue company officer notifies the IC of the helicopter transportation mode after determining that weather and other conditions are conducive to helicopter transportation and that the incident meets criteria for automatic helicopter transportation. Automatic helicopter transportation without prior approval of the IC is generally limited to one or more of the following situations:

- Marine emergencies and disasters—USAR/rescue company personnel perform helicopter-based rescue swimmer operations
- High-rise fires—USAR or rescue company personnel inserted onto the roof to conduct helo/high-rise operations, including rooftop evacuation of victims, topside ventilation, fire attack, SAR, and other assignments.
- Swift-water rescues—USAR or rescue company personnel conduct helo/swift-water rescue operations.

- Ice and snow rescue—USAR or rescue company personnel conduct ice and snow rescue operations using ice axes, crampons, and other appropriate equipment not carried on helicopters.
- Submerged victim rescues—USAR or rescue company personnel conduct surface rescue and/or dive rescue tactics.
- When the helicopter and the USAR or rescue company are responding from the same location, and other pertinent parameters are in evidence.

Guidelines for selecting USAR or rescue company personnel for helicopter transportation. The USAR or rescue company officer should determine the appropriate personnel to be transported via helicopter. This is done by consulting with the pilot (via radio prior to the helicopter's arrival) to ascertain how many members can be accommodated by the helicopter, given the weather, wind, altitude, the mission and destination, and other pertinent considerations. In nearly every case, the USAR/rescue apparatus will respond on the ground while crewmembers are transported via helicopter.

USAR task force equipment to be transported via helicopter.

For trench and excavation rescue:

- Electric fresh-air blower
- Fresh air ducting
- Electrical pigtails
- Quick shores (2 sets) with pump, hose, and wand
- Dirt vacuum
- Air knife
- Hazmat monitors
- Two biopacks
- Consider Arizona vortex for vertical entry
- HTs
- Life detector
- Search camera
- PPE

For confined-space rescue:

- Electric fresh-air blower
- Fresh air ducting
- Electrical pigtails
- Consider dirt vacuum and air knife (situational)
- Hazmat monitors
- Up to 4 biopacks with coolant canisters
- Arizona vortex
- Confined-space harnesses
- Lock-out/tag-out kit
- Haul-safe 4:1 rope system
- Cearley strap
- Homan rescue harness
- Miller half-back confined-space harness
- King pelican lights
- HTs
- Thermal-imaging system
- Search camera (situational)
- PPE

For swift-water rescue:

- Cearley straps
- Rope packs and hardware (situational)
- Line thrower (situational)
- King pelican lights (situational)
- Victim PFD and helmet
- Thermal-imaging system
- Night-vision system (for nighttime operations)
- PPE

For cliff rescues and other high-angle rescues:

- Two 300-ft rope packs or personal mountain packs
- Speed gear/hardware pack or personal mountain packs
- Picket anchor pack
- Homan harness and victim helmet
- Arizona vortex
- King pelican lights (situational)

- HTs
- Thermal-imaging system (situational)
- Night-vision system (nighttime operations)
- PPE

For vehicles over the side:

- Two 300-ft rope packs or personal mountain packs
- Hardware/speed gear packs or mountain packs
- Homan harness and victim helmet
- Arizona vortex
- King pelican lights (situational)
- HTs
- Thermal-imaging system
- Night-vision system (nighttime operations)
- Dewalt sawzall pack
- PPE

For high-rise operations:

- Biopacks with coolant canisters
- Drop bags
- Two 300-ft rope packs
- Hardware/speed gear pack
- High-rise hose pack
- Haligan tool
- 4-ft pike pole
- Axe
- Rescue saw with metal blade
- Thermal-imaging system

For structural collapse:

- Life detector
- Search camera
- Thermal-imaging system
- Exothermic system
- Air bags, hoses, regulator (situational)
- Paratech rescue system box
- Paratech shores, hose, and regulator
- Structural marking kit
- PPE

For marine emergencies:

- Two Switlick rescue platforms
- Cearley straps
- HTs
- PPE
- Thermal-imaging system
- Night-vision system (for nighttime operations)
- Consider Dewalt sawzall and other tools for capsized boats, ship collisions, planes down in the ocean
- Consider additional PPE for shipboard fires and other ocean-going transportation emergencies

For transportation accidents (plane crashes, train derailments, big rig accidents, etc.):

- Thermal-imaging system
- Night-vision systems (situational)
- Search camera
- Trapped person locator (situational)
- Fiber optic scope
- ArcAir exothermic system
- Paratech rescue kit box
- Electric confined-space Amkus system (situational)
- Air bags, hoses, regulators (situational)
- King pelican lights (situational)
- PPE

For mine and tunnel rescue:

- Biopacks and Coolant Canisters
- Hazmat monitors (including pole for above-ground readings)
- Electric fresh-air blower
- Fresh air ducting
- Electrical pigtails
- Arizona vortex
- Haul-safe 4:1 rope system
- Cearley strap
- Homan rescue harness
- Miller half-back confined-space harness
- King pelican lights

- HTs
- Thermal-imaging system
- Search camera (situational)
- Night-vision system
- Picket anchor pack
- PPE

For industrial entrapment (victim trapped in machinery, etc.):

- Dewalt sawzall
- Alternate sized air bags, hoses, regulators
- Paratech rescue kit box with wizzer saw
- ArcAir exothermic torch and rods
- Fiber-optic scope
- Search camera
- Come-along
- PPE

For ice and snow rescue:

- Personal mountain packs, including ice axes, crampons, avalanche poles, cold weather gear, survival supplies
- Consider extra ropes and hardware (situational)
- Picket anchor pack
- Homan harness and victim helmet
- Arizona vortex
- King pelican lights (situational)
- HTs
- Thermal-imaging system
- Night-vision system (nighttime operations)
- PPE

For planes down in mountainous terrain:

- Two 300-ft rope packs or personal mountain packs
- Speed gear/hardware pack or personal mountain packs
- Picket anchor pack
- Arizona vortex
- King pelican lights (situational)
- HTs
- Thermal-imaging system (situational)
- Night-vision system (nighttime operations)
- DeWalt sawzall
- Paratech rescue kit and SCBA bottles
- Air bags, hoses, regulator
- Axe
- Haligan tool
- Pry bar
- PPE

For submerged victim/dive first responder incidents:

- Wetsuits and other PPE
- First responder dive packs
- Thermal-imaging system
- Night-vision system (nighttime operations)
- Inflatable rescue boat (situational)
- Rescue board (situational)
- Swift-water throw bags or marker buoys
- HTs
- Dive gear

For landslides, mudslides, mud and debris flows:

- Cearley straps
- Rope packs and hardware (situational)
- Line thrower (situational)
- King pelican lights (situational)
- Victim PFD and helmet
- Thermal-imaging system
- Night-vision system (for nighttime operations)
- Life detector
- Search camera
- Thermal-imaging system
- Air bags, hoses, regulator (situational)
- Paratech rescue system box
- Structural marking kit
- PPE

Appendix XIII: Tribute to a Helicopter Rescue Pioneer

Some of the helicopter rescue tactics discussed in chapter 5 were developed in the wake of the deadly storms that swept across southern California in 1992. The success of the new rescue methods is evidence that properly trained firefighters, pilots, lifeguards, and police officers—supported by appropriate equipment, planning, and strategy—can save people from flood-related predicaments that would have been characterized as *hopeless* just a decade ago.

Most victims rescued by helicopter crews using these tactics will never realize that one of the people who helped develop the improvements that led to their survival was Jeff Langley. Langley was a 28-year-old firefighter/paramedic who perished in a fall from a fire/rescue helicopter just minutes after he and his crew rescued a mortally injured hiker from the bottom of a cliff in Topanga Canyon.

Langley's death came just one year after he helped lead the charge to develop the HSW rescue techniques discussed in this textbook. He was known as a perfectionist. A more eager and dedicated student of the science and tactics of firefighting and rescue work could not be found. He was intensely committed to a philosophy of giving more than he took. While this is a common theme in the fire service, it was especially evident in Jeff Langley's case. He volunteered to serve on the LACoFD's water rescue committee, whose members devote a great deal of off-duty time to develop innovative flood rescue methods. Later, he became a member of the USAR task force, where he was recognized as an innovator and a natural leader.

Perhaps even more impressive than these tangible contributions were other traits that marked him. His coworkers remember a day when Langley scaled a cliff after a resident of Box Canyon reported hearing the wailing of what he thought was an injured child at the base of a waterfall during the storms of 1992. Actually, the source of the noise *was* a kid. When Langley emerged from the waterfall, he was carrying a young goat that had become trapped in the rocks and was being drowned by torrents of cold water.

Langley also displayed a willingness to push himself to the very limit when necessary, and to take smart, calculated risks to save innocent lives. He implicitly understood the hazards associated with such a personal commitment to public safety.

After transferring to LACoFD air operations as a firefighter/paramedic, Langley showed a knack for independent thinking and the ability to come up with new ways to reduce risks and improve the effectiveness of rescue operations. Some of the research conducted by Langley and his partners was inherently hazardous, in part because the techniques had never been tried before, and there was little margin for error. By 1991, they were hard at work developing new ways to pluck people from swift-water rescue situations with helicopters.

Several of Langley's coworkers followed his lead in devoting their time, effort, and money to attend specialized rescue training courses across the United States. Their goal was to bring the best of what they learned back to their department. Soon, Langley and his coworkers were introducing new innovations to the department's helicopter operations, improving the likelihood of survival for victims and rescuers alike.

Then came the deadly floods of 1992, a series of events that proved to be pivotal in the development of local swift-water rescue capabilities. In February of that year, dozens of people were trapped by the Sepulveda Basin flood, an event that even trapped a dozen firefighters and swamped several fire engines and a ladder truck. Helicopters were used extensively to pluck people from high ground and from the water. The Sepulveda Basin flood was quickly followed by the Ventura River flood in which several people died, and a firefighter (wearing turnouts) had to be rescued from rising floodwaters after he became stranded.

Then a boy named Adam Bishoff fell into a flood-control channel and was swept to his death in the roiling waters of the Los Angeles River, despite valiant efforts to rescue him. It was an incident that agonized rescuers and citizens alike and brought the problem of swift-water rescue to the forefront of public concern. In March 1992, the previously mentioned Rubio Creek incident threatened to end in similar tragedy.

These incidents galvanized the determination of Langley and other members of the air operations and USAR units, including Senior Pilot Rick Cearley, Captain Wayne Ibers, and Firefighter Layne Contreras. After Ibers wrote a proposal to the LACoFD's administration, Fire Chief P. Michael Freeman named a working group to develop new methods of rescuing flood victims with helicopters. The group included Cearly, Ibers, Contreras, Captain Roger Wilhelm, this author, and Langley. With the support of LACoFD air operations and the USAR unit, the group began a series of high-risk experiments to develop and test innovative new helicopter-based swift-water rescue methods and equipment.

The two biggest questions were (1) how to secure a victim in the water for helicopter extraction in fast-moving water, and (2) how to place the rescuer in a position where he could effect the extraction in fast-moving water.

Ibers and Contreras solved this problem by adapting certain high-angle rescue techniques to the swift-water problem. Later, Cearly would improve on these methods by developing two new rescue tools in his own garage. The group also adopted certain existing high-angle rescue methods to the helicopter environment. They combined these disparate elements into a cohesive system that included five innovative new rescue evolutions.

Langley played a significant role in field-testing the newly devised methods and equipment. To ensure a realistic testing environment, Langley and the other group members volunteered to take turns acting as victims in the fast-moving water of rugged and fast-flowing Azusa Canyon, as well as actual flood-control channels during major storms. The others attempted to rescue them while hanging from ropes attached to fire department helicopters. It was dangerous duty because one wrong move could have lethal consequences.

After months of experimenting, the group settled on a series of rescue evolutions that would serve to raise the standard of HSW rescue to a new level. Since that time, these evolutions have been successfully used in simulated rescue situations (to rescue firefighters posing as victims in fast-moving water) literally thousands of times by helicopter-based LACoFD rescuers. Various other agencies across the United States have also developed variations of these evolutions.

The newly devised HSW rescue techniques are based on long-held swift-water rescue fundamentals, reinforced by lessons learned at Rubio Creek and the Los Angeles River. Early on, the group recognized that successful rescue from fast-moving water is dependant upon the ability of rescuers to use the power of the water to their advantage. Simply put, when victims are being swept downstream in fast-moving water (including those dangling from helicopters), it is far more effective to snatch them from peril while they match the victim's speed in the water than to fight the current by remaining stationary. Remaining in a stationary position in the water (i.e., while tied to a rope or cable) exposes the rescuer not just to the fierce current but also to impact with debris carried by the current.

One of the methods developed by the group and field-tested at significant personal risk by Langley and Firefighter Eric Fetherston is a good example of adherence to this principle. First, a specially equipped rescuer is lowered from a rope attached to a helicopter. The rescuer is suspended beneath the helicopter while the pilot flies it to a point downstream of the victim. He hovers in a position where he can see the victim approaching in the current, with the rescuer hanging just above the surface of the water.

A firefighter stands on the landing skid (connected to the helicopter by a special strap) and uses a hot microphone to verbally guide the pilot to exactly the right position over the channel. As the victim reaches the rescuer's location, the helicopter moves downstream and descends slightly until the rescuer is immersed in water.

The rescuer then quickly swims to the victim, a difficult maneuver in water that may only be 2-ft deep and moving more than 30 mph. The rescuer then gains control of the victim and quickly applies a special device called the Cearly strap. The strap was named after Rick Cearly, who conceived the idea and developed the prototype in his own garage. The Cearly strap allows the rescuer to capture the victim and secure him to the rope. All the while, the helicopter follows overhead, matching the speed of the rescuer and victim. Once the Cearly strap is applied and connected to the rope, the rescuer signals that he is ready, and the pilot slowly ascends, lifting rescuer and victim from the water. The helicopter then moves laterally to the shore, where the victim and rescuer are gently deposited.

Variations of this and other helicopter rescue methods that owe their origins substantially to Jeff Langley and the other members of the group have been used to rescue many victims since 1992. For example, on a single day in February 1993, four people were saved from floods by HSW teams using these rescue methods.

Several weeks later, Langley's helicopter was dispatched to rescue a hiker who fell 100 ft to the bottom of Eagle Rock in Topanga Canyon. The hiker lay gravely injured at the base of the rock, and he required an immediate MedEvac. While responding across the 101 freeway in USAR-1 (the LACoFD's central USAR company at the time) to the reported cliff rescue, this author listened intently to the fire radio as the pilot reported that they were committing themselves to attempt the rescue. There was an uneasy feeling because the site of the rescue was a remote canyon, with little access available to ground units. This meant that the air squad would be operating alone, without immediate ground support. This is a familiar predicament for the air squad crews who are frequently first on the scene of difficult rescue problems in the most remote regions of the mountains.

The pilot came on the air and reported that they would be able to handle the rescue and for the dispatch center to cancel all responding ground units. Against my better judgment, I complied without suggesting to the pilot via radio that perhaps the ground units should continue responding to ensure some level of support and rapid intervention capabilities in case something unexpected happened.

Then came a moment that every firefighter dreads. The pilot reported an emergency situation. A firefighter had just fallen from the helicopter. He didn't have time to elaborate because he was attempting an immediate rescue of the fallen firefighter. Not knowing the exact situation at the time, or which member was stricken, we could only hope that it was just a minor mishap. Unfortunately, that was not the case, and Langley died that day.

Since that time, the list of successful helicopter-based swift-water rescues using techniques that Langley helped inspire continues to grow. Recently, the NASAR named a national award after Langley and Earl Higgins, a local man who lost his life while attempting to rescue a boy from the Los Angeles River in 1980. The Higgins/Langley Memorial is now awarded annually to individuals and groups who make outstanding contributions to the field of swift-water rescue.

Jeff Langley would have been pleased to see the success of recent rescue operations. For Langley, there may have been no greater reward than the knowledge that his work (and the work of his friends) continues to play a part in saving innocent lives.

Appendix XIV: Two-point Tether with Life Float

Table A–3 Two-point Tether with Life Float

Operation:	Key points:
1. Communicate your intentions to the victim if possible.	a. Have him remain in the safest place possible. b. Ascertain injuries and his ability to don a PFD and helmet, and to hold onto the life float.
2. Line the river.	a. Use safest method available
3. Assure sufficient personnel are on both sides of the river.	a. Varies with size of float, width of river, and strength of current. b. Normally, one or two rescuers on each side of the river is sufficient for a two-point tether.
4. Attach tether lines.	a. Wrap the end of the lines to life float around a section of the float and clip the running end to the standing part. b. If the life float is an innertube or other inflatable device, use a method which will not pinch the tube when the rope becomes taut. (Place a loop in the standing part. Make a loop in the running end. Take running end loop around the device, and clip it to the standing part loop).
5. Attach PFD and helmet to the float if applicable.	a. Use a carabiner to clip to the float if applicable.
If the victim appears able to assist in his own rescue:	
6. Shuttle the life float to the victim.	
7. Instruct the victim to remove the PFD and helmet from the line.	
8. Instruct the victim to don the PFD and helmet.	
9. Instruct victim to lay on float, facing upstream, with his torso draped over the float.	a. Do not allow victim to clip himself to the line or the float.

10. Shuttle victim to shore.	
11. Downstream safety should be prepared to attempt rescue if victim becomes separated from the float.	
If the victim is unable to assist in his own rescue (small child, elderly person, injured person, etc):	
12. Determine hazards to rescuers entering the water on a float.	a. This is a judgment call which must be left to the discretion of personnel at scene.
13. If danger to personnel outweighs the chance of successful rescue, alternative methods should be attempted. Prepare to attempt a shore-based rescue in the event the victim is swept into the current.	
14. If personnel decide to attempt contact rescue, have one or two rescuers don PFD's and helmets.	
15. Clip PFD and helmet to float for victim.	
16. Have rescuer(s) lay on float.	a. If two rescuers, have the first drape his torso over the float, facing upstream. Have the second rescuer get behind (downstream of) the first. The second rescuer should drape arms and torso around the first rescuer.
17. Shuttle rescuer(s) to victim.	
18. If safe to do so, place a PFD and helmet on the victim.	a. Conditions may be too hazardous to place PFD and helmet on the victim. Our purpose is to get the victim to safety with or without the PFD and helmet. b. This equipment is a safety precaution for the victim; don't risk losing a rescuer (or the victim) by placing yourself in a precarious position just to put the victim in a PFD and helmet.
19. Position the victim for the ride to shore.	a. If one rescuer, there are two choices: 1. Place the victim on the float. Get behind the victim and drape arms and torso around the victim, securing him onto the float. The victim will take the force of the water, but you can assure that he stays on the float. You will be in the eddy created by the victim. This technique is recommended if you are concerned that the victim cannot hold onto you from behind.

2. Lay on the float and instruct the victim to wrap his arms around you (and grab the float if he can). This method will place the victim in the eddy created by your body. However, the victim may become separated from you.

b. If two rescuers, have the first rescuer lay on the float. Have the victim wrap arms and torso around the first rescuer. The second rescuer can now wrap his arms around the victim and grab the float or the first rescuer. This may place quite a load on the first rescuer, but it allows the victim and the second rescuer to have a relatively smooth ride.

20. Shuttle the rescuer(s) and victim to shore.

21. Downstream safety should be prepared to attempt rescuer if the rescuer(s) or victim become separated from the float.

Appendix XV: Two-point Tether for Lowhead Dam Rescue

Table A–4 Two-point Tether for Lowhead Dam Rescue

Operation:	Key points:
1. Do not send rescuers to assist in the hydraulic of a low head dam. Keep all personnel completely clear of the boil line and the hydraulic. Keep all personnel out of the water upstream of low head dams.	
2. Shuttle the life float to the victim, or to his estimated position.	a. The victim may be pulled under repeatedly, and rescuers will have to estimate where he is.
3. If victim can grab the float, pull him along the face of the dam to shore *or* through the boil line into the downstream current.	

Appendix XVI: Two-point Tether for Foot Entrapment Rescue

Table A–5 Two-point Tether for Foot Entrapment Rescue

Operation:	Key points:
1. Line the river.	
2. Attach float or weighted drop bag to middle of line.	a. If victim's head and/or torso are on or above the surface of the water, use a life float. b. If victim's head and/or torso are submerged, weight a throw bag with rocks. The weight will allow the bag to drop beneath the surface to snag victim.
3. Starting downstream of the victim, shuttle the float or drop bag out to the victim.	a. The float or drop bag must be directly downstream of the victim.
4. Move upstream until the float or drop bag is at the victim's chest or waist.	a. If victim is submerged, a momentary slack in the line may help the drop bag to sink under the surface. b. As soon as the bag sinks, move the line upstream to snag the victim.
5. Snag the victim, and move upstream, bringing the victim's head and chest out of the water.	
6. It may be possible to continue moving upstream, freeing the victim's foot. If this does not work, it may be possible to use a weighted line to snag the victim's legs.	a. If moving upstream with the first line is unsuccessful in freeing the victim, you have at least assured that his head/torso are out of the water.
7. If the second line fails to free the victim, the next step depends on the severity of the victim's predicament. If the victim is able to breathe, you have time to try other options. If the victim's airway is still compromised, more drastic measures may be necessary (i.e., mechanical advantage systems, etc.).	a. Depending on conditions at scene, more drastic measures may include the use of mechanical advantage to free the victim's legs. For other alternatives, see page ___________.

Appendix XVII: Filling 2½-in. Fire Hose with Hose Rescue Device

Table A–6 Filling 2½-in. Fire Hose with Hose Rescue Device

Operation:	Key points:
1. Pull adequate 2½-inch hose off apparatus.	1. Length varies with the evolution being used and the width of the river. 2. 50′ section is suitable for most applications. 3. Longer sections may be needed to reach victims trapped in the hydraulic of a low head dam. 4. Bypass section is useful for narrow waterways.
2. Move the hose, one SCBA bottle, adequate rope, two spanner wrenches, and the hose rescue device to the rescue site.	
3. Lay the hose out.	
4. Connect the brass cap to the male coupling.	a. Hand tight is usually adequate; spanner tight will assure good seal.
5. Connect the brass plug to the female coupling.	a. Hand tight is usually adequate; spanner tight will assure good seal.
6. Connect the pig-tail air hose to the quick-connect fitting on the brass plug.	a. Test the connection by attempting to pull the pig-tail hose from the brass plug.
7. Connect the other end of the pig-tail air hose to the quick-connect fitting on the regulator.	a. Test the connection by attempting to pull the pig-tail hose from the regulator.
8. Connect the regulator to the SCBA bottle.	
9. Assure that the one-way valve is shut.	
10. Open the SCBA bottle, charging the regulator.	
11. Check the air pressure gauge on the regulator.	a. The pressure has been pre-set at 100 psi. The pressure gauge reading should not exceed this pressure.
12. After assuring all connections are tight, *slowly* open the one-way valve, charging the fire hose.	a. It is not always necessary to inflate the hose to 100 psi. Depending on the use, lower pressure may be desired.

13. Shut the one-way valve.	
14. Shut the SCBA bottle.	
15. Step on the pig-tail air hose to prevent any whipping action when it is disconnected from the brass plug.	
16. *Carefully* disconnect the pig-tail air hose from the brass plug.	
17. While stepping on the loose end of the pig-tail air hose, *slowly* open the one-way valve to bleed pressure from the regulator.	
18. The inflated hose is now ready for use. While using the inflated hose, avoid dropping it or banging the couplings against rocks and cement.	

Deflating 2½-in. Fire Hose with Hose Rescue Device

OPERATION:	KEY POINTS:
1. Remove the inflated hose from the water.	
2. A 125 psi pop-off valve is tapped into the brass plug and protected with a shield.	
To deflate the hose, pull the ring connected to the pop-off valve. The air will bleed through the valve.	
3. When pressure is sufficiently low, unscrew the brass cap and the brass plug from the hose.	

Appendix XVIII: Lowhead Dam Rescue with Hose Rescue Device

Method 1: Looped Hose

Table A–7 Lowhead Dam Rescue with Hose Rescue Device—Method 1: Looped Hose

Operation:	Key points:
1. Remove adequate hose from the apparatus.	a. The hose length must be at least double the distance from the victim to the near shore (because you will be folding the hose into a bight).
2. Move the hose, SCBA bottle, rope, spanners and hose rescue device to the rescue site.	
3. Make necessary connections and inflate the fire hose.	
4. Disconnect the pig-tail air hose from fire hose.	
5. Bend the hose back on itself to form a bight.	a. If the hose is too rigid, bleed off extra pressure through the pop-off valve.
6. For extra control and safety, connect one end of the rope to the brass cap, and the other end of the rope to the brass plug.	a. This forms a loop consisting of hose and rope. The rope can be used to help control the hose. b. If there is a danger of personnel slipping in due to steep banks or slippery surfaces, extra manpower can be used to help tether the hose from a point further from the shore.
7. Push the "bight" portion of the hose out parallel with the face of the dam.	a. The hydraulic action will keep the hose between the boil line and the dam. This is where the victim is trapped.
8. If the victim can grab the hose, pull him back along the face of the dam to shore.	
9. If a vertical wall prevents a direct exit to shore, it may be necessary to pull the hose (and the victim) through the boil line into the downstream current.	a. An alternative might be the use of a roof ladder suspended with the hooks over the vertical wall. b. If access permits, an aerial ladder or aerial platform may be utilized to extricate the victim.

Method 2: Straight Hose with River Lined

Line the river, connect the line to the hose rescue device, and pull the hose out across the face of the dam. Now the hose can be controlled from both shores. If the victim is able to hang onto the hose, shuttle him along the face of the dam to the nearest shore.

Equipment needed

- One 4,500-psi SCBA bottle (one bottle fills approximately 200 to 300 ft of 2½-in. hose)
- Two sections of rope or two drop bags. Drop bags are adequate in many cases. However, longer lengths may be necessary. If 300-ft sections are not available, drop bag lines can be clipped together for added length.)
- Two spanner wrenches
- Hose rescue device

Table A–8 Lowhead Dam Rescue with Hose Rescue Device—Method 2: Straight Hose, River Lined

Operation:	Key points:
1. Line the river.	
2. Position the line.	a. Just downstream of the dam.
3. Remove adequate hose from the apparatus.	a. The hose length should be longer than the distance from the victim to the near shore. b. If feasible, the hose length should equal the width of the river.
4. Move the hose, SCBA bottle, rope, spanners and hose rescue device to the rescue site.	
5. Make necessary connections and inflate the fire hose.	
6. Disconnect the pig-tail air hose from fire hose.	
7. Connect the rope lining the river to one end of the inflated hose.	a. Clip to ring with carabiner or tie to the ring.
8. Connect the second section of rope to the other end of the hose.	a. Clip to ring with carabiner or tie to the ring.
9. Begin feeding hose into the water while personnel on the far shore pull their line.	a. The hydraulic action will help keep the hose between the boil line and the face of the dam. This is where the victim is.
10. If the victim can grab the hose, shuttle him along the face of the dam to the nearest shore.	
11. If a vertical wall prevents a direct exit to shore, it may be necessary to pull the hose (and the victim) through the boil line into the downstream current.	a. An alternative might be the use of a roof ladder suspended with the hooks over the vertical wall. b. If access permits, an aerial ladder or aerial platform may be utilized to extricate the victim.

Hose Rescue Device: Bridge-based Rescue

Many rivers and flood-control channels are crossed by bridges. Bridges can be used to perform safe, efficient rescues without excessive danger to firefighters. Bridge-based rescue evolutions from swift-moving water require close coordination of personnel. It is important that firefighters have the opportunity to practice these methods prior to use in actual rescue situations.

Inflated fire hose makes an excellent rescue tool when working from a bridge. This is particularly true when the channels are very wide.

Equipment needed:

- One 4500-psi SCBA bottle
- Three drop bags
- One long section of rope (a 300-ft section is preferred, but a drop bag will work in a pinch)
- One 50-ft section of 2 ½-in. hose *or* a bypass hose
- Hose rescue device

Personnel needed:

A minimum of four rescuers. However, six to eight rescuers are the preferred minimum during storm conditions when the water is moving 20 to 30 mph in flood-control channels. This personnel requirement must be considered when assigning units to set up this evolution. An engine should be assisted by a truck, squad, patrol, or combination thereof when possible.

Strategy:

One firefighter is assigned to the upstream side of the bridge to act as a spotter. His job is to look for the victim coming downstream in the river. Once the victim is sighted, the spotter must key on the victim's position in relation to the shores. Remember that personnel working on the downstream side of the bridge will not be able to see the victim. They must watch the spotter to place the hose where the victim will emerge from underneath the bridge.

If the victim moves river-right in the water, the spotter must mimic his position by moving river-right. If the victim moves river-left, the spotter must move river-left to line up on him. Personnel on the downstream side will follow the lead of the spotter.

The engine should be spotted on the downstream side of the bridge, blocking the far right lane of traffic. Safety of personnel may dictate the blocking of more than one lane or the complete shutdown of traffic on the bridge.

Law enforcement can assist with traffic control. In fact, law enforcement should be active in the preplan for rescue on each waterway. Under ideal conditions, law enforcement can preplan to provide traffic control at all strategic bridges.

Adequate 2 $^1/_2$-in. hose is pulled off the engine and stretched out along the downstream side of the bridge (watch for traffic). The use of a single 50-ft section of inflated hose allows rescue of victims from even the widest rivers. For very narrow channels, a bypass hose is effective. The hose is inflated, and belay lines are attached to the hose. For 50-ft sections, one belay line should be placed at each end and one in the middle.

When only four firefighters are available, two belay lines can do the job. However, three belay lines are preferred for maneuvering the hose. For a bypass section, only two belay lines are required. Rescuers tending belay lines are called *belays*.

If drop bags are used, simply run the end of the rope around the hose and clip the carabiner around the standing part of the line. For lines without carabiners, use a running loop, lark's foot, or a girth hitch.

Attach the long section of rope to a ring at one end of the hose. Run this section of rope along the bridge to shore. The hose is suspended from the downstream side of the bridge so that it dangles just above the surface of the water. This places the hose in clear view of the victim as he is swept under the bridge.

A firefighter is positioned on the shore with the long section of hose. This position is called the *pendulum*. The duty of the pendulum is to swing the victim to shore once the hose is dropped into the water.

As the victim goes under the upstream side of the bridge, the spotter will throw his arms up and yell "He's under!" Now the belays must look straight down to the water from their side of the bridge. As the victim emerges, the belays should drop the hose right into his lap, or within easy reaching distance. The belays *immediately* drop the belay lines into the water with the hose and the victim. *Do not hang onto the belay lines!* The hose must float with the victim in the current, or he may not be able to grab it.

At this point, the spotter and belays should make their way to the shoreline to assist the pendulum. The pendulum will be moving downstream while pulling the victim toward shore. Just as in the floating drop bag rescue method, the pendulum will be more effective remaining downstream of the victim. This may require the pendulum to run along the shore.

During this time, the firefighter assigned to downstream safety must remain downstream for the entire operation. The spotter and belays must assist the pendulum. During drills on the Los Angeles River during storms, it was found that at least two persons were required to pull the victim to shore. Under severe conditions, it may take four or five persons to overcome the force of the current in a wide river.

Extra personnel are needed to pull the victim from the water once he/she is at the shoreline. The downstream safety can do this in a pinch. However, the downstream safety should remain out of the operation if possible. If the safety falls in while trying to pull the victim out, who will help him? This dramatizes the importance of assigning other units along the entire river. If something goes wrong, we want to assure that our people will be rescued.

Note: a variation of this evolution involves the use of rope in place of the inflated fire hose for units without a hose inflator. The evolution is performed virtually the same as just described for hose.

Table A–9 Bridge-based Rescue Using a Hose Rescue Device

Operation:	Key points:
The spotter position:	
1. Find advantageous location.	a. On *upstream* side of bridge.
	b. Be aware of traffic hazard.
	c. Find location which offers the best vantage point for spotting the victim.
2. Observe for victim downstream in the current.	a. It may be helpful to utilize coming police or other personnel to help watch for victim.
	b. Be aware that the victim may not be easily seen. He may be floating with debris, and may be submerged for long periods.

3. Notify personnel when victim is spotted.	a. Yell, use whistle or other sounding device.
	b. Remember that traffic and the roar of rushing water may drown out voice communication.
4. Key on victim's position.	a. Continue to line up on the victim's location with reference to the sides.
5. Attempt to notify the victim that he should grab the hose on the downstream side of the bridge.	a. Bullhorn or other public address system is helpful.
	b. Law enforcement will normally have public address capability.
6. Notify personnel when the victim goes under the bridge.	a. Throw both hands up in the air and yell "he's under!"
7. Assist the *pendulum* position at the shore line.	

The belay position:

1. Spot apparatus on *downstream* side of bridge.	a. Blocking the right lane of traffic.
	b. Block additional lanes if necessary for safety.
	c. Shut down all traffic on bridge if necessary for safety.
	d. Use law enforcement for traffic control.
2. Remove adequate hose.	a. Place on sidewalk on *downstream* side of bridge.
3. Attach hose rescue device, inflate hose.	
4. Detach pig-tail from brass plug.	a. Shut off one-way valve and bottle first.
	b. Step on pig-tail to avoid whipping action.
5. Attach belay lines.	a. Carabiners, "lark's foot" knot or girth hitch.
6. Attach pendulum line to either end of hose.	a. Clip into ring at either end, or tie knot to ring.
7. Stretch pendulum line to shore line.	a. Assure no tangles or snags.
8. Lower hose off *downstream* side of bridge.	a. Allow hose to dangle just above the surface of the water.
9. Adjust position to line up on spotter.	a. Follow movements of spotter.
10. When spotter signals that victim is under bridge, look straight down to water.	
11. When victim appears, drop hose into the water.	a. Attempt to drop the hose into the victim's lap (or within easy reach).
	b. Once the hose is dropped, drop all belay lines *immediately* to prevent dragging the hose back upstream of the victim.

12. Assist the pendulum position.	
13. Assist removing the victim from the water.	

Pendulum position:

1. Take pendulum line to selected position on shore.	a. Stretch the rope out and move as far downstream as possible. This will help keep the victim upstream from you and provide a nice pendulum arc. b. Watch for obstructions along the shoreline. c. Do not anchor yourself to stationary objects. The victim is likely to pass you before you can get him to shore. If there is danger of falling in, move back from the river bank. d. An alternative is to tie a line to the *pendulum* and use it as a belay line with another firefighter providing a body belay. This will allow both rescuers to move downstream.
2. When hose is dropped in with the victim, take up all slack and begin to pull victim to shore.	
3. Stay downstream of victim.	
4. If victim gets downstream of you, begin to give slack to avoid forcing him under the water.	
5. "Hold" victim at shore line as other rescuers remove him from the water.	

Downstream safety position:

1. Remain downstream of the rescue operation with floating drop bag.	
2. Be prepared to rescue victim if he loses the hose.	
3. Be prepared to rescue firefighters.	

Appendix XIX: Bridge-based Rescue Using Life Float

Table A–10 Bridge-based Rescue Using Life Float

Operation:	Key points:
Spotter position:	a. On *upstream* side of bridge.
1. Find advantageous location.	b. Be aware of traffic hazard.
	c. Find location which offers the best vantage point for spotting the victim.
2. Observe for victim coming downstream in the current.	a. It may be helpful to utilize police or other personnel to help watch for victim.
	b. Be aware that victim may not be easily seen. He may be floating with debris, and may be submerged for long periods.
3. Notify personnel when victim is spotted.	a. Yell, use whistle or other sounding device.
	b. Remember that traffic and the roar of the rushing water may drown out voice communication.
4. Key on victim's position.	a. Continue to line up on the victim's location with reference to the sides.
5. Attempt to notify the victim that he should grab the hose on the downstream side of the bridge.	a. Bullhorn or other public address system is helpful.
	b. Law enforcement will normally have public address capability.
6. Notify personnel when the victim goes under the bridge.	a. Throw both hands up in the air and yell "he's under!"
7. Assist the pendulum position at the shoreline.	
Belay position:	
1. Spot apparatus on *downstream side* of bridge.	a. Blocking the right lane of traffic.
	b. Block additional lanes if necessary for safety.
	c. Shut down all traffic on bridge if necessary for safety.
	d. Use law enforcement for traffic control.
2. Remove adequate rope from apparatus. This rope will serve both as the belay line and the pendulum line.	a. 300-ft. section is desirable for extra flexibility on the pendulum.
	b. Drop bags can be clipped together to get adequate length.

Step	Key Points
3. Attach one end of the line to the life float.	a. If a carabiner is available, clip to a loop tied in the rope, run around the life float, and clip back to the standing part of the line. b. To prevent excessive stress to the life float, avoid tying the line directly to the float. The line should be clipped back to itself, forming a loop around the life float. This places the stress on the rope, which is designed to handle the drag which will occur.
4. Stretch the rest of the line to the pendulum position on the shoreline.	a. Assure no tangles or snags. b. Keep enough slack on the bridge to lower the float to the water.
5. Lower the life float to the water on the *downstream* side of the bridge.	a. Allow the float to lay on the water. It will "plane up" on the surface.
6. Adjust the position of the life float to allow maximum rescue potential.	a. Allowing the float to move further downstream in the current will give you more time to react when the victim appears from underneath the bridge.
7. Line up on the spotter.	a. Follow movements of the spotter.
8. When spotter signals that victim is under bridge, look straight down to the water.	
9. When victim appears on the downstream side of bridge, make lateral adjustments to place the float in his path.	a. *Important*—keep bystanders and others out of the way. You may have to make swift lateral movements to get the float to the victim. Use law enforcement to clear the area if necessary.
10. When the victim contacts the float, drop the rope immediately, allowing the float to drift with the victim.	
11. Assist the pendulum position on the shoreline.	

Pendulum position:

Step	Key Points
1. Take pendulum line to selected position on shore.	a. Stretch the rope out and move as far downstream as possible. This will help keep the victim upstream from you and provide a nice pendulum arc. b. Watch for obstructions along the shoreline. c. Do not anchor yourself to stationary objects. The victim is likely to pass you before you can get him to shore. Also, the river can rise dramatically during storms. This might cause tethered rescuers to be trapped at the end of a taught line in moving water. d. If there is a danger of falling into the water, move back from the river bank or utilize a dynamic belay (with the belayer working from a flat surface).

e. An alternative is to tie a line to the pendulum person and use it as a belay line with another firefighter providing a body belay. This will provide safety for the pendulum position while allowing downstream movement.

NOTE: *Never* tie a rope around a rescuer if there is any possibility that he might fall into moving water.

2. When the belay releases the float in the current with the victim, take up slack and begin to pull victim to shore.
3. Stay downstream of the victim.
4. If victim gets downstream of you, begin to give slack to avoid forcing him under the water.
5. "Hold" the victim at shore with appropriate tension on the line as other rescuers remove him from the water.

Downstream safety position:

1. Remain downstream of the rescue operation with a floating drop bag.
2. Be prepared to rescue victim if he is unable to grab the float.
3. Be prepared to rescue firefighters.

Appendix XX: Single-line Self-rescue System

Table A–11 Single-line Self-rescue System

Operation:	Key points:
1. Assure downstream safety is in position, and other units are setting up for rescue downstream (if possible).	a. In isolated areas it may be impossible to get any units downstream in time. b. During major floods, there may be no other units available.
2. Explain plan to victim with bullhorn or other public address system (or by yelling if he can hear you over the sound of the water).	a. If possible, and if time permits. If the victim is about to be washed into a raging river, you may just have to go for it and hope he picks up on the idea. Most victims in this situation will grab anything you throw to them.
3. Get line to victim.	a. Use floating drop bag if possible. b. If the victim is out of throwing range, use whatever method is available.
4. Make a hand loop in the line and attach a helmet and PFD.	a. Only if victim is in a position to don this equipment. If he is about to be swept away, forget about the helmet and PFD.
5. Shuttle the helmet and PFD to victim (if applicable).	
6. Have victim don helmet and PFD (if applicable).	
7. Move downstream of the victim's position as far as possible. This will set up a good pendulum arc for the victim.	
8. When ready, have the victim lie on the life float (or hold the loop in the rope) and enter the water.	a. Don't let the victim place the rope around his body. b. If using a life float, have victim lay on the float prone and facing upstream. c. If using the loop in the rope, the victim should enter the water with his feet pointing downstream (river swimming position). d. Watch for dangerous debris before signaling the victim into the water.
9. Pendulum the victim to shore.	a. Stay downstream of victim if possible. b. If victim becomes separated from the float or the rope, be prepared to make a rescue from shore.
10. Assist victim from the water.	

Appendix XXI: Double-line Self-rescue System

Table A–12 Double-line Self-rescue System

Operation:	Key points:
1. Assure downstream safety is in position, and other units are setting up for rescue downstream (if possible).	a. In isolated areas it may be impossible to get any units downstream in time.
	b. During major floods, there may be no other units available.
2. Explain plan to victim.	a. If possible, and if time permits.
3. Get line to victim.	a. If line is long enough, toss the middle of the rope to victim, keeping both ends on the near shore.
	b. If one line is too short, connect two with a carabiner or Figure of 8 follow through knot.
4. Make a hand loop in the line, and attach a second line, a helmet and a PFD.	a. Include the PFD and helmet only if the victim is in a position to don this equipment. If he is about to be swept away, it is unlikely he will have a chance to do anything but grab the rope.
5. Shuttle the second line, helmet, PFD and life float to the victim.	a. Keep in mind that in some cases, the victim may not be able to assist in shuttling operations. He may just have to grab the rope and go.
6. Have the victim don the helmet and PFD (if applicable).	
7. One line tender moves downstream, judging the angle necessary to get the victim out before he hits the hazard.	
8. The other line tender moves upstream.	
9. When ready, have the victim lie on the float (or hold the loop in the rope) and enter the water.	a. Don't let the victim place the rope around his body.
	b. If using a life float, have the victim lie on the float prone and facing upstream. This will help him plane up in the water.
	c. If using the loop in the rope, the victim should enter the water supine with his feet pointing downstream (river swimming position).
	d. Watch for dangerous debris in the water before signaling the victim into the water.
10. Pendulum the victim to shore.	a. The downstream rope tender pulls as hard as possible, attempting to bring the victim directly to shore.
	b. The upstream rope tender pendulums the victim.
11. Assist victim from the water.	

Appendix XXII: Tripod Method for Shallow-water Crossing

Table A–13 Tripod Method for Shallow-water Crossing

Operation:	Key points:
1. Find appropriate tool.	a. Pike poles, wooden handles or similar tools are ideal.
2. Place the tool in front of you, leaning into it to get a tripod effect.	
3. Start across facing upstream.	
4. Use the tool to "sound" the river bottom ahead of you.	a. This is similar to a firefighter "sounding" a roof as he walks across. You are "feeling" for unseen drop offs and other unexpected hazards.
5. It may be helpful to carry a rescue line for assistance in the event you lose your footing.	a. Personnel on shore should be prepared to pendulum the rescuer to shore if he is swept off his feet. b. The line can also be used to assist in shore-based rescue evolutions.

Appendix XXIII: Static Line or Belay for Shallow-water Crossing

Table A–14 Static Line or Belay for Shallow-water Crossing

Operation:	Key points:
1. Tie a rescue line to a tree or other stable sturdy object on shore. Have manpower on shore set up a static belay. (see page ________)	
2. Wade out, holding the line for support.	
3. If you are swept off your feet, hold onto the rope. You will automatically be pendulumed back to shore by the static line (or by personnel tending the rope).	a. If swept off your feet, assume the same position you would use when being rescued with a floating drop bag (river swimming position holding the rope over your head).
4. This method utilizing a static line is ideal when you are working alone, and there is no one on shore to belay you.	
5. If there is only one other rescuer on scene and a good anchor is available, the second rescuer may be better utilized in the downstream safety position.	

Appendix XXIV: Line Astern Method for Shallow-water Crossing

Table A–15 Line Astern Method for Shallow-water Crossing

Operation:	Key points:
1. Line up single-file along the shoreline.	a. Any number of personnel can be involved in the type of a crossing.
2. Those in the rear should find a suitable handhold in the PFD of the rescuer in front.	
3. The upstream rescuer may choose to use a pike pole and/or rescue line.	a. Hand loop in the rescue line if used.

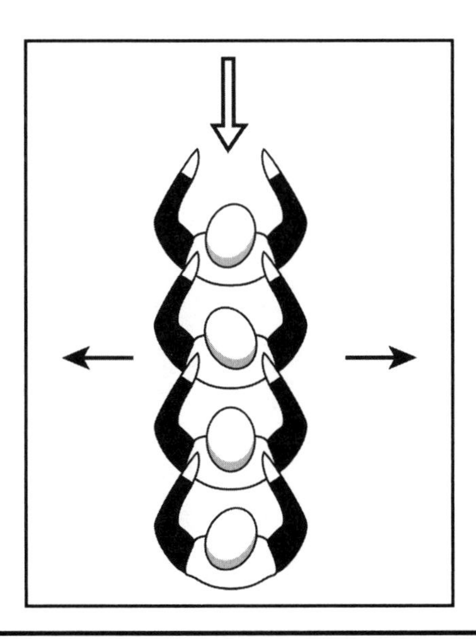

Fig. A–9 Single-file Line

Operation:	Key points:
4. Start across the river.	a. Be aware of unseen drop-offs or foot entrapment hazards.
5. Those in the rear should lean into, and hold down, rescuers in front.	
6. If for rescue, an ambulatory victim can be placed in line for the return trip.	

Appendix XXV: Line Abreast Method for Shallow-water Crossing

Table A–16 Line Abreast Method for Shallow-water Crossing

Operation:	Key points:
1. Line up side by side on the shore.	a. Normally two to four rescuers are used.
2. Place a pike pole or similar tool across chests of the rescuers.	

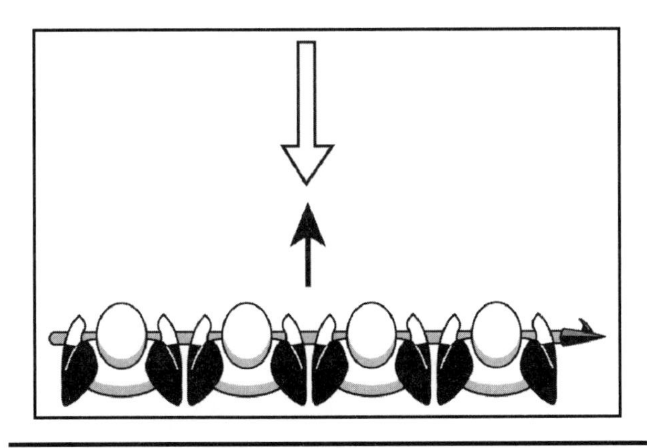

Fig. A–10 Side by Side

3. Interlock arms and hands, grasping the pike pole.	
4. Start across the river.	
5. A rescue line can be used.	
6. For return trip, it is not necessary to change positions. Simply lift the pike pole over your heads, turn around, and replace the pike pole at chest or waist level. Interlock arms and grasp pike pole.	
7. If for rescue, a victim can be placed in the middle and supported from both sides.	
8. A non-ambulatory victim can "sit" on the pike pole with support from each side.	

Appendix XXVI: Circle of Support Method for Shallow-water Crossing

Table A–17 Circle of Support Method for Shallow-water Crossing

Operation:	Key points:
1. Make a circle and interlock arms around the torso of the next rescuer.	a. Three rescuers are ideal.
2. Start across the river.	a. The current will tend to spin the group as they move laterally in the river. This is caused when the current deflects off the rescuers. The effect is similar to a water wheel in an old mill. b. Because of the spinning effect, this method is not recommended for taking a line across (for the obvious reason that the group will become wrapped up like a ball of yarn).
3. A victim can be placed between two rescuers for the return trip.	

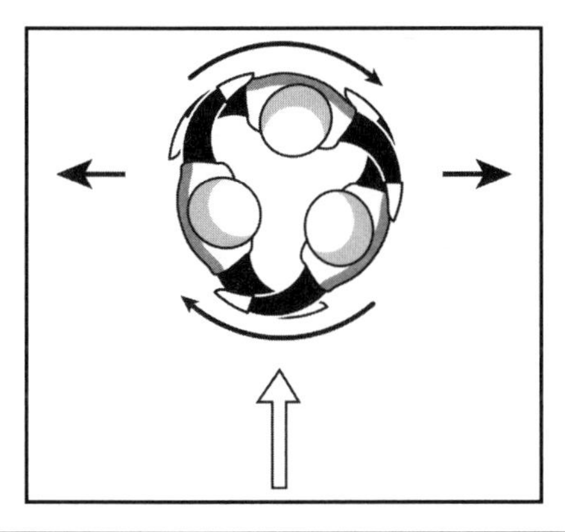

Fig. A–11 Interlocking

Appendix XXVII: Shallow-water Crossing with Victim on Backboard

Table A–18 Shallow-water Crossing with Victim on Backboard

Operation:	Key points:
1. Place victim on backboard.	a. The Miller board is excellent for this purpose due to its shape and the numerous handholds.
	b. *Do not secure the victim to the board.* If the river overpowers the crossing party, it is likely to scatter them in the current. If the victim is strapped to the board, it may be very difficult to retrieve him. It will be easier to rescue the victim if he is free of the board. (In a life-threatening situation, like this, cervical spine precautions must take a back seat to getting out of the water alive.)
2. Pick up board.	a. A minimum of five rescuers should be used. Additional rescuers may be added as necessary.
	b. The rescuer at the head of the board will have a tough time holding the board while fighting the current. If this is the case, place an additional rescuer in front of him to break the current. Now the rescuer at the head can hold the board with one hand and the PFD of the upstream rescuer with the other hand.
	c. The other rescuers should hold the board with one hand and the PFD of the rescuer in front with the other hand.
3. Make crossing.	a. The "point man" may want to use a pike pole for additional support. All personnel must be aware of possible drop-offs and submerge objects.
	b. Move one step at a time in unison. The leader should call out "Step" (or some other direction) to initiate each movement. Communicate with the others to assure coordinated movement.

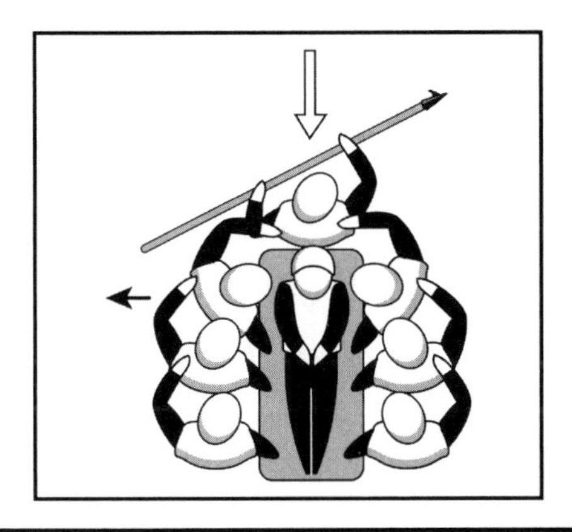

Fig. A–12 Victim on Backboard

Appendix XXVIII: Continuous-loop Rescue System

Table A–19 Continuous-loop Rescue System

Operation:	Key points:
1. Choose proper length of rope.	a. Rope should be about three times as long as the distance from the victim to the shore. b. Two lines can be clipped together if necessary to obtain desired length.
2. Form a continuous loop.	a. Use a carabiner or b. Tie a square knot or c. Tie a figure of eight follow through
3. Take rope to edge of river.	a. Assure no trees or other large obstacles between the rope and the water. Any obstacles along the shore may interfere with the operation of this system if they end up in the middle of the continuous loop.
4. Rescuer #1 takes a position upstream of the rescue site.	a. To act as upstream belay point.
5. Rescuer #2 sets up downstream of the site.	
6. Rescuer #3 will make crossing from rescuer #1's position.	
7. Rescuer #3 makes a hand loop and crosses to the victim.	a. Using a shallow water crossing method. b. Swim if necessary. c. Rescuer #1 and rescuer #2 belay the line. Be prepared to pendulum rescuer #3 to shore if necessary.
8. Rescuer #3 checks victim's condition and explains rescue procedure to victim.	
9. Rescuer #3 finds a safe place to belay from the victim's position. He is now the third belay point.	
10. Rescuer #4 clips a PFD and helmet to the line.	
11. Rescuers #1,#2 & #3 shuttle the PFD and helmet to the victim.	a. Victim dons PFD and helmet.

Step	Key Points
12. Rescuer #4 takes a hand loop and crosses to join rescuer #3 and the victim.	
13. Rescuer #4 assists victim to safety using a hand loop.	a. Use shallow water crossing method if possible. b. Swim if necessary. c. Do not attach rope to victim.
14. Rescuers #1,#2 & #3 belay the line.	a. Be prepared to pendulum rescuer #4 and the victim to shore if necessary.
15. Rescuer #3 takes a hand loop and returns to shore.	
16. Rescuers #1 & #2 belay the line.	a. Be prepared to pendulum rescuer #3 to shore if necessary.

Endnotes

1. www.firescope.oes.ca.us.gov; Incident Command System USAR Operational System Description (ICS-USAR-120-1).

2. Hail searches are often most effective when combined with technical search operations to detect vibrations and noise emanating from within the collapse zone. Devices like trapped-person locators, life detectors, and search cameras (with the microphone/camera head inserted into void spaces from the surface) may be placed around the collapse zone before the hail search begins to detect any return signal from trapped victims as the hailing is performed around the collapse.

3. *County of Los Angeles Fire Department Air Operations Manual, Rappel Insertion Operational Guidelines.* These sample guidelines are intended for Bell 205 and Bell 214 model helicopters as indicated. They are intended only as sample procedures.

4. *County of Los Angeles Fire Department Air Operations Manual, Rappel Insertion Operational Guidelines.* These sample guidelines are intended for Bell 205 and Bell 214 model helicopters as indicated. They are intended only as sample procedures.

5. *County of Los Angeles Fire Department Air Operations Manual, Rappel Insertion Operational Guidelines.*

Index

A

B

C

D

E

F

G

H

I

J

K

L

M

N

O

P

R

S

U

V

W–Z